Violence Through Environmental Discrimination

Social Indicators Research Series

Volume 2

The titles published in this series are listed at the end of this volume.

VIOLENCE THROUGH ENVIRONMENTAL DISCRIMINATION

Causes, Rwanda Arena, and Conflict Model

by

GÜNTHER BAECHLER

Swiss Peace Foundation,
Institute for Conflict Resolution,
Berne, Switzerland

KLUWER ACADEMIC PUBLISHERS
DORDRECHT / BOSTON / LONDON

A C.I.P. Catalogue record for this book is available from the Library of Congress.

ISBN 978-90-481-5144-8

Published by Kluwer Academic Publishers,
P.O. Box 17, 3300 AA Dordrecht, The Netherlands.

Sold and distributed in North, Central and South America
by Kluwer Academic Publishers,
101 Philip Drive, Norwell, MA 02061, U.S.A.

In all other countries, sold and distributed
by Kluwer Academic Publishers,
P.O. Box 322, 3300 AH Dordrecht, The Netherlands.

Printed on acid-free paper

TO MAREN AND CLARA ANNINA

CONTENTS

PREFACE

Since all-out interstate wars for the time being seem to belong to the past, conflict studies focus more and more on domestic conflicts. This is a broad field, not only because the arbitrary line between war and sub-war violence disappears and the analyst is confronted with phenomena reaching from criminal violence and clashes between communities to violent conflicts of long duration and civil wars with massacres and genocides as their characteristics. It is also because there are so many different types of conflicts to be analyzed, so many different types of behavior to be studied, whereas there is often little information available on what is really going on.

Against the background of internal conflicts, which tend to be as protracted as diffuse in terms of time, intensity, actors, and their goals, this study aims to follow a specific pathway through the current thicket of violent circumstances. It focuses on causation patterns by exploring the causal role of the environmental factor in the genesis of violent conflicts occurring today and probably even more so tomorrow. This approach, which for once does not focus on a specific level of the conflict system, on one area in the conflict geography, or on a specific category of actors, analyzes causation dynamics. Concentrating on one causal chain in a syndrome of factors may not only contribute to a better understanding of organized violence, which is mainly violence in developing or transitional societies. It may also help to explain why certain conflicts escalate. Today many analyses as well as media news and reports restrict themselves by highlighting the *how* and *when*. Concerning the question *why*, only thin explanations about ethnic, religious, or nationalist cleavages are offered. However, if one fails to find out the root causes, therapy will be either wrong or superficial, in any case not long-lasting and sustainable. Therefore, the premise in this study is that the conflicts have major *material* roots. These roots are related to the socioeconomic and sociopolitical transition of developing economies, the transformation of the society-nature relationship being only a crucial part of it.

The study focuses on the African continent. Not only because it is the forgotten continent, studying it may bring no major benefits to the scholar in the North. It is because African countries have to find their way out of devastating structures and the vicious circle of economic decline, environmental degradation, livelihood insecurity, and political power struggle. And it is because the author became deeply engaged in the affairs of this beautiful continent during his stays in Africa - not least as an observer of the first free elections in the Republic of South Africa.

This study could not have been undertaken without the help and encouragement of many people. First, I would like to thank the Africans from different countries who so trustfully shared with me their knowledge and their experience. Together with my friends and colleagues - to name only Mohamed Suliman and Naigzy Gebremedhin - I hope that the environment they grew up in and used to live in will again become a place of fun and exciting discussions. This thanks include the many colleagues who contributed to the International Environment and Conflicts Project (ENCOP) from all over the world. Without their knowledge and the stimulating discussions we had during several meetings and workshops, this study would not have been possible. Particular thanks go to Volker Böge, Stefan Klötzli, and Stephan Libiszewski for their creativity and solidarity during the last five years. Kurt R. Spillmann, who - as a co-director of ENCOP - always encouraged me to proceed with this study and to go even beyond it, is gratefully included.

I am also indebted to many colleagues in the community of conflict and peace research. Many of them provided me with new ideas, provoked new thoughts and supported me in publishing preliminary results. I especially want to express my gratitude to Dieter Senghaas whose extremely rich contributions have always been of exemplary value for my own work.

It was in the second half of 1996 in Cambridge, Massachusetts, when I found both time and *ambience* to do the writing. Special thanks go to the Center for Science and International Affairs (CSIA) at the J.F. Kennedy School, which kindly accepted me as a research associate during that time. I profited very much from the discussions, especially during the brown-bag lunches when Steven Miller gave me the chance to present my arguments on the causes of conflicts in the Horn of Africa. I want to thank Michael Brown for being a good colleague and partner during this time as well as the staff of the institute for its kind support, especially of program manager Dawn Opstad. Being a »Harvard official« also allowed me to use the Widener Library as a huge source of knowledge.

The sabbatical leave to Cambridge would not have been possible without the generous support of the Swiss Peace Foundation in Berne. It was not easy to leave the office for quite a long period of time. However, the Foundation Council as well as the staff encouraged me to undertake this endeavor. I am particularly indebted to the Foundation's president, Fritz R. Staehelin, who during my stay in Cambridge took over responsibility and carried out a lot of everyday work on my behalf. I am also grateful to my secretary, Franziska Bacher, who did a great job during all these years that both ENCOP and this study went on.

She was responsible for the close-to-everyday communication between Berne and Cambridge. And she skillfully supported the typing of the tables and figures in the appendix. I am grateful to my colleague Laurent Goetschel, who read an earlier version of the manuscript and made helpful comments. I also want to thank Lyn Shepard, an American journalist living in Zurich, who carefully corrected my English manuscript. Finally, many thanks go to my friend and colleague Catherine Schiemann Rittri in Stockholm. She did a great job indeed with both the time-consuming final reading as well as the careful preparation of the camera-ready copy of this study.

I am grateful to all these persons and institutions for their support. This encompasses the Swiss Development Cooperation and the Swiss Foreign Ministry, Division for Peace Promotion, the financial contribution of which made ENCOP possible. However, responsibility for carrying out the research and the views expressed here, of course, remains my own.

Particularly important has been the help of my family. I am deeply indebted to my wife Maren; not only did she give generous support in all phases of the work; she was also concerned that our daughter Clara Annina got all the loving care she required. It is to Maren and Clara Annina that I dedicate this study.

Cambridge, MA and Berne
May 1997

SUMMARY

(1) This study aims to elaborate on the violent transformation of society-nature relationships (*see: 1, 2*). Various methods adapted to examine environment-related conflicts provided similar and complementary results: the cultural-ecological background of dryland and mountain systems transformation has shown that a correlation does exist between environmental degradation and violent conflicts (*see: 3*). Moreover, as the synthesis of 40 case studies carried out by the ENCOP team indicates, there are different causal pathways of present violent conflicts and wars that could be traced to the environmental roots of the conflict genesis (*see: 4*). However, not all conflicts became violent. Quite a number of them - although acute - remained below the threshold of organized violence.

(2) There is *ample evidence* for *correlations* between rural poverty, environmental stress in rural arenas, and discriminated access to resources. These factors coincide with political instability, *i.e.*, non-democratic change in regimes and/or violent conflicts and wars. These findings primarily apply for the least developed countries, secondly for countries with medium development characterized by poor state performance and unstable political conditions (*see: app. 5*). If one compares regional aggregates of human development indices, the trend of violence being characteristic of low human development is even more significant; Sub-Saharan Africa being at the peak concerning war-torn societies.

(3) Environmentally caused violence is hardly found in interstate relations. Although climatic change, ozone layer depletion, and loss of biodiversity have created global conflict lines, which did not exist until recently, no major catastrophe threatens the world immediately. »Climatic world wars«, mass flight due to rising ocean levels, military crusades by northern countries against destruction of tropical rain forests, and Atlantic »fishery wars« may rather remain threatening blusters than lead to virulent conflict situations with a high risk of escalation.

(4) Arenas of environmental conflicts are most often domestic ones. They encompass distinct strategic groups using and overusing one or more ecoregions, subdividing ecoregions through political boundaries, or

migrating across such boundaries. Acute conflicts that tend to become violent constitute a kind of small-scale north-south conflicts within the countries concerned. Countries with high percentage of labor force in agriculture, little cropland availability per capita, and low commercial energy consumption per capita belong to the group of countries with both high environmental stress as well as violent conflicts or wars going on. There are only few countries with a low resource availability index (RAI) and no violent conflicts occurring. In these cases a low political freedom index (PFI) indicates that a repressive regime is capable to suppress any organized violence successfully.

(5) Transformation of landscape leads to violent conflicts and wars if and when it accentuates structural heterogeneity, which tends to discriminate chiefly against those rural producers who are the victims of bad resource allocations, unequal resource distribution, high dependence on natural capital, and bad state performance outside the federal district or national capital areas. The widening gap between the modern sector, where state performance concentrates on, and the marginal sector, which encompasses poverty clusters in sensitive environments, constitutes a fault line. In many places this has become a real front line of environmentally caused violent conflicts.

(6) Conflicts about degraded renewable resources manifest themselves as transboundary, center-periphery, interregional, and group identity struggles exacerbated by migration and displacement, and in some cases accelerated by population dynamics. Actors with access to state power typically have access to the most productive arenas, whereas identity groups facing environmental discrimination are forced to use and degrade marginal arenas with low productivity, thereby perpetuating impoverishment. In addition, groups against which environmental discrimination works are confronted with environmental deterioration beyond their control: *e.g,* deforestation by loggers destroys the livelihood of indigenous forest dwellers, dam building degrades land both upstream and downstream, mining leads to widespread contamination of the landscape.

(7) As a rule the resort to violence only occurs if and when some of the following *five key situations* coincide: *Inevitable environmental conditions* - group survival is dependent on degraded resources for which no substitutes are apparent; *scarcity of regulatory mechanisms and poor state performance* - a political system is incapable of producing certain social and political conditions, whereas sustainable resource use is far from being attainable; *instrumentalizing the environment* - using the environment (and its destruction) in order to pursue specific group inter-

ests, so that environmental discrimination becomes an (ideological) issue of group identity; *opportunities to build up organizations and find allies* - actors organize and arm themselves in political settings, often behind a strong leader, and gain allies. Rural actors capable of waging conflict need powerful coalition partners from different social levels to support their goals, for instance, part of the *intelligencia*, members of the middle class, or a charismatic leader of an ethnic minority at risk; *spill-over from a historic conflict* - environmental discrimination occurs within the context of an existing (historic) conflict structure and, as a result, the conflict receives new impetus. These five contexts constitute the arena for actors who choose violence as their best alternative to nonviolent activities.

(8) There are significant cases where individuals prove to be unwilling or unable to join a group with strategic goals; various communities and identity groups have virtually no capacity to go to war. There are different reasons: marginalization, lack of means and organizational skill, lack of leadership, state repression, poor health conditions, wide-ranging apathy, fatalism, or defeatism.

(9) As the case study of the Rwanda arena in particular indicated, environmental conflicts are related to socioeconomic, ethnopolitical, interregional, migrational, and demographic problem syndromes (*see: 5*). In the Rwanda arena two factors cumulated and thus accentuated the conflict potential of environmental scarcity. Firstly, demographic dynamics led to high density clusters of small rural producers which - due to lack of financial and human capital - were unable to escape the deteriorating society-nature relationships. Secondly, social stratification led to discrimination against the majority of small rural producers. However, these facts alone may not explain the escalation of violence sufficiently.

The crucial point in Rwanda is that the rural-urban gap was overarched by clear-cut ethnopolitical segregation used by the ruling elite to exclude the other group almost completely from any resources, whether power, education, natural capital, or even the entire country. With the emergence of absolutism, further exacerbated and strengthened by the colonial powers, a steep hierarchy developed with firm role assignments in which a participant was never supposed to be subordinate in one area and dominant in another. The dissolution of traditional mechanisms and kinship relations as well as the transfer of the *social hierarchy* to *politics* led the expanding central state to become an arena where ethnicity became a predominant channel to either hold power or compete with it.

The central state itself became an important tool in this struggle: the winner could easily turn the state into an agent of his own interest, which was supposed to be ethnic interest - the latter following the logic of a winner-takes-all system in a fragmented society. Hence, despite the relative cultural unity of the country in the long period from the 15th century to colonial times, the ethnic amalgamation had not progressed far enough to avert accentuation of ethnic identity once and for all.

(10) The environmental conflict model, which was checked by a number of case studies, basically confirmed the results found in the previous chapters. The linked hypotheses based on a set of different types of indicators turned out to be valid and significant for the cases examined. Environmental degradation plays different roles depending on the specific conflict: in many conflicts it is a systemic background reason; in some other conflicts it is a catalyst leading to channeling along lines between distinct groups; the avoidance of degradation turned out to be a target of parties to a conflict, too. However, passing the threshold of violence definitely depends on *sociopolitical* factors and not on the degree of environmental degradation as such. Environmental discrimination against actors is a strong entry point in order to explore the genesis of conflict escalation in the arenas analyzed. Other key factors are the high dependence on natural capital coinciding with a low resource availability index, lack of institutional capacities for peaceful conflict settlement, the readiness and/or capacity of authorities and leaders to organize and mobilize collective actors, and the (mis-)perception of alternatives to resorting to violence.

(11) Environmental conflicts indicate that transition from agricultural to industrial urban societies is at stake. The »silent revolution« of the peasants discriminated against has become a noisy battlefield in many rural arenas, chiefly in African countries. The *Great Transformation* from rural to industrial societies started not only with remarkable delay, it coincides - *horribile dictu* - with an era of accelerated human-ecological transformation.

Peasants, smallholders, landless people, rural workers, pastoralists, and nomads - the major victims of imperfect penetration through modernization - are permanently fighting structurally and actively against impoverishment, marginalization, and displacement. Decreasing productivity, lack of innovation, and migration or flight belong to a syndrome of key factors, which - if there are triggers and strategically organized actors - cause endemic violence of changing intensity. This is to be understood as struggles of the land against the town, as competition between endangered rural livelihoods and invasive modernity.

(12) The solution of land issues in developing countries is enormously important. Unclear property laws led to disappearance of biological and human resources. Nationalizing land, legal uncertainty, and development by large farming producers are a considerable potential for conflict,

which may even increase through further degradation. Much more land must be privatized in order to give small farmers the legal security to increase productivity, to test alternative methods, and to maintain genetic resources. Governments can play a considerable role in this process. Instead of policing environmentally dependent actors or enacting new laws with little consultation in advance, they can offer or at least facilitate an exchange of equivalents between the actors involved, according to the *formula*: land for reasonably paid work, resource use limitations for compensation payments or agricultural credit, ban on forest use for offering cheap alternative energy, etc.

(13) Thus to achieve sustainable solutions for both security and development policy, economic uniformity in rural sectors must be replaced by *socioeconomic diversity* adjusted to environmental conditions, population dynamics, and modernization attempts. Only if and when economic off-farm alternatives for the rural population are established - be it in new sectors of sustainable farming, in commerce, in small rural industries, in the service sector, or in administration of revalued small and provincial towns - can local production pressure be reduced on biological resources. And only in this way is there an intermediate- and long-term chance to maintain biological wealth and renewable resources for future generations. To protect biodiversity socioeconomic diversity has to be strengthened.

CHAPTER 1

INTRODUCTION

The Transformation of Society-Nature Relationship

Armed conflicts, low human development, and degradation of the environment belong to the most severe problems human beings are confronted with at the end of 20th century. Not surprisingly, many analysts suggest those phenomena being closely intertwined. The Indian ecologist Vandana Shiva calls this threefold syndrome *maldevelopment*, which means »the violation of the integrity of organic, interconnected and interdependent systems, that sets in motion a process of exploitation, inequality, injustice and violence« (Shiva 1994:275). In the early 1990s both war and endemic violent conflicts have indeed been more widespread than during any other decade since World War II, almost all of them were located either in the South or in the East (*see: app. 5*). What at first glance is striking: the arena of present regional conflicts consists of widespread poverty and misery in politically stressed countries, in sometimes highly militarized but nevertheless weak states with poor performance, and in fragmented societies with endemic competition between ethnic or religious groups. Only recently it has been acknowledged in scholarly literature that natural resource scarcity and environmental degradation may also be reasons for intergroup violence and anti-regime struggles (*see: 2*). Resource competition is not only seen as related to development projects but more and more to poverty-related issues, such as overuse of scarce land resources, high population growth, and lack of technical competence or financial means in order to deal with resource degradation. In addition, societies throughout the developing world are confronted with natural disasters - drought, floods, earthquakes, etc. - weak states can hardly cope with. Natural catastrophes and environmental disruption - both being closely intertwined - aggravate if not trigger internal tensions, social competition, power struggles, deadly quarrels, or civil war. In comparison to classic all-out wars between states present domestic violent conflicts seem to be of relatively low intensity; however, this does not mean that their costs in terms of victims, economy, and infrastructure are generally low. On the contrary, many small-scale wars are protracted conflicts. They have no clear boundaries, neither in terms of time concerning the beginning or the end of hostilities nor in terms of dynamics concerning the level of violence or the number of parties involved. They are stop-and-go-wars, rumbling conflicts erupting from time to time, viciously affecting one region or another. Clear-cut victories of one side over the other - the latter declaring total capitulation - are seldom.

Resource conflicts - over land, water, energy, and minerals - have historically lent themselves to military solutions. But what was once a zero-sum game - to the victor belong the spoils - has been transformed into a non-win situation as environmental threats have become more prominent and global in scope; not the *common goods* but the *common bads* are the casus belli, and increasingly so. Thus states will never be able to restore ecologically lost territory with military means.

The supposed correlation of violence and poverty-driven environmental disruption is, however, only the upper side of the coin, whereas the reverse consists of complex causal relationships. Analyzing them requests more detailed area studies. With quite some effort recent conflict studies have been focusing on the question if »environment matters«. Nonetheless, there is still a lack of straightforward explanations on questions such as: why does environment matter? What role does the environment play if it matters? How are the different factors of the man-nature problem syndrome related to each other? What is a major cause? What is only an effect? There is no easy answer to any of these questions. Neither short definitions of »eco-conflicts« nor thin explanations provide satisfying answers. The dilemma between definitional clarity and factual multidimensionality is especially challenging in this field of research to some extent still being a carte blanche despite the increase of scholarly contributions in the last few years (*see: 2*).

This study aims at overcoming both data constraints and methodological dilemmas through a step-by-step approach which combines the advantages of different methods in a cumulative sense. It deals with the syndrome of factors involved in environmentally caused violence as thoroughly as necessary and as reductionisticly as possible. The notion syndrome is designed to cover the problem adequately. A syndrome is »a running together«; in medicine, it constitutes a number of symptoms occurring together and characterizing a specific disease (Webster's 20th Century Dictionary). In our context syndrome means a running together of socioecological key factors in a certain politico-geographical arena. Such an arena encompasses different conflict parties or actors with contradictory interests and preferences »which mobilize their cultural, social and economic capital in view of playing an active role in policy design and receiving a significant part of the cake« (Sottas 1996:216; Evers/Schiel 1988). *Cake* in the context of the present study means the arena's natural capital or - as pointed out further down - the arena's landscape with its available renewable resources.

In the past three centuries science, technology, and industry have opened opportunities to replace scarcity by abundance and wealth on a scale never known before; in certain arenas the cake has grown enormously. The price for this exceptional growth is high. On the one hand, only relatively few people in the industrialized countries and newly developed countries can enjoy the privileges of abundance of resources available. The high standard of living of a privileged group of people cannot be generalized; nor is it foreseeable if and how the pattern and level of consumption can be maintained worldwide. On the other hand, in several arenas abundance and wealth coincide with environmental disruption inducing new and severe forms of resource scarcity. It is one of the working hypotheses that fundamental change - or better transformation of the society-nature relationships in a given arena - provokes a running together of several key factors, the constellation of which causes violent reactions in many ways. Land, wood, water, and other scarce renewable resources have become objects of conflict among actors who, in order to survive, struggle either to preserve their endangered traditional forms of life or to lift their standard of living.

It is of basic importance at present as well as vital for survival at the threshold of the 21st century that these conflicts be better understood from both their proximate and ultimate aspects in order to defuse them, to resolve them peacefully, or at least to avoid violent excesses. Before highlighting correlation and causation I will put the concept of environmental conflicts into the historical context of violent changes in society-nature relationships.

1.1. Environmental Conflicts: Historical and Contemporary Phenomena

1.1.1. SOCIETY-NATURE RELATIONSHIPS

Historically it is the »instability of human relations with the environment« (Cronon 1983:13) that shapes cultural-ecological transformation processes. This instability has two dimensions: the first one consists of environmental *change*, the second one of environmental *transformation*.[1] Transformation

---

1 The term transformation is used in a number of different contexts. To transform originally meant quite simply to change one form into another form (to remodel, to transpose), *e.g.*, a circle into an oval or one organism into another organism. To take another example, the metamorphosis of a caterpillar into a butterfly may be understood as a transformation in exactly the same way as transformation in the aggregate condition of water when it becomes steam or ice. In social sciences the idea of transformation is also used and applied specifically in conflict research with reference to »transformation of conflicts«. In revolution and democracy research it is »transformation of society« which is being studied. Polanyi comes very close to the human-ecological transformation concept. He states with regard to creation of the industrial system that large-scale transformation consists in the fact that societies which were previously feudal and agrarian were completely transformed. This transforma-

differs substantially from environmental change. Regardless of its popularity the term *change* creates quite some confusion because it supposes untouched nature to be a »timeless wilderness in a state of perfect changelessness« (Cronon 1983:11). The human-centric standpoint widely ignores that ecosystems have their own history of both slow and sudden changes. Forests, for instance, have been changed by disease, drought, and fire, species have become extinct, and landscapes have been drastically altered by climatic change without any anthropogenic intervention (*ibid.*). Humankind is confronted with environmental change on an enormous scale. When social entities cultivate or modify their environment by their own means, they only induce additional changes that differ in both quantity and quality from non-anthropogenic ones.

However, these changes added by human action are tremendous: »To transform is to effect fundamental change in appearance or nature« (Kates et al. 1990:1).[2] Over and above that it is the qualitative side, including sudden and chaotic changes within the transformation process, which matters: »The ability of our species to change landscapes or physical environments at the local and regional scales is ancient (...) Some early activities may have led to wide and lasting transformations«. However, »during the past 300 years (...) the scales, rates, and kinds of environmental change have been fundamentally altered as humanity has passed through an era of rapid population growth and the development of a fossil-fuel-based industrial society. The results are unprecedented« (*ibid.*). Climate historians call the conglomerate of social, economic, and cultural developments based on cheap and overall fossil energy supply »the syndrome of the fifties« (Pfister 1995) because the early 1950s are seen as the starting point of accelerated global transformation by human action. Still in the 1960s and 1970s the finiteness of nonrenewable - that is fossil, geological, and mineral - resources were at the center of the emerging environmental consciousness. Remember only the first Club of Rome Report or the »oil crisis« of 1974, whereas more recent debates of the 1980s and the early 1990s have dramatically shifted the focus on transformation of renewable resources, such as water and land (Meadows/Meadows/Randers 1992).

---

tion was caused by the development of both the market economy and the nation-state. As a consequence, it created a subsystem within itself, *i.e.*, the industrial system. Since this was basically an expansionist system it began to *dominate* or to colonize its societal and natural environment (Polanyi 1995:87-101, 243-260). Concerning the term *domination* see further down.

2 At the 1987 symposium *The Earth as Transformed by Human Action* at Clark University, Worcester, for the first time a large-scale interdisciplinary attempt was made in order to create an inventory of transformation and to comprehend the theoretical significance of the term. This was done with the support of a number of scientific institutions (notably IIASA and the Academy of Sciences of the USSR).

Analysts suggest therefore: »The biosphere has accumulated, or is on its way to accumulating, such a magnitude and variety of changes that it may be said to be transformed« (Kates et al. 1990:1). It has not only »changed« and it has been more than »modified« or »degraded«. Transformation is used to indicate the domination of non-human nature by human beings but also the human-human aspects of society-nature relationships (Merchant 1994:1).

As far as the *human-nature* relationship is concerned, domination can be conceptualized with *degradation* of renewable resources. Degradation is used as a technical term in order to describe introduction of tracer gases into the atmosphere, conversion of energy into heat (according to the law of entropy), and changes in soils through washing away, drying up, overgrazing, overuse, deforestation, etc. Degradation is exclusively used as an indicator of the degree of environmental transformation.

Concerning the *human-human* relationships the concept of *discrimination* against actors in gaining access to productive natural capital may best serve to analyze the sociopolitical impacts of domination of nature.[3] *Environmental conflicts* are to be seen as rebellions against discrimination within the context of the transforming society-nature relationship[4] (*see: 2; 6; 7*).

A discipline that specifically addresses transforming relationships between human beings and the environment is *human ecology* (Glaeser 1989:28).[5] Being divided into numerous areas of study its subject is the structural network of

3 The concept of domination stems from the critical theory of the early Frankfurt school (Leiss 1972). Even though Adorno, Horkheimer, and Marcuse did not directly influence the ecological debate of the last two decades or so, the concept itself is at the core of modern approaches, such as deep ecology, social ecology, ecofeminism, environmental justice (environmental racism), and postmodern science (Gaia) (Merchant 1994).

4 In an earlier contribution (1992) I used the term »human-ecological transformation» almost identically. However, »transformation of society-nature relationship» is closer to the socio-political focus of this study, whereas »human ecological transformation» describes more the overall historical background.

5 In everyday language the word *ecology* has a double meaning. On the one hand, since Haeckel (1866) it has meant the science (*logos*) of the »economy of nature». On the other hand, ecology also refers to the object being examined, *i.e.*, complex events in the biosphere (plants and animal ecology). The transition from plant and animal ecology to *human* ecology took place in the 1920s and 1930s. Human ecology was first used by the Chicago geographer H.H. Barrow in a speech in 1922. At first it was applied by Paul Ehrlich mainly in epidemiology, social medicine, ethnology, and anthropology, and then in graphical and geographical population ecology (Huber 1989:58). The social science side of ecology then took a diversion through economics. K. William Kapp, working together with Barry Commoner and Kenneth Boulding, used the expression »political economics of the environment». In introducing »natural resources and environment» as a production factor, it was stated that »economic and production methods used so far are responsible for increasing and even irreversible damage to the natural and human-social environment, whose costs will have to be paid by the community or by future generations« (Huber 1989:60).

relations between human beings and nature, between societies and their environment. Political ecology as a subdiscipline is directed towards social sciences in the framework of which the human environment is seen as part of the societal reproduction sphere. Thus ecology is understood as the political economy of nature, agriculture and industry are only the most important interactive fields in both qualitative and quantitative terms.[6] The analysis of society-nature relationships not only implies the study of opportunities and limits in the use of natural capital. It also implies the study of opportunities and limits in the demands made and burdens laid on the human body and on the social life of a community. Political ecologists put emphasis on the social nature of human beings and their ways of acting and behaving towards natural capital as collective actors motivated by economic and political considerations.

A crucial role plays *scarcity* determining the »economy of nature« (Haeckel). Scarcity indicates a conflictual relationship between those dependent on the use of natural capital available in a certain place at a certain time. The *common* interest of two or more actors to use a plot of land automatically includes the *competing* interests of who uses it, why, how, and probably when. Both common and competing interests lead - as a consequence - to the transformation of nature which, in turn, affects and transforms the relationship between the users as well. Nonetheless, only in exceptional cases or at some turning points in history the competing interests dominated the common interests to the degree that the »running together« of key factors intentionally led to violent struggles over resources in defined arenas. This happened because the natural capital available was considered to be too scarce and/or because social and political regulations over access to resources established in the arena broke down.

Transformation is ill-understood if it is only seen as destruction and devastation of a once »healthy« environment. Analysis of transformation processes requires a differentiated approach with regard to new developments taking fully into account both the ecological benefits *and* the costs of human development and modernization. The extent, type, speed, and reversibility of environmental transformation depend on the specific interweaving of economic, technological, political, demographic, and time factors. Transformation can lead to either degradation or upgrading of the environment. It makes a difference to consider an arena where biodiversity has been enhanced through methods of cultivation or where sustainable resources together with technical

6 The former International Organization for Human Ecology in Vienna used to define its subject in the following comprehensive terms: »Consideration of human life in all its physical, chemical, biotic, spiritual, social and cultural manifestations from an ecological point of view; this includes consideration of the interactions between a human being or a number of human beings and the world surrounding him or them» (I.O.H.E. 1981. In Glaeser 1989:32).

improvements have contributed to higher productivity of natural capital compared with arenas where dirty and energy-intensive industries led to deterioration of a suburban environment, where a copper mine caused the pollution of a river and land in a densely populated agrarian arena, or where overgrazing of land in an arid zone contributed to soil erosion and desertification.

If peace and conflict research is to be seen in the light of human ecology or political ecology respectively, it should basically be concerned with violent transformation of society-nature relationships by focusing on power struggles, hierarchical structures, resource distribution, maldevelopment, and security issues. Moreover, it should design sociopolitically as well as ecologically sustainable concepts for peace on a regional and global level (*see: 8*).

1.1.2. HISTORICAL EVIDENCE

Against the background of transformation analysis it is appropriate to recall that scarcity has been an elementary and ever-present condition of existence throughout human history. There have always been common as well as competing interests of actors concerning use of the natural environment in their arena. Hence, conflicts over scarce environmental goods are not at all a new phenomenon; they form an intrinsic part of dialectical interactions between human beings and nature. People in all cultures have tried to overcome resource scarcity, each in their own way by using their societal and instrumental possibilities (Cohen 1977).

Regarding the pre-Islamic Arabian world, Henri Bruno noted that »(a)mong the settled populations as well as among the nomads, water was scarce and often disputed. It was a prize above all others, and bloody struggles were waged for its possession« (Bruno 1913:10-11). For the selected arenas of European agrarian societies during the 12th and 13th century Caroline Merchant suggests:

> »Through force and the need for military security, a hierarchical structure of landlord domination had imposed itself on the communal structure of agrarian society, extracting surplus value in the form of labor, services, rents, and taxes. The amount landlords exacted was regulated by long-established tradition, and medieval landlords did not, under stable conditions, strive to maximize their gains. But the built-in pressure for seigniorial privileges - the rights traditionally accorded a feudal lord over his domain - was in constant tension with the peasant pressure for community control of common rights and resources. When combined with other interacting destabilizing forces - population pressure and technological innovation - this tension could produce sharp conflict, altered relations between landlords and peasants, and significant changes in the ecosystem as a whole« (Merchant 1980:44-45).

In medieval Europe, conflict was already acute concerning use and control of technology for energy production. The energy of the pre-industrial economy was drawn from renewable sources - wood, water, wind, animal, and human power. Simultaneously, population growth all over Europe resulted in an era of social change combined with political unrest and protest. Forests, fens, and marshes were converted to arable lands and marginal wastelands turned to pasture for stockbreeding (Braudel 1995 I). Quite a number of regulations over private use of meadows, pastures, and woodlands were implemented. Relatively small family holdings were already divided and subdivided. The growth of towns as centers for trade and crafts induced by the slight change in rural structures enhanced the pressure to increase cropland areas. As a result, by the early 14th century most wasteland was being cultivated. Forests had shrunk dramatically over much of Europe (Merchant 1980:46-47; Braudel 1995 I).

The European economic takeoff combined with demographic trends led to both the transformation - and even breakdown - of medieval agrarian economy and to anthropogenic transformation of the ecosystem as two sides of the same coin. Population pressure, soil erosion, and landlord exactions led to a dearth of fertile land per family to ward off starvation in a poor harvest year - such as in 1315. Thus »in the 14th century, European ecosystems reached their saturation point: Western Europe had become a truly full world. The hour of the great ecological crisis had struck, heralding the so-called Industrial Revolution - a revolution which spanned several centuries« (Deleage/Hemery 1990:25).

The human-ecological transformation of the European ecoregions induced a change in the relationship between landlord and peasant as well. On the one hand, in those arenas where peasant collectivity and self-determination were strong and landlords relatively weak, mainly in Central Europe, socioecological balances could be more or less maintained. In East Prussia, for instance, violent conflict broke out in the so-called Peasants' War of 1525. The peasantry rebelled against the power of the landlords because the latter had taken over regulation of grazing, hunting, and fishing rights. Peasants rose in armed rebellion to demand the return of self-regulation of common resources and control over village officials. The result of this success story of the rural population and perhaps of the European modernization process as such can be studied centuries later: by the 17th century, in the west of Germany peasants controlled up to 90 percent of the agricultural land. On the other hand, in those arenas where princes, landlords, or other institutions were powerful enough to continue to levy high taxes, to control natural resources, and to attack traditional land-owner rights, soil quality degraded tremendously. One of the reasons was that the peasants could not invest in enough animals to maintain soil fertility over a long period of time (Merchant 1980).

Europe is by far not the only place to demonstrate early transformation that led to violent conflict, upheavals, and revolutions. Soil erosion, salinization, and disorganization of water supply in Mesopotamia around 2400 B.C. (Goudie

1990:127), scarcity of wood in the Middle East in the 7th century, or destruction of the primary forests in China during the centralist Tang dynasty in the 7th/8th century (Deleage/Hemery 1991:25-26), the degradation of forest and water resources caused by the Maya civilization during the 10th century, or the decline of the empire of Aksum in the Ethiopian Highlands (Hurni 1990:52) are all said to have led to upheavals and unrest, endemic violence, major shifts of power of dynasties, fall of civilizations, etc.

The physical nature of a certain environment may shape the economic choices human beings make in their arena during a certain era. Culture reshapes the environment, which »reacts« to the economic choices made by its users. The reshaped environment, in turn, provides a new set of opportunities for societal reproduction, etc.

The »economy of nature« consists of dynamic interactions. It is indeed obvious that in the history of humankind the density of these interactions - in terms of the necessity for technological control of the environment - has dramatically increased. Every technological change has been accompanied by a change in social organization as well. Simultaneously, influence on nature has covered an ever greater amount of territory because of socioeconomic and demographic developments. Nature and its subcomponents have been transformed into resources in both mental and substantial terms since historical times, however increasingly so after the so-called Industrial Revolution.

1.1.3. DEVELOPMENT OF MALDEVELOPMENT

The historical transformation of the natural environment through changing resource use patterns, coupled with the change or destruction of either rural self-regulation mechanisms or the rise of the early modern industry, combined with social change, population growth, and accompanying violent struggles bears striking parallels to present environmentally induced social change and political conflict in many developing countries. As in early Europe, the transformation of prairies, marginal lands, lakes, rivers, etc. has a major impact on agriculture and production systems, on human health and nutrition, on welfare and technological innovations. Moreover, it induces societal and political change often accompanied by social uprisings, civil strife, clashes between different groups, and regional wars (Godelier 1984). Hence, should present maldevelopment in the Third World be seen just as a continuation of maldevelopment in Europe before the Industrial Revolution? Are the violent conflicts in those countries nothing else than modern peasants' revolts? Are they integral part of an ongoing process of modernization, which - as empirical studies suggest - mostly »begins with peasant revolutions that fail« (Moore 1966:451)?

Indeed, this study aims at putting the widespread phenomenon *environmental conflict* into the broader context of transforming Third World's agrarian societies. However, despite the supposed and more or less obvious similarities of transformation phenomena in two completely different eras as well as in different parts of the world, one must be aware of fundamental differences which are not negligible at all.

In comparison to medieval Europe both the quality and quantity of transforming society-nature relationships have changed tremendously. As mentioned above, transformation is determined by a severity, speed, and globality unprecedented in history. Adequate indicators in order to illustrate the accelerated and globalized process no individual actor can neither ignore nor escape from are the exponential growth rate of commercial energy use in the last 50 years, the high rate of fertile land under production which leads to cultivation and overuse of marginal lands, the deforestation rate, the declining quality of the humus layer, or the scarcity of freshwater (*see: 1.2.1.*).

Another point is the fundamental difference between genuine development and development of maldevelopment. Generally speaking, this is the difference between consecutive development where one era follows another, as it has been the case in Europe before and during industrialization, and simultaneity of historically different periods of time in one and the same era, as in today's developing societies of the Third World. Whereas the medieval European economy had been based on organic and renewable energy sources, the emerging market economy was based not only on a nonrenewable energy source - coal - but also on inorganic matter, such as iron, copper, silver, gold, tin, and mercury. There were socioecological transformations going on which induced modernization conflicts affecting and revolting almost all of society as one era followed another one. A genuine civilization process took place without the intervention of third parties or a transfer of foreign cultural values (Siegelberg 1994:151-153).

In contrast to Europe, in colonized countries natural resources have been transformed on the basis of both the traditional use of renewable as well as industrial exploitation of nonrenewable resources (drilling for fossil energy, large-scale strip mining, minerals, etc.) from the very beginning of the modernization process initiated by foreign intervention. The confrontation of modern or high-energy societies of the North and traditional or low-energy societies of the South is still engraved on each developing economy. Against the heterogeneous background of a dynamic modern sector and a traditional sector becoming more and more unstable, a politically powerless and economically ever more marginalized rural society is forced to overuse the renewables available and, as a consequence, to destroy the living space rural producers depend on.

1.1.4. SOCIOECOLOGICAL HETEROGENEITY

The co-existence of a modern and a traditional sector and their interrelatedness have been highlighted with the help of the concept of *structural heterogeneity*. When first introduced by *dependencia* analysts as a critical answer to the predictions made by modernization theory, explanations were focused on socio-economic disparities and differentiations in underdeveloped societies. Dieter Senghaas defines structural heterogeneity as simultaneity of interrelated but distinct modes of production being separated by completely different productivity rates (Senghaas 1974:22-24). The internal disparities in Third World economies were seen as induced by external forces depending on domination of the world market and on interests of highly productive »centers« creating sub-centers or modernized segments in developing countries. Those internal sectors not forming part of the dynamic economic modernization will either be integrated and subordinated to the interests of the modern sector or, if not, socio-economically marginalized. The approach basically questioned a major prediction of modernization analysts according to which the traditional sector of developing economies would have been absorbed sooner or later by a constantly expanding modern sector - to the benefit of all people.

Beyond this, with persistent poverty, high population growth rates, and environmental discrimination new dimensions put additional pressure on already twisted structures. Therefore, the ecological dimension is more than a contributing factor to political economy. The transforming natural environment has to be conceptualized as an independent or an exogenous variable (*see: 2.3.5.*).

There are various aspects of transformation, which have to be seen as consequences of structural heterogeneity. Those, for instance, who depend more on degrading renewables may be further marginalized by those who are able to mobilize substitutes through access to centers related to the world market. The lack of feasible alternatives may accelerate widespread processes of marginalization and deepen regional, social, and ethnic fault lines within fragmented societies. Adverse impacts on the environment, in turn, may aggravate structural heterogeneity, for instance, through a decrease in the already low productivity in the traditional rural sector due to degradation of land and water resources. Deterioration of the environment in the marginalized sector may also have negative impacts on the modern sector if and when it depends on the use of scarce renewables as well, such as freshwater and fertile land for large-scale irrigation schemes and/or cash-crop plantations. To further illustrate this: dam building often leads to upstream destruction of renewables and forces local producers to retreat to more marginal arenas where the pressure on and the risk of marginal soil overuse is much higher than at the original site. A dam also leads to the transformation of the downstream landscape. The effects may contribute to enhance the productivity of the modern agricultural sector, whereas other users downstream may suffer from seasonal water scarcity because of storage, pollution, or deviation.

The traditional rural sector is most affected by socioecologically heterogeneous development. First of all, in the rural sector, which still hosts most producers in Third World countries, dependence on natural capital is extremely high. Degradation of badly managed resources means that the natural capital itself is shrinking - and not accumulating whatsoever. Only areas with agroindustry, mechanized farming, or *latifundia* systems get access to resources in order to enhance productivity. In some countries, as a result of the green revolution and programs established by international and/or state agencies, a new sector is emerging between the two poles of traditional smallholders and agroindustry. However, this remains a rather small niche and does not absorb landless, migrants, livestock breeders, smallholders, etc.

Furthermore, developing societies are fragmented through regionally bound economies dominated by ecogeographical boundaries due to a lack of means of communication, legal regulations, and viable political institutions. The consequences of such *poor state performance* are numerous: *e.g.*, competing land-use and land-tenure systems leading to confusing property rights, subdivision of already small plots leading to overuse of scarce land resources, overcentralization and poorly developed regional centers, lacking off-farm opportunities, high taxes combined with low capital investments leading to a lack of financial input in rural areas, high dependence on interventions of the central state leading to failures in rural development if not of the state itself.

Poor state performance is a lack of state outputs regarding civil and political rights, welfare expenditure, livelihood security, resource management, income and job creation. The state may not produce good outputs for two reasons. Firstly, the decisions and actions of the state are correct in terms of publicly stated legitimate goals, but their impact is not strong enough to reach the goals. Secondly, the rulers, although proclaiming that the state enhances the public interest, may pursue ends that are actually in its own interest. Both reasons apply especially for regions outside the capital area.[7]

Figures provided by the Human Development Report (UNDP 1995; 1996) or World Bank statistics (1991; 1992; 1995) indirectly support the critical approach to modernization theory. These sources have shown that in maldeveloped countries the traditional sector is not only still there, but the modern sector itself induced more and more poverty-driven sectors, which could not be absorbed by the market. 60 to 95 percent of Third World populations still

7 Adopted and modified from Lane/Ersson (1994:82-83).

depend heavily on the traditional low-profile rural sector, whereas only small elites benefit from the dynamic sectors most often concentrated in the capital districts of their countries.

Even though there are still relatively few disaggregated data available, which measure internal disparities, there is important work in progress carried out by the World Bank, the World Resources Institute, UNDHA, and UNDP. The World Bank is experimenting with measuring different degrees of dependence on natural capital. High dependence means low human development, limited technical means, a rural-based economy, and the like (World Bank 1995). UNDP provides examples of disaggregated data about regional human development disparities within states. In the Philippines the National Capital Region has a relatively high Human Development Index (HDI) of 0.871; the index reaches from 0.01 to 0.99 (*see: 6.1.2.*). West Mindanao, on the other hand, with 0.410 is at the bottom of all Philippine provinces. The same disparity occurs in many regionally fragmented countries, such as Brazil, India, Malaysia, Mexico, Turkey, or South Africa. Disparities often follow societal fault lines, especially concerning the ethnic fragmentation of countries. In Mexico the results show that the HDI of non-indigenous people is 0.887 or 1.3 times as high as that for indigenous people who average 0.700 (UNDP 1996:31). In Venezuela the HDI value of Delta Amacuro is the lowest in the country (0.659), whereas the Federal District tops the list with an HDI value of 0.887. Such examples lead to the conclusion that there is a resource flow bias towards the capitals or federal districts. They also reflect regional disparities which quite often correlate with ethnopolitical disparities: »These disaggregation results signal the need for timely, appropriate measures to address disparity - whether provincial or regional, between ecological zones or between ethnic groups - to avoid explosions down the road« (UNDP 1996:32).

Internal disparities are visible expressions of what recently has been called environmental injustice or what is called here socioecological heterogeneity leading to such different things as a widening gap between privileged and marginalized land-tenure systems, unequal land distribution, inequitable use of water resources, resettlement of smallholders in unproductive areas because of dams and mines, establishment of protected forest zones in areas where people highly depend on fuelwood.

Taking all these sociopolitical factors into consideration, one may speculate about where the transformation of the Third World's society-nature relationship is going to lead in political terms. According to observations made by Barrington Moore (1966), history provides three different possibilities: struggles for democracy, rebellions based on communist ideology, or fascist mobilizations. Moore has no general answer on the probability of which option will occur. However, referring to India as one of his major studies of reference, he ends up with a rather pessimistic prediction:

> »If the agrarian program of the present Indian government fails to solve India's food problem, and there is substantial evidence for a pessimistic evaluation, a political upheaval of some sort will become highly likely. But it will not necessarily take the form of a communist-led peasant revolution. A turn to the right or fragmentation along regional lines, or some combination of these two, seems much more probable in the light of India's social structure. The situation in India leads one to ask whether the great wave of peasant revolutions, so far one of the most distinctive features of the twentieth century, may not have already spent its force. Any attempt to consider the question seriously would require detailed study of Latin America and Africa, a huge task that must be left to others. Nevertheless, one consideration is worth pointing out. By and large, during the process of modernization the circumstances of peasant life have seldom made peasants the allies of democratic capitalism, a historical formation that in any case is now past its zenith. If the revolutionary wave continues to sweep through the backward world in the years to come, that is scarcely the form it is likely to take« (Moore 1966:483).

One may struggle with Moore about several aspects of predictions he made in the 1960s. However, his prognosis on India is closer to reality today than two or three decades ago. Communist-led rebellions are not to be expected in the foreseeable future. The former Soviet Union experienced a turn to the right, political turmoil, and ethnic as well as nationalist fragmentation. China is probably closer to similar developments than ever before in history; some other regionally and/or ethnically fragmented developing countries - for instance Sudan and Nigeria - are at stake as well. Thus the question arises whether environmental conflicts have to be seen in the light of peasant rebellions against modernization and democratization that fail or whether they have to be regarded as a new category combining goals which were seen by Moore as mutually exclusive, such as the successful struggle of peasants for democratic capitalism.

1.2. Landscape as Threatened Life-Support System

In the context of rural maldevelopment the term *environment* may be somewhat misleading and cloudy because it means everything surrounding us: land, forests, urban and industrial areas, recreation zones, individual habitats, etc. Basically, the environment this study deals with is the landscape or, better, the concept of *landscape* as designed by analysts. It is the culturally shaped landscape above all developing societies heavily draw their wealth from. Landscape might be defined as the biophysical system that provides human beings with its life support by its function of hosting the natural resources water, biomass, minerals, energy, etc. (Falkenmark/Subrapto 1992).

1.2.1. WATER AS PART OF THE LIFE-SUPPORT SYSTEM

It is foremost freshwater which life depends on. The amount of water that a particular region receives, in combination with the amount of water that evaporates determines the aridity or humidity of the landscape and therefore its plant production capacity. Together with the seasonality and erratic nature of rainfall and the quality of soils such hydroclimatic preconditions clearly constrain the population-supporting capacity in terms of rain-fed agriculture. These and the following arguments are drawn from the Swedish hydrologist Malin Falkenmark (1995:28-42).

The incoming precipitation is divided into two main branches: the vertical branch returns water to the atmosphere, the quasi-horizontal branch feeds aquifers and rivers. In order to stress the production functions of water, Falkenmark found it useful to distinguish between »green water production« in the former branch based on water passing through the root zone and operating biomass production, which stops when *green water* is lacking, and »blue water production« in the latter branch, involving socioeconomic production activities based on the water available in aquifers and rivers (Falkenmark/Rockström 1993). The water is accessible for societal use while it passes through the landscape both above and below the surface.

The potential for green-water dependent biomass production and blue-water dependent socioeconomic production related to health, industry, and urban activities shape the constraints on the landscape as a life-support system; polluted water adds further constraints. The industrialized North experienced that most societal sectors in some way depend on water. In order to manage the problem in arid climate zones, one must realize that water is a complex resource: vital, non-substitutable, and finite; where there is no water available, people cannot survive but must leave. Water has basically four main life-support functions; these are the

- *health function*, which indicates that safe drinking water is a basic precondition for socioeconomic development;
- *habitat function*, hosting aquatic ecosystems, which are easily disturbed when water in the water bodies gets polluted;
- *carrier function* for dissolved and suspended material picked up by the mobile water along its pathways through atmosphere, landscape, and water courses, and carried along. This function plays a central role in land degradation processes (leaching of nutrients; erosion and sedimentation);
- *production function* in the sense that water should be seen as an economic good. There are two production functions to distinguish: (a) biomass production, operated by a flow of *green water* entering through the roots and leaving through the foliage. In the absence of green water photosynthesis

stops altogether and the vegetation wilts; (b) societal production in households and industry based on *blue water* withdrawn while passing through the landscape and delivered to cities and industries through water-supply systems (Falkenmark 1995:14-15).

1.2.2. SOIL AS PART OF THE LIFE-SUPPORT SYSTEM

Another crucial dimension of landscape transformation is declining soil fertility and - as a consequence - decreasing soil productivity. In developing countries the main reasons for soil degradation are overgrazing and overcultivation. These may lead to nutrient deficiency, salinization, and devegetation. However, one major aspect of deficient soil productivity is nothing else than deficiency in soil water. This may impede crop growth in the sense that there is not enough *green water* available in the root zone to meet the crop's water requirement. Soil deficiency may be caused in two main ways: (1) through drought caused by no precipitation or erratic rainfall or (2) through a lack of soil permeability needed to allow the water to infiltrate into the root zone. »Both causal factors, when combined with droughts (temporary rainfall deficiency), result in local soil productivity deficiencies and a lack of 'green water' needed by the plants to allow photosynthesis« (*ibid.*:16). When drought is combined with other eco-geographical factors, such as short growing seasons, recurrent drought as part of the climate, dryland degradation, and scarcity of surplus from rainfall (rivers, aquifers) it tends to become a factor of extreme societal vulnerability in terms of crop failures and loss of livestock. This is best illustrated by the situation in sub-Saharan Africa where during the famine years of 1984 and 1985 drought passed through a region where all four modes of water scarcity were superimposed.

The multiple environmental vulnerability in the dry-climate tropics can be expressed in terms of a whole set of water scarcity modes: lack of *green water* manifested as a short growing season due to limited rainfall and a long dry season; intermittent droughts becoming part of the climate and generating disturbances in water vapor circulation between the Pacific and other regions; dryland degradation of the landscape due to soil vulnerability and adding to the scarcity of *green water*; scarcity of *blue water* originating from aquifers and rivers recharging rainfall surplus (*ibid.*:18).

A country may receive its water resources from either the net precipitation over the territory (endogenous availability) or entering water in multinational aquifers and rivers bringing water from upstream countries into the same river basin (exogenous availability). These two complementary *blue water* sources may exist in different constellations: in one country there may be scarcity of endogenous water (*e.g.*, Egypt), in another one scarcity of exogenous water or both (*e.g.*, Israel). The typical problem pattern terms of water allocation vary in principle according to this relationship: water availability differs between countries due to different relationships between runoff produced from rainfall within the territory (endogenous water) and imported water in entering aquifers and rivers fed by rainfall over upstream territories (exogenous water).

Concerning these direct water scarcity and quality aspects further conflicts may be analyzed according to societal needs to mobilize a greater fraction of the water available. A first type of dispute relates to seasonal problems and the question of how to compete with it, for instance by construction of water storage facilities behind dams (*e.g.*, Eastern Turkey). Second, irregularities of human settlement patterns - *e.g.*, transhumance, migration, urbanization, etc. - call for redistribution of water from water-richer regions to poorer ones (*e.g.*, Central Asia). Such measures are designed to provide water for activities in arid areas focusing on urban areas and industrial production as well as irrigation for cash-crop production. The side effects of both types of water projects tend to make them controversial. For instance, in case of dams disputes arise from the unavoidable side effects of flooding the area behind the dam. Disputes are also reflected in the accumulation of avoidable and unavoidable side effects of either a dam project or downstream use of water made available by the dam owners (Dixon/Talbot/Le Moigne 1990).

Falkenmark suggests: conflicts induced by problems of livelihood security will chiefly emerge in the following areas:

- land degradation resulting from soil/water mismanagement, driven by problematic land-tenure and land-use systems, accelerated by population growth, and triggered by recurrent drought;
- escalating water scarcity provoking controversies, which may be structured into three categories: (a) conflicts related to *blue water* competition either between upstream and downstream *blue water* uses or between *green water*-consuming biomass production and *blue water*-dependent societal uses; (b) conflicts related to water quality, where upstream water users pollute the lifeline of downstream users along the same river; and (c) conflicts related to water projects, which may be related either to coping efforts (reservoirs to cope with seasonality; transfers to cope with spatial distribution problems) or to hydropower production (Falkenmark 1995:24).

1.3. Summary and Overview

Transformation indicates a variety of processes, which may run together in certain patterns. These very patterns - or specific syndromes of factors - have to be analyzed if one aims at explaining and understanding serious environmental conflicts. Instead of suggesting a comprehensive and therefore meaningless definition of environmental conflict, one can do no more than formulate a plausible research strategy applied to those problems considered relevant.

The landscape as transformed by human action is the basic category for analyzing environmental conflict potential this study focuses on. It comprises the structural pattern on the basis of which conflicts occur in the sociopolitical network dependent on the resources a given arena's landscape provides. This means that - aiming to explain environmental conflicts - we must concentrate on transformation of the renewable resources water and land as threatened life-support systems. Resource scarcity is therefore seen in an ecogeographical and politico-territorial context; territory not defined in terms of national sovereignty but as culturally shaped landscapes transcending political frontiers. Landscape provides livelihood security. Where landscape is transformed livelihood security is at stake and environmental conflicts are likely to escalate. These conflicts have to be seen in the light of dominating both men and nature. The best indicator of domination is the extent to which actors expose the natural capital to degradation as well as the extent to which they discriminate the access of other actors to natural capital.

Against this background the working thesis is as follows: transformation of landscape leads to violent conflicts and wars if and when it accentuates structural heterogeneity, which tends to discriminate chiefly against those rural producers who are the victims of bad resource allocations, unequal income distribution, high dependence on natural capital, and bad state performance outside the federal district or national capital areas. This thesis will be further elaborated on in the following chapters.

After discussing the state of the art concerning environmental security issues in the first sections of *chapter 2*, an overview over the various methods involved will be provided in the last section of the same chapter. The working program and the research questions are to be found there as well. In *chapter 3* correlations between transformation and violent conflicts will be explored with the tools of a cultural ecological approach. *Chapter 4* provides a typology of environmental conflicts. The empirical material is drawn from a set of case studies carried out by a team of researchers in the framework of the Environment and Conflicts Project (ENCOP). *Chapter 5* deals with causation and causalities. The case of *Rwanda* is chosen in order to explore at greater depth the complex causal network leading to violent conflicts. In *chapter 6* a causal model on environmental conflicts, based on seven linked hypotheses, will be proposed. At the core of the model is environmental discrimination, which provides an entry point to the socioecological and sociopolitical part of the model. *Chapter 7* is dedicated to the model check against the background of six case studies and six control cases taken from different research teams. The term

»check« indicates that the model will be confronted with empirical facts in order to see how robust it is. However, it is not empirically testable at this stage of research due to the lack of disaggregated data on the state of the environment at subnational levels of the countries concerned. Finally, *chapter 8* will be concluded with generalizations regarding the patterns of environmental conflicts as well as with answering the working thesis. The *appendices* encompass background material, such as a list of violent conflicts and wars, as well as tables with figures available concerning the main indicators used.

CHAPTER 2

DISCUSSION

State of the Art

A basic assumption made in this study is that environmental degradation under certain conditions may cause violent conflict or war. Indeed, recent years have seen an ever growing discussion about »ecoconflicts« or green wars caused by the global environmental crisis. While environmentalists among peace and conflict researchers argue that ecoconflicts will be the very wars of the foreseeable future, the traditionalists refuse to accept that this phenomenon really does exist. Others argue that throughout human history there have always been conflicts over natural resources, thus »ecoconflicts« are hardly a brand-new type of conflict on the threshold of the 21st century. As a matter of fact, resource conflicts are not at all specific events of the modern world (*see: 1.1.*). And »green wars« do not exist, as there seldom have been wars with one single cause. In most cases wars have a complex and multilayered background.

However, the repeated appearance of transformation in the genesis of violent conflicts is striking. The discussion about *environmental security and environmentally induced conflicts* has kept on expanding in the last few years (Dokken/Graeger 1995; Gleditsch 1997; Research Panel 1996, author: Steven Lonergan) despite the fact that the interest of governments in environmental foreign policy issues has been shrinking during the same time almost worldwide. The time lag is perhaps not surprising for an issue, which emerged only at a very late stage of the debate on global environmental change. And it is typical for a debate, which has been going on among members of such different scientific communities as environmentalists, security experts, development sociologists, and conflict researchers, mainly since 1989, that is, since the end of the Cold War. There is, on the one hand, a growing amount of literature available that either concentrates on more theoretical issues or deals with methodological questions and criticism of existing studies (Gleditsch 1997; Matthiew 1997). On the other hand, extremely little emphasis is put on empirical work, case studies, or field research. The somehow ideologically colored debate on *environmental security* has led to new attempts in redefining security over and over again. Simultaneously, attempts to reject the concept as either too narrow or too encompassing are not less persistent. It is indeed unusual that first of all scholars throughout the world try to define what a new security threat is, whereas military and defense ministries try hard to keep the leadership conceptualizing what they suppose security should be about.

Furthermore, there is the twin bias in the debate on environmental conflicts among scholars and experts. The first bias consists of »the discovery of a new issue« bias: whenever there is a new issue emerging, there is a tendency to cut off history. This leads either to rather alarming statements about future events, such as »water wars« and similar disasters, or to an unhistorical approach neglecting the distant root causes of present developments. Concerning the issue of environmental transformation and violent conflict or war, these shortcomings are critical insofar as environmental history should not and indeed cannot be separated from socioeconomic and political history. The second bias is supposed to be »environmental determinism«, which tends to overestimate geographical structures, demographic data, resource dependence by simultaneously underestimating cultural, socioeconomic, and foremost political capacities or shortcomings respectively in order to deal with environmental degradation and discrimination. Even if a correlation between resource degradation and the resort to physical violence does exist (*see: 4*), there is no such thing as a linear cause-effect relationship between the environment and the collective decision to use weapons.

There is no intention to provide an exhaustive survey on the growing field of literature dealing with environmental security in one form or another (Bächler/Böge/Klötzli/Libiszewski/Spillmann 1996:10-17)[1] but rather to concentrate

- on an early contribution that conceptualized well the interrelationship between environmental transformation, underdevelopment, and sociopolitical conflict right at the beginning of the global environmental change discussion, namely the UN Founex Report of 1971;
- on recent studies, which focus on resource scarcity and environmental degradation as a major security issue or as a trigger of various types of serious conflicts.

2.1. Founex Report on Environmental Change and Underdevelopment

The United Nations Conference on the Human Environment, which took place in Stockholm in 1972, for the first time created broader awareness of international and global environmental problems (Kilian 1987). This conference - or better, the previous discussions - were the starting point of debates on »environmental foreign policy«, »international environmental cooperation«, and »poverty as the biggest polluter«. In the forefront of the UN Conference a group of

1 Herein further referred to as Bächler *et al.* 1996.

experts chaired by Maurice Strong was addressing the issues environmental change and underdevelopment from a global perspective. The group provided the so-called *Founex-Report on Environment and Development*, prepared by Mahbub ul Haq (Founex Report 1971). It formed a cornerstone for decisions by the United Nations, international agencies, and member states later taken on in Stockholm and beyond. In the report the point is made for the first time that the environmental problems of the developing countries are predominantly problems that reflect both rural and urban poverty and the very lack of development in their societies. »In both towns and in the countryside, not merely the 'quality of life', but life itself is endangered by poor water, housing, sanitation and nutrition, by sickness and disease and by natural disasters« (*ibid.*:9). While most environmental problems of the industrialized countries were supposed to be related to development, environmental disruptions in the South were seen as those that could be overcome by the process of development itself (*ibid.*:10). However, many developing countries were experiencing problems common to those of the industrialized countries. Therefore, development itself would have to be broadened by new aims and qualitative measures in order to deal with the »environmental side effects that have been known to accompany, in varying degrees, the process of development in agriculture, industry, transport and human settlement« (*ibid.*:16-17). The panelists acknowledged the role of socioecological variables, recognizing that deviant social behavior - and, hence, social conflict - emerge from a loss of community and social organization. »Many developing societies«, the report states, »display a high degree of social organization and a considerable sense of community, even in urban settings, as a result of the transplantation of traditional social structures in the process of rural-urban interactions« (*ibid.*:23). The Founex Report continues with discussing specific problem areas, addressing the world economic order and providing detailed policy recommendations. However, what is important here are two statements made by the panel.

Firstly, the members were well aware that poverty-related environmental change can induce heavy degradation of socioecological structures, which - in turn - leads to social conflicts in developing countries. If we observe the growing trend towards internal violent conflicts and wars in the South - mainly in the least developed countries (LDCs) - over the last two decades or so, it seems that the suggestions of the panelists can be confirmed today (*see: app. 5*). Secondly, as a consequence, the panelists called for a broader and redefined concept, not of security - but of development. They assumed a causal linkage between environmental and sociopolitical deterioration and therefore put the emphasis on a broader concept of development. If this assumption holds, environmental conflicts may be understood as a failure of both socioeconomic development and sustainable resource use over the last decades or at least over the period of time between the UN conferences of Stockholm in 1972 and of Rio de Janeiro in 1992. As we have seen, the problem was obviously fully recognized by analysts already in 1971.

Nonetheless, mainstream discussions took another direction, while poverty issues were broadly addressed again only twenty years later during the UN Conference on Environment and Development in Rio.

2.2. Linkage between Environmental Change, Security, and Conflict

It was only in the late 1980s and early 1990s when the debate on war-related environmental destruction shifted away from classical military security topics. Previously »environmental destruction« was mainly considered either as a side effect of military training (Krusewitz 1985), as a means of warfare (Westing 1976, 1984), as a by-product of wars (Weizsäcker 1971; Westing 1980, 1985), or as the catastrophic outcome of a future war waged with weapons of mass destruction (Westing 1977; Ehrlich *et al.* 1985). The debate moved on to »environmental factors in strategic policy and action« (Westing 1986a,b; Brown 1990), focused on »redefining security« bringing into the picture new threats such as global environmental change (Ullman 1983; Mathews 1989, 1991; Gleick 1991) and results in the concept of environmental degradation as a major cause of violent conflict and war (Lipschutz 1989; Renner 1989; Bächler 1990a,b; Brock 1991; Homer-Dixon 1991a,b).[2]

Meanwhile there are myriad arguments and activities associated with »environment and security« (Simmons 1995:2) and sharp disputes on how to define the linkages between environment and security (Brock 1991; Deudney 1991; Gleick 1991; Bruynincks 1993).[3] As has been stated in a report quite recently, »one of the difficulties in assessing the nature of the linkage is caused by the ambiguity surrounding the term 'security'« (Research Panel 1996:2). For Westing, for instance, comprehensive security consists of the two intertwined components *political* security with its military, economic, and humanitarian subcomponents, whereas the second one, *environmental* security, includes protection and use of nature (Westing 1989). The United Nations Development Program, on the other hand, simply lists seven components of »human security«, namely economic, food, health, environmental, personal, community as well as political security (UNDP 1994). Some skeptics completely reject the

---

2 See for an extraordinary early contribution on »redefining security» Brown (1977), and on environmentally caused conflicts Earthscan (1984).

3 The PRIO Report on »The Concept of Environmental Security» (Dokken/Graeger 1995) provides an excellent overview. The ENCOP team refers to the debate in more details in Bächler *et al.* (1993:101ff).

association of environment and security for several reasons. Daniel Deudney claims that it is analytically misleading to think of environmental degradation as a national security threat. First, he points out that the traditional focus of national security - namely interstate violence - has little in common with either environmental problems or their solution. Second, he criticizes the effort to harness the »emotive power of nationalism« to help mobilize environmental awareness and action as being counterproductive because it would undermine global political sensibility. And third, he maintains that environmental degradation is not very likely to cause interstate wars (Deudney 1990:461).

2.2.1. LINKAGES BETWEEN ENVIRONMENT AND SECURITY

Redefining security, it seems, is not the problem at stake but rather the question of *how* to define it adequately. This is not surprising as the notion of security shares this kind of difficulties with other normative terms or general social categories, such as development or peace. There is no way out of the normative trap of being either too broad and therefore vague or too narrow and, hence, necessarily exclusive. If one defines security in the most general way it means absence of any threats in the foreseeable future (Kaufmann 1973). If we want to break this general statement down to a specific type of threat in a given period of time, we must identify the linkages we are talking about.

In principle, four sets of linkages between environment and security have to be distinguished (*see: fig. 2.1.*). In the first line (environmental transformation) are events with empirical evidence; the impact of the various linkages can be measured, at least in principle. In the second line (environmental security) there are normative statements on ideal situations or concepts that should be achieved; it means either wishful thinking, strategic planning, or practically oriented theory. Since the notion of security - narrowly or broadly defined - will remain in dispute, because it is only *one* normative approach to manage threats and dangers concerning both societies and individuals, some critics prefer to replace it with other terms which indicate alternative or even more appropriate problem-solving concepts, such as peace or sustainable development. Therefore, a false hope underlies attempts to redefine security, namely to use environmental concerns as a means of overcoming the controversies both within and among different communities concerned with national security issues. Even *redefined* security may be designed to deal with non-conventional threats in a very conventional manner and with traditional means. It does not necessarily induce major shifts from curative (end-of-pipe) thinking to preventive actions and to therapies of the root causes of problems.

	global linkage	functional linkage	intentional linkage	causal linkage
environmental transformation	as a threat to human security	as a consequence of war	as a means of warfare	as a cause of violent conflict and war
environmental security / peace with nature / sustainable development	absence of threat to human security	no environmental damage through war	environment is not/should not be used as a means of warfare	environmental transformation does/should not cause violent conflicts/wars

Figure 2.1. Taxonomy of environmental transformation linkages

Of course, the same is true for *causal linkage* as well. If environmental disruption does lead to violence, it is not sufficient to concentrate on how to mitigate or stop violence. In an integrated approach one must rather provide concepts connected to *global linkage*. The absence of a threat to human security or better to human habitat is a necessary prerequisite for sustainable conflict resolution and resource use.

This study concentrates on the upper segment of causal linkage (*see: fig. 2.1.*), namely environmental transformation as a cause of current violent conflict and war. A cause is the meaningful connection between independent (*explanans*) and dependent variable (*explanandum*) in terms of cause, in the sense of primary source and effect. Only in the conclusions (*see: 8*) there will be examined if environmental conflicts have to be seen more with regard to the *security* dilemma between states or rather in view of deepening of the *development* dilemma within states. A third option suggested here will be the integration of security and development issues in one concept.

2.2.2. DISCUSSION OF THE CAUSAL LINKAGE

Why is causal linkage so important? First, if it is possible to identify environmental transformation as a major cause of present conflicts in many of the poor countries, this might help to redirect both political awareness and financial contributions to address a root cause of domestic violence ill-understood so far. If this happened, conflict management could and would be more successful than it is today. To take only the case of Rwanda/Burundi, where the third parties involved totally failed to address root causes of the protracted conflict (*see: 5*). Secondly, »ecoconflict« is supposed to be or at least to become a major threat to human kind at the threshold of the next century. Therefore, it is crucial for policy makers to deal with »global linkage« in order to get precise analytical tools and to highlight causal relationships as well as to improve both predictability and preventive measures.[4] Both reasons require some major research efforts in the domain of transformation, poverty, and violent conflicts.

A first major step in exploring causal linkage is the »Studies in Environmental Security«, which resulted from a joint venture of the United Nations Environment Program (UNEP) and the Oslo Peace Research Institute (PRIO). Part of the program concentrated on the Horn of Africa where causal linkage was analyzed in great detail: »A degraded environment has, in the context of the Greater Horn of Africa, frequently led to war and competition over scant resources. (...) Taking up the challenge of promoting the resolution of the complex problems of the region would require identifying the linkages that undoubtedly exist between environmental stress, the clash of economic interest and the tensions inherent in political conflicts that escalate to armed confrontations« (Imru 1990:15,16).[5] Another participant in the Horn project states: »Indeed, the environment itself is among the causalities of war. (...) Nevertheless, it is generally recognized that competition over scarce resources is frequently a contributing cause of conflict because it generates insecurity. War itself is very wasteful of resources and destroys the environment that produces them, thereby increasing scarcity and insecurity and aggravating conflict« (Markakis 1990:1). The same analyst finds it legitimate »to examine the role scarcity and competition for basic resources play in the generation of conflict« even though »this role may, or may not, be clearly perceived by the embattled parties« (*ibid.*).

As a consequence, while highlighting the environmental dimension of the »tragedy in the Horn«, the analyst suggests including socioeconomic and political factors that have environmental impact. The Horn study group perceived the task as not easily resolved - despite availability of a lot of information on environmental change. Therefore, a member of the study group pointed out clearly: »It may be difficult or impossible to prove, in a scientific way, that wars or conflicts or hostilities between nations are caused by environmental degradation« (Molvaer 1989:2). Indeed, even though the literature widely accepts that such a connection or linkage often exists - not as a sole factor but as a contrib-

---

4 Since the latter are part of a four-year project called Ecological Conflict Management (ECOMAN), the normative approach to causal linkage is excluded in this study with exception of chapter 8 dealing with conclusions and outlook (Bächler/Bellwald/Suliman 1996).

5 Most of the papers on the Horn are unpublished. I quote here from manuscripts gathered in six volumes and owned by the person formerly responsible for the program with UNEP, Ato Naigzy Gebremedhin, Asmara/Nairobi.

uting one that may trigger unrest - there have been few attempts to »prove« the existence of this linkage. The study group provided empirically rich area studies and identified fields of environmental cooperation. However, it failed to sort out causes that may trigger environmental conflicts in a rigorous or systematic way.

It is not that the UNEP/PRIO program would not have worked on the basis of a widely defined environmental security concept, stating, for instance, that »(w)hile military activities can have a severe and debilitating effect on the environment, environmental factors can, in turn, have a profound effect on security and prospects for peace« (Ulrich 1989:3). However, the concept has its benefits as well as its shortcomings on empirical terrain. One definite advantage is that it sheds light on an emerging issue, considering simultaneously data on wars as well as on victims (refugees), on political events, and on devastating effects of warfare on the environment. On the other hand, the socioeconomic, institutional, and actor-oriented variables, which intervene between the causal relationship linking environment and conflict, tend to be underestimated. The Horn project, as other projects listed by Gleditsch in his critical overview of the scholarly literature, suffered a lack of clarity about the concept of environmental conflict, single-factor reasoning, case selection on the dependent variable (violent conflict), and thus a lack of control cases. Beyond this there is no clear distinction between internal and international conflict and therefore no clarity about the level of analysis (Gleditsch 1997:0). This criticism applies for most academic approaches in this domain. It is still not clear, for instance, if one can infer immediately the *intensity* of conflict behavior of actors involved from the degree or *severity* of environmental degradation in a certain area. Some studies presume a linear relationship; others do not. Quite often socioeconomic effects of environmental change are cited as responsible for violent outbreaks of conflict, but it is not indicated which effects. Hume's question *same cause - same effect?* yet remains to be answered.

The concept of environmental conflicts has often been criticized as being unable to really *prove* the interrelationship between environmental degradation and violent conflict. The main problem that arises repeatedly seems to be the request to »prove« causality (Molvaer 1989:2).[6] However, this is neither a

6 Many skeptics of the Environmental Change and Acute Conflict Project (ECACP) (Homer-Dixon 1994) and the Environment and Conflicts Project (ENCOP) (Bächler *et al.* 1996) argued that both projects failed to explain how environmental change »really« matters in terms of conflict. This is why Dessler's caveat concerning explanatory relativity, considering the example of explaining a *lunar eclipse*, is worth noting here. He argues that scientists cannot explain the eclipse *tout court*. They explain its timing, its duration, its visibility from certain parts of the earth, etc. They only select some aspects of it. These are the aspects which really matter. However, the quality of these aspects or indicators is important for understanding the *explanandum* (Dessler 1994:99).

shortcoming of social science in general nor of the environmental conflict program in particular. It is not *proven* whatsoever that other factors than environment, such as economy, ethnicity, or contention for state power cause violent conflict. Even for medicine as a »hard« discipline it would be very demanding to request proof that a small number of precisely described factors triggers cancer or heart attacks. If successful methods were developed in distinct domains, then nothing would be easier than adopting them. However, the problem at stake »in fact is that there is little ongoing empirical research that has led to testable hypotheses« (Lane/Ersson 1994). Theories claiming to provide testable hypotheses, such as rational choice, for instance, have not carried out many empirical tests so far (Green/Shapiro 1994).

The term *proof* raises methodological expectations that cannot be met and therefore lead to rationalizations and defensive research strategies; grapes that cannot be grasped are sour. That is why, even after careful exploration of environmental disruptions, conclusions quite often remain weak and somehow impressive. Causes of conflicts are often concluded rather vaguely to be complex and multifaceted or that the environment »contributes« to this complexity. More differentiated statements contend that environmental change either contributes to instability, heightens political tensions, aggravates existing conflicts, or is a catalyst for hostilities (Ulrich 1989:5). Such observations on trigger mechanisms are much more helpful; however, they are rarely empirically grounded or systematically related to specific events. So far, little emphasis has been put on tracing causal pathways accurately. The myriad arguments get often stuck in the marshy ideological battlefield, where people fight for definitions. Moreover, scholarly literature tends to ignore the distinction between conflict and conflict intensity or ways of conflict resolution. At first glance a conflict is no more than a difference in position between two or more actors perceiving this difference as mutually exclusive. Conflicts happen every day, everywhere, and between everybody. Settlement of conflict, on the other hand, can either occur peacefully or result from coercion, organized violence, or war. Thus linkage between environment and conflict must clearly be distinguished from linkage between environment and *violent* conflict or *war* in particular. A clear-cut threshold lies between the two different kinds of linkages, the analysis of which requires specific tools.

Another problem is that many studies hedge their conclusions by failing to distinguish clearly between *past* and *future* developments. A common statement and indeed a truism is that population growth and migration may lead to conflict. Instead one must argue that both population growth and migration always lead to some sort of conflict. Yet the interesting questions in terms of predictability would be: how, where, and why do the two factors lead to serious social and political problems, which tend to become violent? Where does this very problem syndrome also exist but not lead or contribute to violence? And why not?

The environment is often seen as a background variable that only indirectly influences or intensifies an already existing and often more comprehensive conflict situation (Dokken/Graeger 1995:39). Although it is widely acknowledged that research has addressed existing linkages between environmental degradation and conflict, it holds that »establishing one causal element in a conflict situation, *e.g.*, the environment, does not preclude other elements from being equally important« (*ibid.*: 43). Here again analysts seem to question whether environment really matters by suggesting that it contributes to some event or state of affairs rather than serves as a necessary precondition. An overview concludes that »the threshold for when environmental degradation becomes intolerable and erupts into violent conflict is difficult to assess in general terms and must be decided on a case by case basis« (*ibid.*:49). Against this background, a widely shared view holds that the environmental conflict approach cultivates rich case studies, which help to justify the environmental dimension of security empirically and therefore make it more applicable for the political community and the general public.

To conclude: there must be something between the two poles *scientific proof* and *case-by-case studies*. If proof of linkage indeed overstates the claim made by (social) science and myriad case studies, why give up the search for common denominators, for well-selected indicators going beyond single case description, and for allowing some generalizations, which could be integrated into more general approaches in conflict research? By considering the state of the art and methodological constraints several distinct options exist in which the results of general evidence can be gained without ignoring empirical evidence. How to deal with this is the focus of the next two sections.

2.3. Methodological Dilemmas in the Study of Causal Linkage

2.3.1. EVALUATION OF TAXONOMIES AND TYPOLOGIES

Several analysts provide differing taxonomies and typologies of the linkage between environment and security in general and the *causal linkage* in particular. In the following section I will examine the relationship between some of the most important systematic orders to learn more about the specifics of the approaches under consideration.

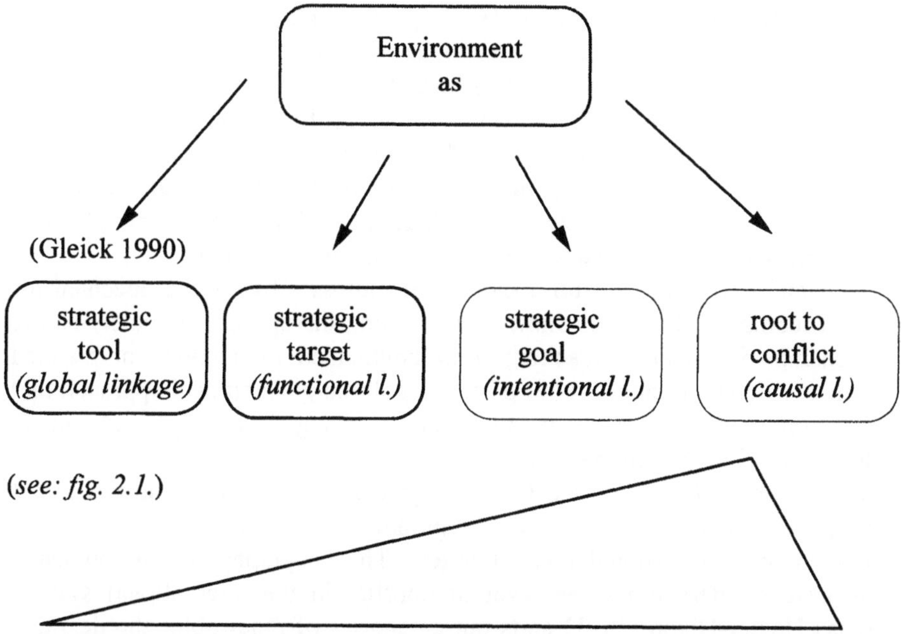

trigger mechanisms (Lonergan 96)	geographical boundaries (Renner 89)	renewable resources (Homer-D. 91)	anthropogenic transformation (ENCOP 92)
1. natural disasters 2. cumulative changes 3. accidental disruptions 4. development projects 5. conflict and warfare	1. global environmental effects 2. regional effects 3. bilateral effects 4. nation-wide effects 5. subnational effects	1. climate change 2. ozone layer depletion 3. degradation of land 4. degradation of forests 5. water depletion 6. depletion of fisheries	1. overuse of renewables 2. overstrain of sink capacity 3. impoverish-ment of space of living

Figure 2.2. Environment, security threat, conflict setting

Peter Gleick (1990) concentrates more on general linkages, distinguishing four ways of using or manipulating the environment in terms of security issues; the environment may be a strategic tool, a strategic target, a strategic goal, and a root to conflict. Gleicks taxonomy (*see: fig. 2.2.*) is more or less at the same level of generalization as the four linkages mentioned earlier (*see: fig. 2.1.*). Since this study focuses on *causal linkage* it only refers to the four different taxonomies drawn from the category »root to conflict« (*see: fig. 2.2.*). This does not mean that they mutually exclude each other. The opposite is the case: they are highly complementary because all of them deal with the same syndrome but from a different angle. They highlight a series of aspects of transformation, which is supposed to induce environmental conflicts.

Steven Lonergan (Research Panel 1996) addresses five trigger mechanisms which encompass almost all dimensions causing disruptions in society-nature relationships in an arch, reaching from unintended to intended or planned activities. Michael Renner (1989) prefers a politico-geographical approach that allows him to analyze the effect environmental change has on political units at different levels of the international system.

Both approaches could easily be integrated. For instance, one can ask if cumulative changes (Lonergan) affect regional commons as a whole or only an ecoregion on a subnational level (Renner). The geography of environmental transformation determines the level of conflict in the international system. Thomas Homer-Dixon (1991) sorts out categories of renewables the degradation of which is supposed to induce violent conflict. ENCOP (Libiszewski 1992; Bächler *et al.* 1996) finally focuses on the historical dimension of changes in society-nature relationships, addressing the transformation of renewables as a cause of environmental conflicts and wars.

As can be demonstrated by case studies having been carried out mainly by the third and fourth group mentioned in the taxonomy, it is impossible to focus on one classification exclusively because each deals with specific aspects of the problem as a whole. For instance, overuse of a renewable resource (ENCOP) can consist of the degradation of land (Homer-Dixon). The degradation of land may affect a subnational or transboundary area (Renner), the transformation of which may be caused by cumulative anthropogenic changes, natural disasters, or by both, *i.e.*, drought (Lonergan). The result of this comparison is not at all surprising: every renewable resource (Homer-Dixon) is geographically bound (Renner), forms part of the historical transformation of societal-nature relationship (ENCOP), and belongs to a broader societal environment (Lonergan). Although each single taxonomy has its own advantages, it is obvious that empirical studies have to draw elements from all of them.

The three issues of the ENCOP-taxonomy can be formulated in an encompassing way, integrating the crucial aspects of the other classification schemes. The *first* dimension of the synthesis consists of transformation in both physical form and chemical substance. To these processes belong: (a) deterioration of the sources, *i.e.*, of renewable resources based on industrial exploitation of nonrenewable resources. This may happen through drilling for fossil energy and strip or underground mining or through large-scale industry and extensive or intensive agricultural use of land. Increasing demands put upon resources by rapidly growing populations accelerate transformation of nature's sources; (b) overuse of nature's sink capacity, namely through introduction of anthropogenic pollutants, *i.e.*, pollution of water, deposits of hazardous waste, deterioration of soils by chemicals, emission of greenhouse gases, etc.

The *second* dimension consists of transformation in an ecogeographical setting. This dimension refers to destabilization of well-adjusted living orders through a syndrome of transformation processes. For instance, soil erosion and drought, which deprive both farmers and stockbreeders of their means of production, and rapid growth of peri-urban settlement are expressions of comparable processes leading inevitably to impoverishment of society-nature relationships. Impoverishment also expresses itself in resettlement of population forced to abandon its original living space caused by industrial development projects, accidental disruption, or natural disasters.

The inherent tendency of local transformation to have consequences in a wider ecogeographical area may easily cause environmental deterioration throughout the world. Simultaneously, globalized transformation may induce serious regional or local damage. Greenhouse gas emissions are blamed for global climate change inducing regional as well as local transformation such as erratic precipitation patterns. Human action affects resources either due to home-made environmental deterioration or due to the impact of externalization caused by third parties. There are those affecting ecoregions within a single country (*e.g.*, soil erosion, deforestation, mining activities); those affecting ecoregions touching two or more neighboring states (*e.g.*, transboundary waterways, mining activities); those not directly affecting the ecoregion within one state nor shared by two states but regional commons in a wider geographical area (*e.g.*, acid rain, air pollution, dumping of hazardous waste); finally, those affecting global commons which, in turn, have an impact on both ecoregions within states and extraterritorial ecoregions (*e.g.*, stratospheric ozone layer depletion, global warming, rising sea level).

Whereas the taxonomies deal with the environment as the *independent* variable, there are typologies concerning the *dependent* variable, namely environmental conflict and war. The typologies on conflicts, however, are far less complementary than the taxonomies on the independent variable (*see: fig. 2.3.*).

Peter Wallensteen's (1992) concept aims at identifying the different stages during which ecological concerns emerge as a topic of political competition in democratic societies rather than a typology of distinct conflicts. However, it indicates that the environment matters in terms of inducing changes in both political institutions and in the setting of priorities of political actors.

	Types of environmentally caused conflicts
Wallensteen 92	1. Reduced resources available result in more contention in society at large; 2. shift of power between already existing parties; 3. formation of new parties as a reaction; 4. environment becomes more important for established parties; 5. environment becomes central in political affairs; 6. involvement of environmentally based groups.
Homer-Dixon 91; 94	1. Simple scarcity conflicts; 2. group identity conflicts; 3. relative deprivation conflicts; 4. environmental scarcity conflicts (1994) (population growth + resource distribution).
Bächler et al. 96	1. Center-periphery conflicts; 2. ethnopolitical conflicts; 3. interregional migration conflicts; 4. transboundary migration conflicts; 5. demographically induced conflicts; 6. international water conflicts; 7. long-distance conflicts (cumulative effects).

Figure 2.3. Typologies of environmentally caused conflicts

Thomas Homer-Dixon (1991, 1994) distinguishes environmental conflicts in Third World countries with the help of hypotheses provided by three general theories well-introduced in social science and psychology, namely economy dealing with scarcity of commodities, group identity theory, and relative deprivation theory. Although it seems quite plausible to explain three types of conflicts with three theories, the approach is too linear. The three theories do not necessarily explain three distinct types of conflict but rather three different *stages* in the genesis of one and the same conflict. An example may help to illustrate this. Environmentally induced migration does not exclusively and necessarily cause typical group identity conflicts (Homer-Dixon 1994:20). A group identity conflict may initially have been caused by a simple scarcity conflict, *e.g.*, about access to fertile land. In a second phase violent reactions may have urged people to flee to another area, where they got into trouble with people belonging to another ethnic group. Migration could also have resulted in clashes between refugees and border police or armed forces of a neighboring state. Thus by tracing a specific conflict, we recognize one and the same conflict passes through different stages, probably with different actors involved. At the same time, the pathways may be quite complex.

ENCOP (Bächler et al. 1996) provides a typology, which is inductively drawn from a set of about 40 area studies. It indicates the amalgamation of actor-oriented analyses and structurally defined arenas of conflict in terms of sociopolitical, ethnopolitical, and regional geography of environmental conflicts and wars. ENCOP did not start with complex theories and models but aimed at applying what is called »a gradualist approach to model-building« (Gleditsch 1997:8). Only in a second step did examination of each of the seven types lead to explanations that referred to hypotheses drawn from several theories, such as *dependencia*, political geography, anthropology, or development sociology. In order to understand and explain complex cases the configuration of several approaches was stressed rather than the linear connection between an individual type and one single theory.

2.3.2. A CONFLICT STORY CAN BE TOLD FROM DIFFERENT ANGLES

Some critics use shortcomings or »methodological flaws« in order to neglect the environmental conflict approach as such, maintaining that there are no interesting mechanisms that are purely environmental because environmental factors interact with a variety of other factors to spawn violent conflict (Levy 1995:45). According to Levy, one has to focus on the »contention« for state power as a fundamental factor of present violent conflicts. It is evident that power contention increased sharply after 1987 and »that power contention was and is the source of much more severe conflicts than ethno-nationalism or indigenous rights« (Gurr 1994:355). However, only 18 out of 50 ethnopolitical conflicts fall in the category »contention for power« (*ibid.*), which means that there are other important issues to study as well. After Ted Gurr a strong case can be made by linking ecological stress, inequalities, and conflict affecting indigenous peoples. It would therefore be »unwise to discount the present or future significance of economic factors, however, because conflicts tend to be more numerous and intense in regions and countries where systemic poverty is greatest« (*ibid.*:358-359).

The suggestion made by analysts (Levy 1995) to concentrate on events on »how nations get into war« - and not on structural root causes - is somewhat biased. Stories of proximate events are tremendously important in order to understand the dynamics of a conflict, but there is a tendency to confuse the outward appearance of a conflict with its root causes. Focusing, for instance, on political or military leadership without understanding the deep motivation of the actors one would fail to explain the emergence of disasters as experienced by hundreds of thousands of refugees in the triangle between Eastern Zaire, Burundi, and Rwanda in 1996. Rwanda, but also Somalia, Liberia, and Afghanistan remind us to ask why there are actually no *nations* going to war but, obviously, many ongoing violent conflicts and wars. The struggle for power is certainly a crucial problem above all in the young states of the international system, where the starting point for social, economic, and ecological transformation as well as modern leadership is in the mid-1950s (Lane/Ersson 1994:170). Since these transformation processes began only a few decades ago, it is all the more important to understand and explain them in the light of historically shaped socioeconomic and cultural-ecological factors - both being closely intertwined. Hence, explaining encompassing patterns of current conflicts does not compete with the strategy to trace back particular causes as part of the causation process of violence.

Thus far, relatively little research has been done on violent conflict and war as a chronic development problem, with few exceptions only (Arnold 1991; Matthies 1988, 1994, 1995a,b; Senghaas 1989; Brown 1996). The association of conflict and poverty has often been excluded in development sociology and economics. Only recently Dan Smith provided a study, which focused on this very relationship, recognizing the problem of identifying single causes of current violent conflicts and wars in the South without being accused of »monocausality« (Smith 1994:47). Although dealing with foreign debt and domestic poverty as causes to conflict the approach is nonetheless helpful in many ways to the study presented here. First of all, Smith insists on both the uniqueness of each single conflict as well as their common features. Concentrating systematically on the causal relationship between violent conflict and foreign debt structure shows that there is »in some cases at least, a connection between debt and violent conflict based on the mechanism of response to economic reform and austerity measures« (Smith 1994:61). The problem is not monocausal explanation, but rather that pure statistical analysis is only the very first step in explaining causes; causes that may even change over time. Only dense case studies that focus on certain causes in a multidimensional arena can fill this gap.

Secondly, Smith stresses the fact that the multiplicity of conflict causes underlines the need for »more focused, comparative studies of conflicts, tracing their genesis within the full context - social and political, economic and environmental, domestic and international« (*ibid.*: 64). Even though one may fully agree with this statement, some difficulties remain when dealing with this request in practical research. Full-context studies can only build on an extremely complex model of linear and nonlinear causalities adding to the picture a great number of variables. If the program, for instance, is to trace the »strong association of poverty and conflict« (Smith 1994:64), we can do so by correlating violent internal conflicts and the country ranking of the Human Development Index of UNDP. However, the country rank is determined by five indicators only, excluding many other variables that would indicate the status of a country in a broader framework.

In comparative approaches the full context often turns out to be wishful thinking. It consists of images, shared meanings, and conventional wisdom, and it depends on data provided by agencies, which try hard to get more valid and, above all, comparable data from each single country. This is especially difficult in the hot spots, hence, the very cases conflict research is normally most interested in.

2.3.3. »SOMEONE'S COW ATE SOMEONE ELSE'S CROP«

Another difficulty is that even parties to a conflict may not fully understand the causes of their own struggle or war. This happens because of »a difference between the causes for which they fight and what it is that causes them to fight« (Smith 1994:54). This difference is well understood by Beth Roy (1994) in her rich empirical field study carried out in Bangladesh. To briefly illustrate her case, which fits excellently into the argumentation of this study: according to Roy the causation of the conflict under consideration was that »in a remote village somewhere in South Asia, someone's cow ate someone else's crop« (Roy 1994:1). Was this the very cause that explains why the villagers were ready to fight an almost all-out war on a community level? Of course not. The villagers were ready to fight because they got a green light from both political and traditional leaders to do so. The leaders did not mediate between the struggling parties because they had the impression that the times »have changed« (*ibid.*:64). Indeed, local leaders had information about similar fights going on elsewhere in the country. This meant that »waves of history« splashed over into a normally quiet place. What happened far away was that Bangladesh - where the village is located after separation from first India and then Pakistan - had become an independent state. The Muslims - among them the owner of the cow - started to complain: »The Muslim did not have any existence, the Hindus were all (...) The elderly Mussalmans had all suffered a lot of loss during the pre-partition days. But they could not make any protest during British time«

(*ibid.*:65). The implication is clear: the cohesion of oppression - or of the collectively shared feeling of being oppressed - created a bitterness among Muslims, so they did not care to compromise any longer. They got a strong feeling of »now the state is ours, it is our turn. Now we'll teach them a lesson«. Or, as Beth Roy concludes, »to translate power on a state level into power on a local level was the task of the day« (*ibid.*). Hence, step by step, an atmosphere was created, which made violence acceptable, possible, and even necessary:

> »People knew, however vaguely and inaccurately, that outside their village forces were at work which corresponded to their own experience. Pakistan was formed. Pakistan represented the culmination of other Muslims' grievances. They knew those wrongs had to do with a state unfavorable to their people. Precisely in what way it was unfavorable was only vaguely understood - it had something to do with job discrimination and exclusion from schools. All that part of the story took place in a class universe far distant from the villagers. But it harmonized with their own experience. They too felt excluded and oppressed. Their landlord and moneylenders, the economic powers in their world, were Hindu, and the erstwhile state was somehow felt to be a Hindu state, even though it was known to have been British. In any case, now the state was Muslim. Those Muslims so far away in town, so far away across India, speaking another tongue and wearing a different garb, those Muslims had risen and won a new state. Such monumental events struck chords of harmony with the desires of the villagers in Panipur. (...) Between the winds of history and the storm in Panipur came some process of synthesis and decision« (Roy 1994: 70-71).

Roy provides a good example of how to study both the gap between causation and causes or reasons as well as the genesis of a conflict in a certain arena heavily influenced by events and perceptions stemming from within or outside this arena - one made up of individual and collective actors who aim at mobilizing their cultural, social, and economic capital in the hope of playing an active role in policy design and receiving a significant part of the cake (either material or power). The causal pathway between cow-eats-crop and almost all-out war on a local scale consists of different pieces. There is the economically motivated fight among neighboring farmers about scarce resources; there is an element of social class struggle between top-dog caste Hindu and underdog Muslims; there is a local political component because the leaders - in whom people trusted - tolerated the resort to violence; there is an ethnopolitical dimension induced by the historically developed cohesion of oppression within the Muslim community; and there is, finally, an important shift of regional power in South Asia, which led to the creation of a new state by the underdogs. The proximate trigger which caused them to fight - namely cow-eats-crop - is not difficult to explain.

However, the other pieces indicate that strong distant triggers are at work - causes which peoples found worth fighting for. The distant trigger is found somewhere in history, not only in political history (this would be too reductionistic an approach) but also in social and environmental history. The rural population has played a crucial role throughout Indian history. The cow-eats-crop conflict is just one out of thousands of similar conflicts. This explains why it is not only interesting as an individual case. It is also interesting as an indicator of heavy socioeconomic pressure felt by struggling neighbors when the cow started eating crops. Rural poverty, unproductive agriculture, and land degradation through overuse or mismanagement - and not »simple scarcity« - have induced social and political conflict in India over centuries; the syndrome of factors is therefore supposed to act as a crucial distant trigger. One can further assume that the farmers in our story were not directly struggling for a change of power in South Asia; instead, their ultimate interest concerned land and crops. And we further know that especially in India there were major obstacles to rebellion explained by Barrington Moore within the structure of the predominant type of village society in India (Moore 1966:330-335). However, this is why the intervening variable *foundation of a new state* in Roy's story cannot be overestimated. If state power - far away in the center - had not shifted in favor of the Muslim community, the cow owner would definitely not have summoned up the courage to protest against the Hindu neighbor capturing his crop-eating cow; nor would he have had the collective support of both the Muslim community and its leaders.

Against the background of Muslim farmers *vs.* Hindu neighbors the foundation of Pakistan must also be viewed as a necessary condition. Some crucial observations concerning the environment and violent conflict program can be made in this regard:

- even though full context analysis may be a fiction, it is obvious that single root causes to a conflict must be seen in the multidimensional arena where the play occurs;
- the cause of a conflict is not necessarily identical with the causes the parties involved believe they must fight for;
- the perception of causes can change during the genesis of a conflict; ethical deterioration - driven by external events - can create new fault lines during the escalation period, which - at a certain stage - may be perceived by the actors as the »real« causes they fight for;
- conditions only contributing to a conflict are not interesting in terms of scientific analysis. What counts is to find out whether - in a certain arena - either necessary or sufficient conditions (or both) exist;
- the analysis of causal pathways - and therefore determination of root causes - depends heavily on the subjective decision of the analyst (the »storyteller«) on how to trace back the development of a crisis and on the point of time to let the pathway start in history.

2.3.4. »THE EFFECTS OF PERSISTENT DROUGHT ...«

Let me briefly illustrate the latter point with a well-documented conflict which often serves as an example in the debate on environmental conflict: the Mauritania-Senegal border dispute that broke out in 1989. The effects of persistent drought in the Sahel in the mid-1980s led to the governmental decision to build a dam and a barrage in the Senegal River Valley. The decision, once taken, had unintended consequences (in terms of Lonergan's taxonomy; *see: fig. 2.2.*): land prices in the region increased significantly. Thus the Mauritanian Arab elite did not hesitate to change the land-tenure system immediately after becoming aware of the possible future benefits of denying black African farmers, herders, and fishermen use of their resources, as they had done for ages. So much for the facts. Now to the analysis of the conflict genesis. As perceived by analysts, the genesis of the conflict usually begins with the sentence: *»The effects of persistent drought ...«* The question arises if this beginning suffices. Is it correct to begin the story with persistent drought? If so, does it lead to the conclusion that - in this case - we are confronted with a typical environmental conflict? Was it not the change in the land law by the white elite, which was responsible for the brink-of-war situation? On the other hand, it could also have been ethnic division induced by the expected increase in land prices that triggered escalation of the conflict. Or was it the clear decision to build a dam? Or could it have been the rather unintended consequences of the elite's decision?

Depending on the answers to these questions the conflict might be classified as either a group identity conflict, as contention for power in the state, as a modernization conflict, or as a simple scarcity or environmental scarcity conflict (Smith 1994). If we maintain this to be a typical environmental conflict caused by drought and probably triggered by the decision to build dams, then we are confronted with a definition problem. Since environmental conflicts are caused by *anthropogenic* transformation of the environment - at least in the approach chosen in this study - the question arises if the Sahel drought of the 1970s and 1980s resulted from anthropogenic climate change or »natural« influences upon the climate. If the latter is the case, is it still an environmental conflict? The question is of some significance because the Senegal River Valley served as a case of reference in the analyses of most participants in the discussion on environmental security and violent conflict (Dokken 1995; Smith 1994; Homer-Dixon 1994; Dessler 1994; Bächler/Spillmann 1996III). Since there is still great uncertainty about the regional impact of climate change, the answer to the question is rather pragmatic: causation - both man-made and natural - may be perceived as a natural catastrophe by human beings affected by persistent drought. Therefore, the perception of actors in the arena may be more important in terms of conflict dynamics than the best scholarly explanation. On the other hand, as soon as means of problem solving are found, it makes a difference. In the case of anthropogenic transformation it may be easier to find ways out of the trap of unsustainable resource use than to deal adequately with the challenge of »natural« climate change.

This leads finally to the question of how best to escape the problems of causal analysis addressed with the example of Mauritania *vs.* Senegal. There are three answers: *Firstly*, one must carefully trace causal pathways. This must be done by highlighting the syndrome of factors involved within a given arena and in a historical perspective. Concerning the Senegal example, this means one must look behind the decision to build a dam and to analyze the political and societal processes that led to the decision. Was it really persistent drought? Or was it the expectation of benefits in terms of land price? Perhaps there would have been better strategies than building a dam in order to overcome drought in the Sahel; strategies which might not have fueled an existing syndrome. *Secondly*, one must identify the different but interacting roles played by a syndrome of factors in a given arena. The *third* answer concerns rejection of any attempt at a quasi-functional analysis of causes in the sense that the Sahel drought would have the function of modernizing rural societies, which would have to be seen, *per se*, as a violent transformation process.

Any study of causal linkage must focus on an institutional and actor-oriented approach by dealing with positions and perceptions, interests and decision-making processes.

2.3.5. CAUSATION IN COMPLEX POLITICO-ECOLOGICAL SYSTEMS

Concerning tracing pathways and interacting roles of causes the *Environmental Change and Acute Conflict Project (ECACP)* took a major step forward. It aimed at combining empirical research by carrying out a set of case studies with a systematic approach focusing predominantly on causal linkages (Homer-Dixon 1991, 1994; *fig. 2.2., 2.3.).* The ECACP study group was definitely not interested in the whole range of factors that currently cause changes in the value of the dependent variable »conflict«; instead, it wanted to know whether and how a hypothesized independent variable, in particular »environmental scarcity«, causes conflict (Homer-Dixon 1995:2). The research team identified four social effects of ecological degradation, which may lead to violent conflict, namely decreasing agricultural productivity, economic decline, migration, and disruption of political institutions and social relationships. Against the background of eight case studies they drew the following conclusion: »Scarcities of renewable resources are already contributing to violent conflicts in many parts of the developing world« (Homer-Dixon/Boutwell/Rathjens 1993:38). Environmental degradation is supposed to trigger internal violent conflicts and wars; only under exceptional circumstances does it lead to wars at an interstate level. This is because resource conflicts are mainly related to degradation of renewables, such as land, water, forests, and fisheries in a local or regional context, whereas the hypothesis that ozone layer depletion and climate change induce violent international conflicts was falsified.

Acknowledging that environmental scarcity always operates together with other political, socioeconomic, and cultural causes, the analysts are confronted with the problem that all these causes may be *contributory* or *necessary* conditions, but not necessarily *sufficient* ones. Therefore, multicausality as such is not the most challenging problem, but rather the relationship between the variables, ranging from being fully endogenous to being fully exogenous. David Dessler confirms that »the central epistemological issue does not concern the presence of environmental change in these causal processes. No one denies that environmental change is somehow 'involved'. The issue is rather the *importance* of environmental change in producing violent conflict« (Dessler 1994: 96).

There are three different standpoints concerning the role of environmental scarcity. Firstly, environmental scarcity can be an important force behind changes in the politics and economics governing resource use (*see: fig. 2.4a*; fig. 2.4a,b,c taken from Homer-Dixon 1995:5).

Figure 2.4. Degrees of causal independence

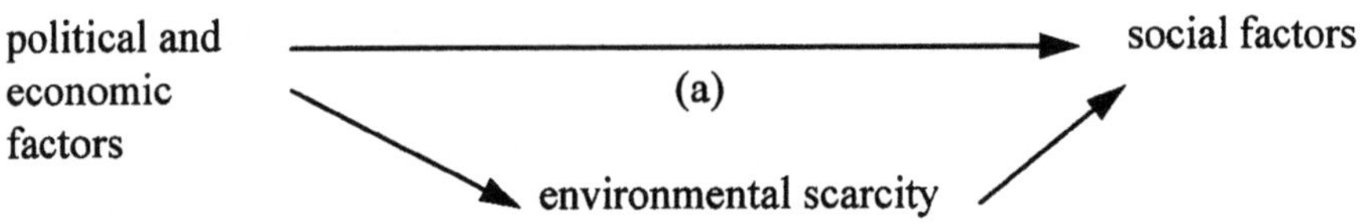

Secondly, ecosystem vulnerability is often an important variable contributing to environmental scarcity, and this vulnerability is, at least partly, an external physical factor that is not a function of human social institutions or behavior (*see: fig. 2.4b*).

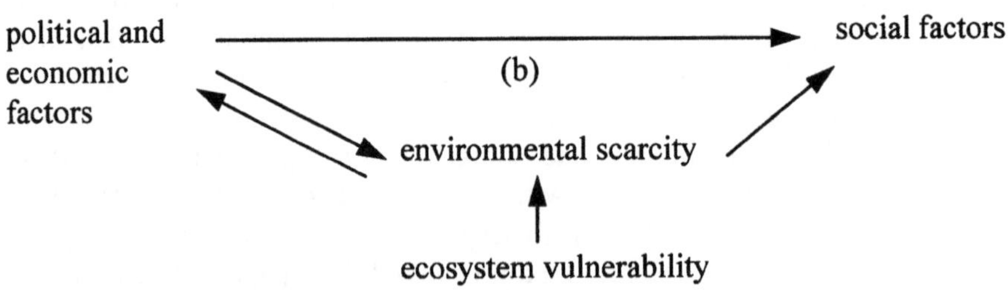

Thirdly, in many parts of the world environmental degradation has crossed a threshold of irreversibility. Even if enlightened social change removes the original political, economic, and cultural causes of degradation, it will definitely be a continuing burden on society: »Once irreversible, in other words, environmental degradation becomes an exogenous variable« (Homer-Dixon 1995:4; *see: fig. 2.4c*).

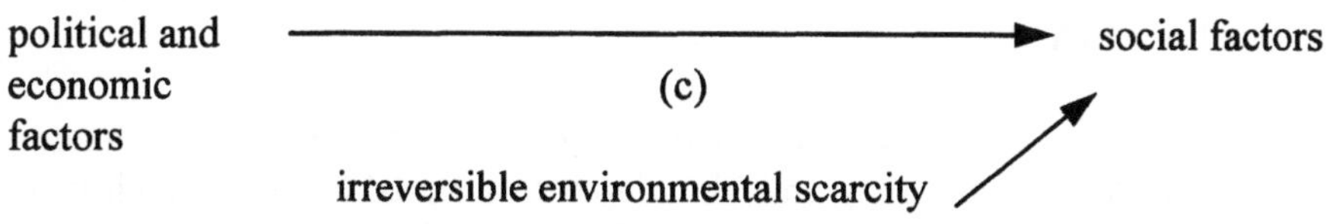

The degree of *exogeneity/endogeneity* of the independent variable is only to be determined by tracing back the causal linkage through rich case studies and/or by model testing. To come back to our Mauritania/Senegal case: what is the degree of exogeneity of the Sahel drought: is it (a), (b), or (c)? One might come to the following conclusion: (a) is to be excluded because each analysis begins with *»persistent Sahel drought«* stressing predominantly the fact of ecosystem vulnerability. There seems to be ample evidence that the drought is an independent variable rather than a contributing factor to the syndrome; it is supposed to play a major role in the arena. However, if it is (b), it could also be (c), namely irreversible environmental scarcity perceived as a fully exogenous variable, the role of which dominates the arena. It is difficult to decide whether it is (b) or (c) because there are no scientific tools available so far to predict irreversibility.[7]

Hence, the choice depends heavily on perceptions of the actors in the arena. If, let us say, a community of farmers perceives drought and its effects on fertile land as irreversible and reacts accordingly - by abandoning the fields for instance - we may chose (c) as the degree of independence we have to deal with. As demonstrated by a huge comparative study in the Shewa-Wello Province, one of the most drought-prone parts of the Highlands of Ethiopia, the farmers firmly believe they suffer »under God's environment«. Indeed, the deeply rooted belief in God as »the fountain of authority and power« insulates those in power from any accountability to the people. »In famine or war, the peasants almost never charge any government or institution with responsibility. Nothing, including their abject poverty and suffering, happens without God's

7 There is a tendency in international discussions to state that environmental degradation is in principle reversible in terms of physical or chemical recovery. However, from the perspective of chaos theory, for instance, which stresses linearity of time, every single impact induced by human action is irreversible; change is always and everywhere cumulative, one reason why the term transformation was introduced (*see: 1.1.*).

will« (Wolde Mariam 1991:181). And since people are prepared to accept responsibility for their conditions, the situation does not lead to social action »because the basis of responsibility is personal rather than social« (*ibid.*). This is a good example of how drought - which is perceived exclusively in terms of a sequence of crop failures - becomes an exogenous variable.

2.3.6. CAUSAL RELATIONSHIP BETWEEN INDEPENDENT AND DEPENDENT VARIABLE

In the above section a scheme was provided, which can help clarify the connection between the independent and the dependent variable. The conditions involved are - according to system theory (Easton 1965) - contributing, necessary, and sufficient conditions. A contributing or *c-condition* (*see: fig. 2.4a*) is a condition, which contributes to occurrence of some event or state of affairs. In comparative politics economic factors, for instance, are supposed to be without exception c-conditions, »which means that other factors need to be recognized, allowing for complexity« (Lane/Ersson 1994:108). A necessary or *n-condition* (*see: fig. 2.4b*) for an event or state of affairs posits that, if the event takes place or the state of affairs holds, the condition will certainly be present or, to put it the other way round: if the condition is not there, the event will not occur that way. A condition is both necessary and sufficient, or an *s-condition*, when in its absence an event or state of affairs would simply not occur (*see: fig. 2.4c*). If we could identify conditions, which are both necessary and sufficient, we would have come a long way towards explaining causes of violent conflicts. Most often we have to be content with identifying c-conditions only (Lane/ Ersson 1994:108).

A reference to the dependent variable (I) of this study, namely violent conflict and war, and to the independent variables listed under (II), may illustrate the difference between c-, n-, and s-conditions:

(I) DEPENDENT VARIABLE:
Violent conflict and war

(II) INDEPENDENT VARIABLES:
- *(a) physical conditions: environmental degradation*
- (b) economic conditions: affluence, structure, growth
- (c) social conditions: homogeneity vs. heterogeneity
- (d) cultural conditions: family structure (lineages)
- (e) political conditions: institutions, rules, actors

Focusing on the question what kind of condition environmental degradation (independent variable, IIa) in a specific conflict consists of, we get to the following alternatives:

- if variable (IIa) is a c-condition, then it necessarily has an impact on either variable (IIb), or on (IIc), or on (IId) etc., or on all other variables (IIb-e). All these factors, being together with (IIa) c-conditions, contribute to the occurrence of the dependent variable (I).
- If variable (IIa) is a n-condition, then it has an impact on either variable (IIb), or on (IIc), or on (IId) etc., or on all other variables (IIb-e). At the same time, it has independently a direct impact on variable (I) as well. It causes changes in the independent variables (II b-e) as well as in the dependent variable (I), regardless of the changes induced in (II b-e).
- If variable (IIa) is a n- as well as a s-condition, than it has a direct impact on variable (I), whereas the impact of variables (IIb-e) on the dependent variable (I) is only minor and thus to ignore.

If there are any interactions between I and II, but the correlation is rather weak, then we are confronted with either c- or n-conditions. Emphasis has to be put on cases of environmental conflicts where degradation is a n-condition. Although it is obviously difficult to identify the *full* set of causally relevant factors in a definitive way, focusing on n-conditions takes multicausation processes into account as encompassingly as possible. Interestingly enough, newer approaches in comparative politics found that physical (geographical, climatic) variables do significantly matter concerning the issues state stability and state performance. There is, for instance, statistical confirmation for a combined socioecological model dealing with climatic impact and family structures. The correlations between dry climate, traditional family structures, and a weak state »are so pronounced that these factors have to be looked into. Even though there is no obvious causal mechanism at work here, these findings require more interpretation« (Lane/Ersson 1994: 147;199). But not only the question to what extent environment matters is crucial (in terms of regression coefficients), but rather how it matters. It is still an open question to what extent the degree or severity of environmental scarcity directly influences the dependent variable. I assume this can only be determined with the help of rich case studies combined with generalizations and model building.

Another problem concerning domestic conflicts is to determine relevant indicators and to quantify them. Mesfin Wolde Mariam's survey in the Ethiopian Highlands was confronted with this very problem when he asked: »How do we measure total environmental quality? And how do we relate the cultivation of crops and the animal and human population to this environmental quality and measure the strain on it? This is something that requires the cooperation of individuals from various disciplines as well as field experience« (1991:172). To complicate matters even more, not only states, but also parapolitical systems, such as local and regional governments, kinship groups, and NGOs play a role of one kind or another in a given arena.

There is a final point to be addressed. There is some criticism of the procedure for selecting the dependent variable. Critics maintain that one should not only select cases in which environmental scarcity and conflict both occur, but also - in controlled testing of hypotheses - cases without conflicts. Homer-Dixon rejected this kind of criticism by justifying the case selection with the complexity of the environmental conflicts program. Since many other issues in conflict research are also complex, a more convincing argument must be found to counter the argument. More important is that there is no such thing as a 0-hypothesis in the sense: »Environmental scarcity does not lead to conflicts«. The reason is that *scarcity* as a notion drawn from economic theory *per se* implies some sort of conflict. If resources are perceived to be scarce by more than one actor, there is conflict. Therefore, in order to fulfill the postulate of introducing control cases, one has to be more precise on what we mean by conflict. The adequate controlling after all is to choose cases, where environmental scarcity is linked either to *nonviolent* conflict (0-hypothesis) or to *violent* conflict. This distinction is designed to help explaining the *why* and the *how* of the difference it makes (*see: fig. 2.5.*).

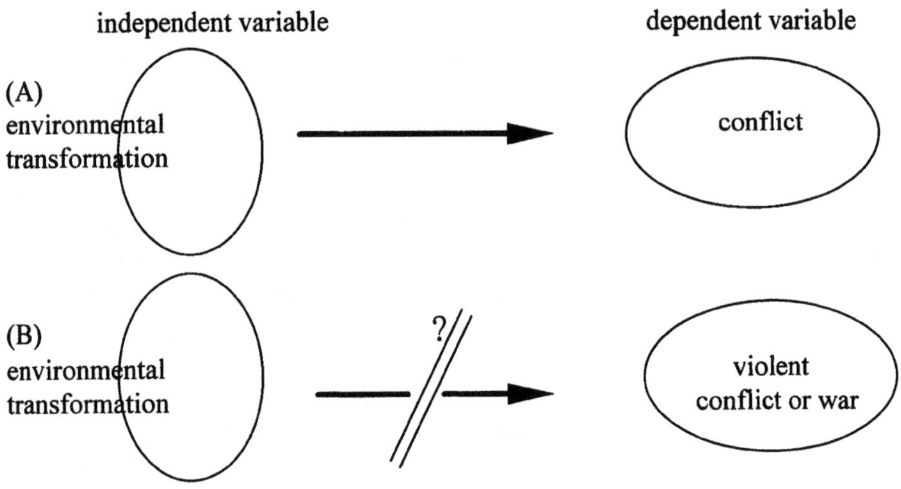

Figure 2.5. Independent and dependent variable

To summarize: in an initial step of a cumulative research program with many projects involved it is methodologically legitimate to focus the first empirical programs of this kind only on causal linkage (B), environmental scarcity, and violent conflict, for two reasons: almost all violent conflicts that can be analyzed now occur in the South. The causation of violent conflict there is ill-understood in general, and this is true for environmental scarcity and violent conflict in particular. Linkage (A), which is a very common one in industrialized countries and to some extent also in developing countries, is far better documented. Concentration on violent conflict is important in order to provide both a better understanding of the dynamics and adequate tools to prevent or mitigate violence. Only sound empirical ground enables analysts to compare linkage (A) as control cases with linkage (B). Based on both case studies as well as a set of indicators we will be able to answer why some countries succeed in peaceful conflict and resource management and others do not (*see: 7*).

2.3.7. ACTION AS A RESULT OF TWO FILTERS

Since the causal contribution of environmental transformation to any given conflict can only be analyzed as part of a syndrome of factors in a given arena, the analytical task is to develop a method to determine the contribution of a particular type of factor, environmental transformation, to a specified type of outcome - such as violent conflict and war.

There are different ways to determine the importance of a factor. For instance, the concept of *probability trajectory* as discussed by David Dessler (1994). This is a causal pathway along which some probability of the outbreak of violent conflict exists. In other words, a causal factor is to be considered important in proportion to its quantitative contribution to the probability of violent conflict breaking out. The problem with this approach is that »we have no way of accurately measuring the probability of conflict at a particular point in time, nor of making fine-grained judgements as to changes in the probability of conflict from one moment to the next« (Dessler 1994:99). Neither trajectories of causal pathways nor accurate constructions of counter-factual pathways would provide a viable basis for sorting causes in empirical analysis (*ibid.*:99). Moreover, *multiple* causation would appear to be a major problem at every point *along* the curve.

There is another trap to fall into, well-known in social science as the *post hoc ergo proper hoc* trap. Why should we believe that an event (a) and an event (b) relate to each other as cause and effect? For instance, how can we know that it was the Sahel drought of the mid-1970s that triggered violent conflicts between farmers and nomads in the Jebel Marra mountains (Sudan) a few years later? It is only theory that enables us to link two events substantially and not only occasionally. We must have a certain knowledge and understanding of the situation and the link must be analytically plausible.

Dessler (1994:103) proposes an intentional-actor model of human behavior in order to get robust and testable results in the environmental conflicts studies. Human action is the product of two successive filtering operations (*see: fig. 2.6.*). The first filter separates what is feasible from what is not; it determines the content of the feasibility or opportunity set of an actor. As such, feasibility indicates a hard separation between what is physically, psychologically, economically, and socially possible for a person or a collective actor from what is not. The second filtering operation results in an action drawn from the opportunity set. In other words, two mechanisms co-react in this process: the first one is rooted in individual and collective rationality, the second one in social norms.

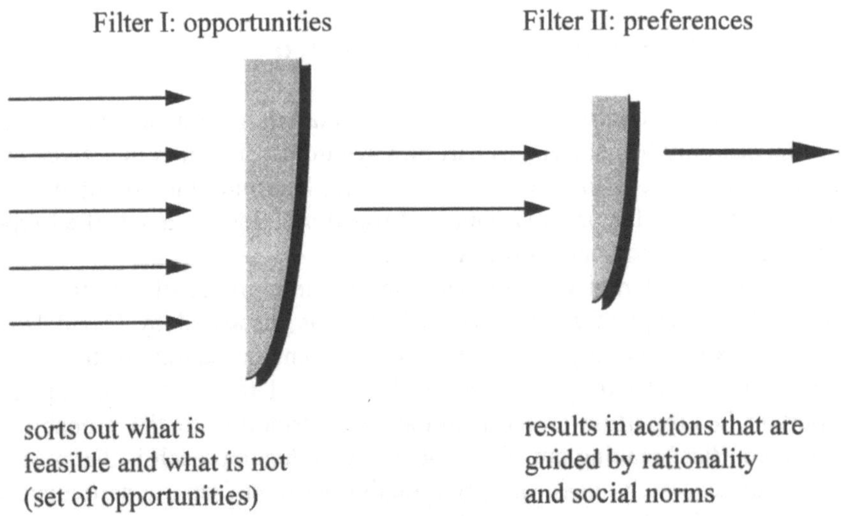

Figure 2.6. Action as the result of two filters
(taken and adapted from Bächler et al. 1996:56)

Through these double-filtering mechanisms an individual or collective actor selects a certain course of behavior. Against this background, a choice is seen as rational when it is perceived as the best means of reaching a given goal and when it is chosen because this goal exists (Dessler 1994:103). The distinction between rationally and normatively guided action seems to be crucial regarding rural actors in developing countries. Belonging to traditional communities they are supposed to share strong beliefs in traditional rights. Since there is some evidence that primary organizations or ascriptive structures such as kinship are more important than secondary organizations of such unions (Lane/Ersson 1994:114), one has to bear in mind that social norms are shared beliefs about proper and improper behavior in defined circumstances: »An action is norm-guided when it is meant to conform to such beliefs, rather than to serve an instrumental goal. Social norms (...) encompass standards of behavior directly relevant to the dynamics of intergroup conflict - such as norms of revenge for instance« (Dessler 1994:103; see also: Elster 1987, 1989). We may call this approach *soft rational choice* (R. Jervis) because it hypothesizes neither totally objective rationality nor complete rationality (Zürn 1992:13). Simultaneously and even more important, it is focused on collective actors and not exclusively on individuals. Regarding a particular case study, we would ask how individual behavior results in collective action and how collectively shared views become social phenomena that urge individual reactions of some kind.

Both independent and dependent variables influence the options of action and the set of preferences in various ways. Actors may understand the social and economic constraints of resource depletion as a major challenge and concentrate the available human capital on development of new technologies to compete with the challenge. On the other hand, they may have a completely different order of preferences and therefore decide to fight against those actors they perceive as responsible for resource degradation, discriminated resource access, or general mismanagement of natural resources. The different options depend heavily on the feasibility set or on the perceived capacities to deal with a problem.

The two filter models of »soft« rational behavior of both individuals as well as collective actors lead us to formulate two pairs of research questions; these are as follows (*see: fig. 2.7.*):

	opportunities	preferences
environmental transformation	how does environmental transformation influence the opportunity sets of the actors?	Why does environmental transformation influence the preferences of the actors?
Violent conflict and war	how do violent conflict and war influence the opportunity sets of the actors?	why do violent conflict and war influence the preferences of the actors?

Figure 2.7. Two sets of research questions

Both pairs of questions lead to complete causal-intentional explanations of action including their effects, be they intended or not (Zürn 1993:13-14).

2.3.8. TO SORT OUT CAUSES BY ATTRIBUTING SPECIFIC ROLES TO THEM

Now, after having formulated the four research-guiding questions, we still miss a systematic approach to trigger mechanisms. Considering the role environmental change plays in causal linkage, there are - as we have seen above - assumptions that the deteriorating environment either contributes to instability, heightens political tensions, aggravates existing conflict, or is a catalyst for hostilities (Ulrich 1989:5). Only recently the fuzzy »folk« concepts of causal verbs, such as aggravate, amplify, trigger, etc. were problematized (Homer-Dixon 1995:6). In order to overcome cloudy concepts, David Dessler provides a helpful system of four different roles causes play in a given arena. I adapt this concept by linking it to a case study on Rwanda in East-Central Africa. This arena study is designed to demonstrate how helpful identification of roles is empirically and to see if it leads to explanations of a more systematic character (*see: 6*).

Concerning the first role - the trigger of a conflict - Dessler's distinction between a *distant* trigger and a later or *proximate* trigger is somewhat problematic. An earlier or distant trigger is considered to be low on the probability curve. It should create situations in which later or proximate triggers lead to events much closer to the outbreak of violence and therefore supposedly higher on the probability curve. If a trigger is seen as »any event causing actions that increase the probability of conflict«, Dessler is unintentionally confronted with the same difficulties that led him to reject the concept of the probability trajectory. As he pointed out correctly, we cannot describe any event that increases probability of violence without having precise tools to measure it (*see: 2.3.7.*).

There is another problem related to the first one: if a trigger is the intended result of a choice made by actors who prefer violence to other means to reach a goal, why then is the distant trigger considered to be very low on the probability curve? In the distant/proximate trigger concept distant triggers are always considered weak, fading away in the clouds of the past, whereas proximate triggers are supposed to be strong in any case because they just occurred. First, this is not necessarily so, and secondly, this concept bears the risk of an endless regression: any later trigger mechanism can be explained by a previous trigger mechanism; it is a never-ending story of triggers becoming ever weaker by growing passage of time. To avoid this one better distinguishes between *reasons* and *triggers* as a time-neutral concept.

Dessler's slightly modified concept consists now of five causal roles, which are defined as follows (Dessler 1994:104-105):

(1) A *reason* is a combination of actions that are perceived by an actor as »historical problems« (traumata, history of oppression, injustice, former

wars, etc.), which influence his preferences in a way he thinks justify the resort to violence historically. The reason - however distant or immediate - could be a strong factor in mechanisms leading to the outbreak of violence. Reasons play the mighty background role, which shape normative behavior more than rational behavior. A reason works on both the content and the dynamics of a conflict.

(2) A *trigger* of an action is an event, which causes that very action to become the most favored alternative in someone's feasibility set. In terms of violent conflict, a trigger causes an actor, who previously preferred nonviolent solutions to a problem, to favor violent action instead. A trigger must always be seen near the outbreak of a violent conflict or war under consideration. It is part of the cause, whereas reasons are, by contrast, the causes actors fight for. Triggers are the outcome of decisions that led to violence - even though the latter is not necessarily intended. Violence could also be the unintended result of a strategic threat by an actor or of a perceived threat by another one. A trigger works foremost on the dynamics of a conflict.

(3) »A *target* is an actor's objective, aim, or goal. The target is what the conflict is about« (*ibid.*), at least in the eyes of the parties to the conflict. If we can define the target(s), we can explain the purpose of the conflict and see why an actor prefers resort to violence. A target works on the content of a conflict more than on the dynamics.

(4) »A *channel* is a line of political, social, economic, or national cleavage (...) To cite a channel is to explain the social, political, economic, and/or cultural structures that cause individuals to fall into the groups they do« (*ibid.*). Channels are designed to form the group identity. They figure primarily in subintentional explanations (why actors have certain beliefs and desires) and supra-intentional explanation (why individual actions have certain collective or cumulative effects), whereas triggers and targets figure in explanation of intentional action. For instance, a reason can be - and often is - a strong motive to stress fault lines that can be traced back to both recent or ancient history and that are ideal to build channels along which groups can more easily be divided. A channel can be strong enough to influence causal linkages, so that new reasons arise during the genesis of a conflict. A channel is working on both the dynamics and the content of a conflict.

(5) »A *catalyst* is any factor that controls the rate or intensity and the duration of a conflict, once initiated« (*ibid.*). A catalyst might serve to lengthen a conflict if it stabilizes opportunities and preferences of violence in a given conflict. It might cause a conflict to become extremely violent. Ethical deterioration in a conflict can itself be a catalyst inducing more violence.

Together with channeling, a catalyst can also transform reasons. Ethnic groups, for instance, that may have had an economic reason to fight each other may - as ethical deterioration goes on - perceive differing ethnicity as a reason rather than the resources they fight for. A catalyst works on both the dynamics of a conflict and its content.

2.4. Research Strategy

The five causal mechanisms give us the tools to explore causal linkage more systematically. In order to counter-check the causal linkage analyzed in the ENCOP case studies (Bächler/Spillmann 1996II, III), hypotheses provided by ECACP are applied. According to this project the independent variable *environmental scarcity* consists of environmental problems related to forests, water, fisheries, and cropland. With the exception of fisheries, which are excluded from this study, these are the constituent factors of the landscape conceptualized in chapter 1 as a life-support system (*see: 1.2.*). Beyond this, ECACP includes scarcity caused by population growth, resource maldistribution, degradation, and depletion (Homer-Dixon 1994:17-18). The team produced the following results bearing on three hypotheses:

(1) There is little empirical support for the first hypothesis that environmental scarcity causes simple scarcity conflicts between states. Scarcities of renewable resources, such as forests and croplands, do not often cause resource wars between states.

(2) There is substantial evidence to support the hypothesis that environmental scarcity causes large population movements which, in turn, cause group identity conflicts.

(3) Empirical evidence partially supports the third hypothesis that environmental scarcity simultaneously increases economic deprivation and disrupts key social institutions. These, in turn, cause deprivation conflicts, such as civil strife and insurgency.

The conclusion is reached that environmental scarcity causes violent conflict; conflicts that tend to be persistent, diffuse, and subnational (*ibid.*).

Causal analysis of political and social events or states of affairs is confronted with a range of difficulties related to multicausality, identification of appropriate indicators, and availability of data. It is therefore legitimate to isolate one cause of a multiple-cause event, if one bears in mind that it is only part of a syndrome analysis in a given arena. The conduct of studies dealing with environment in particular is necessary as it is not interesting *if* environment matters - as part of a multicausation pattern - but rather *how* and *why* it matters and what the specific contribution of the environment is to a given violent conflict or war.

Acknowledging the probability nature of social science knowledge, one should try to find traces of connections between variables, which make sense in a causal interpretation. Since there is no single widely accepted paradigm for conduct of research in this field, a cumulative step-by-step approach may be appropriate, which combines the advantages of each single methodology. It is acknowledged in scholarly literature - especially in comparative politics - that various methods from single-case studies to multivariate modeling should be tested for whatever usefulness they may offer (Lijphart 1971).

The *first step* begins with hypothesizing simple *correlations*, tracing a first clue, which may already bear some circumstantial evidence. This can only suggest where to look for causes. Correlation simply means that there is some kind of interaction between the dependent and independent variable. There is a high possibility that the interaction is spurious, meaning that some other variable accounts for the interaction. Nevertheless, one must interpret the empirical associations so that they make sense from a theoretical point of view, even though co-variation is not a safe tool to justify causal hypotheses (Lane/Ersson 1994:187).

Therefore, this step can be complemented by the conventional approach dealing with broad configurative descriptions (*second step*). One should by no means underestimate the heuristic value of in-depth case studies especially in a relatively new field of research, where major constraints always develop because of a lack of sufficient data to build models and test them. Case studies support behavioral analysis of actors by looking into the action that really occurs behind political ideologies. Political phenomena are substantially affected by other kinds of social phenomena, such as economic development, physical environment, social stratification, and cultural system. However, socioeconomic and ecological conditions do not affect political behavior in a direct way. There is a crucial property of politics, of behavior of political leaders, of their capacity for adaptation and innovation, and of their significance for explaining properties of political systems, such as stability and conflict regulation. Apart from structure we therefore have to cover institutions and actors as well (*ibid.* 1994: 105).

In a *third step*, it is necessary to trace back causal relationships in order to provide insights on a specific conflict history and to generalize the findings of a single analysis. Causal analysis is therefore necessary but not sufficient. In order to avoid the *post hoc ergo proper hoc* trap, tracing pathways of conflicts has to be combined with an actor-oriented approach that conceptualizes positions, interests (aims), and perceptions of both individual and collective actors - hence, actions undertaken by actors in the light of their aims, interests, and perceptions, as well as outcomes resulting in interplay between those actions. This should lead to generation of inductively won typologies drawn from area studies and to some general causal patterns, which may provide ample evidence of the supposed causal linkage.

Thus the step-by-step approach suggested in this study consists of the following sequences:

chapter 3: formulates simple correlation hypotheses by associating current conflicts with data on significant patterns of socioeconomic development as well as on disruption of the environment. In the first part the correlation between environmental conflicts and *desertification* or *soil erosion in arid zones* is analyzed. In the second part supposed correlations in the context of historically transforming society-nature relations in sensitive *mountain ecozones* are highlighted;

chapter 4: re-evaluates case studies provided by ENCOP. It aims at checking the hypotheses formulated by the ECACP;

chapter 5: designs an in-depth case of the conflict syndrome in the arena of Rwanda by using Dessler's roles concept. It aims at identifying more systematically the causal linkages in one single case;

chapter 6: builds the model of environmental conflicts and wars based on results provided by the chapters 4, 5, and 6;

chapter 7: checks the model with the help of six arena studies and six control studies carried out either by ECACP or ENCOP teams;

chapter 8: concludes and discusses both the methods and the final results in the light of different levels of knowledge achieved in the previous chapters.

How and to what extent do environmental causes contribute to violent conflict? And why do they contribute to violent conflict and war? These are the leading questions readers of the following pages should bear in mind.

CHAPTER 3

CORRELATIONS

Environment, Maldevelopment, and Violent Conflict

Conflicts have always existed in human societies. What changes over time are the sources of conflicts, the forms that conflicts take according to available arms technologies, and the types of parties to a conflict, such as tribes, states, ethnic groups, social groups, etc. Finally, conflict behavior has also changed. Today environmentally caused conflicts appear in the light of the development dilemma, that is the specific contradictions in the societal nature relationships produced by maldevelopment. The sheer coincidence of *underdevelopment, transformation of landscape*, and *violent conflicts or wars* in many regions of the South and East leads to the supposition that analysts are not confronted with an isolated phenomenon in one specific area alone. Many local or regional forms of conflict conform to a general basic pattern. Their origins possibly stem invariably from qualitative and quantitative changes in transformed ecoregions, which impose stress upon human habitat. The coincidence of the three phenomena may therefore indicate that meaningful correlations exist.

There are different methods to analyze correlates of violent conflict and war: purely *quantitative* ones and more *qualitative* ones. For instance, a long-term project by David Singer and his team has become a backbone of quantitative conflict research (Small/Singer 1982). An increasing amount of quantitative research is based on different models and catalogues of indicators, which focus on particular crises, *e.g.*, ethnopolitical conflicts (Gurr 1993), genocide (Harff 1992; Gurr/Harff 1989), human rights violations (Jongman/Schmid 1994), or forced migration (Schmeidl 1995). All these studies of different segments of the international conflict geography share the advantage of being able to draw relatively accurate data from different available sources (World Bank, UNDP, UNHCR, Amnesty International, etc.).

Concerning the environmental conflict program, however, it is impossible so far to use any methods involving either metric or non-metric statistics to test correlation or even a complex model of causation. Comparable disaggregated data are hardly available on environmental degradation within countries. It is difficult to measure the impact of human behavior on natural resources on a local level. Beyond this, the sample of conflicts analyzed so far in case studies (Bächler/Spillmann 1996 II; III) is too small for any kind of elaborated statistical analysis. One could, for instance, suggest building country groups from 1 to 10 according to different percentages of arid land or high mountain coverage of the overall territory and correlate the ranking with the degree of violence. However, possible outcomes such as »the more arid land, the higher the conflict intensity« seem to be too crude a result. One would have to find more fine-tuned indicators such as development of soil productivity in certain ecoregions over a certain time - including significant events (such as drought). In addition, one would have to sort out intervening variables on socioeconomic change within subnational regions in order to regress violent conflicts against these variables.

Despite the data constraints quantitative analysis in the environmental domain faces, a simple concept of correlating the relatively small number of present violent conflicts and wars with existing data provided by various UN agencies may at least provide some rough guidelines on where problem areas are and on where analysts have to go deeper into detail (*see: 3.1.*). However, the focus of this study is clearly on qualitative research. In the second part of this chapter pathways are traced and patterns of configurations are identified. It concentrates on two major types of landscapes or life-support systems at risk, namely *arid lowland arenas* and *mountains and highland arenas*. The assumption is that violent conflicts mainly occur between actors depending on the use of these kinds of sensitive landscapes highly exposed to transformation.

3.1. Human Development Correlated with Violent Conflicts and Wars

There are different global statistics based on national social indicators that can be useful here. For instance, the World Bank provides a country-by-country review of over 170 economies with up to 94 indicators reported for each country including determinants of population growth, natural resources, income and poverty, expenditure on food, housing, energy, etc. (World Bank 1994; World Bank: World Data on CD-ROM). The World Resources Institute (WRI) can be used as a supplement to get basic data on environment and landscape transformation (World Resources Institute 1994). UNDP with its Human Development Report (HDR) provides country ranking based on an index composed of five indicators of human development (UNDP 1995, 1996).

To start with I compiled a register of violent conflicts and wars in the last three years (between January 1994 and December 1996). This list will be correlated with the country-ranking system of HDR called Human Development Index (HDI). This step leads to building different groups of war-torn societies in order to determine any significant conflict and maldevelopment correlations.

3.1.1. WAR REGISTER AND HDI-COUNTRY RANK

War is »generally understood to embrace only armed conflicts on a fairly large scale, usually excluding conflicts in which fewer than 50,000 combatants are involved«. For a study of modern wars - for instance on the African continent - this clear-cut definition of a classic war taken from the *Encyclopedia Britannica* is too rigid a demarcation. Carl von Clausewitz sees war as an act of violence intended to compel our opponent to fulfill our will (Clausewitz 1976: 90). This qualitative description applies very well for modern wars, whereas his view that war is an armed conflict between rationally acting states or leaders of states with both a declaration of war and capitulation of the defeated is a classical Eurocentric characterization, which does not hold any longer. This would imply that in today's Africa there are almost no wars going on. After a broader definition provided by Small and Singer, »a sustained military conflict has to array at least one sovereign member of the interstate system, and results in at least 1,000 battle deaths of the system members participant(s)« (Small/Singer 1982:50; see: 51-55). However, even this expanded view of the phenomenon of war is still too narrow. After World War II, but even more after decolonization most wars have been internal with often less than 1,000 battle deaths, while the percentage of civil casualties has been increasing tremendously.

A broader concept originally provided by Istvan Kende will be applied here; *war* is seen as an armed, violent mass conflict following a planned strategy, encompassing the following three constitutive qualitative criteria:

(1) it must be a conflict with a minimum of continuity (rather months than days);
(2) there must be central organizations on both sides (this could also be a paramilitary or guerilla force);
(3) at least one of the parties to the war must be a government with regular or at least government-associated troops (Kende 1982:5; Gantzel 1987:33).

Violent conflicts must be distinguished from war as defined above. Regarding their intensity, duration, degree of organization, and involvement of governmental forces violent conflicts remain below the threshold of war. They are defined as repetitive outbursts of organized political violence intended and committed by collective actors in which quite a number of people are killed (military unrest, rebellions, riots, etc.). However, since almost all current wars are internal ones, it is often difficult to draw an arbitrary line between intrastate war and domestic violent conflict. Some violent conflicts can be seen as the forerunners of a war, whereas some wars can cool down to stop-and-go violent conflicts with no continuing major clashes between regular forces and other groups. But if we adapt too broad a concept of violence, where then should the line be drawn between violent conflict and criminal, police, or domestic (private) violence? However difficult a clear-cut distinction may be, the lists compiled of both wars and violent conflicts are clear and encompassing enough to respond to the purposes of this study (*see: app. 2, 3*).

I will begin with correlating violent conflicts/wars and human development. For this purpose the Human Development Index data (HDI) 1996 compiled by United Nations Development Program (UNDP) will be used. The HDI relies on national estimates reported by the United Nations and its agencies and by other internationally recognized organizations. It is heavily dependent on the standardization and consistency of data produced by these offices. Thus some important limitations remain in the coverage, consistency, and comparability of data across time and countries; for instance, estimates on population size and growth are often derived from old censuses. A general problem is available data aggregated exclusively on a national level. Figures on average national income may enhance the rank of a country with a strong economic center by ignoring absolute rural poverty, which would lower the country rank significantly. UNDP therefore blames »the extremely limited availability of data on such important human development issues as poverty, gender equality, crime, violence, HIV/AIDS, maternal mortality and the environment« (UNDP 1996:132). Disaggregated human development indices at the level of different groups, gender, ethnic groups, regions, etc. would presumably confirm that there is no automatic link between growth and human development: »Growth has been failing over much of the past 15 years in about 100 countries, with almost a third of the world's people. And the links between growth and human development are failing for people in the many countries with lopsided development - with either good growth but little human development or good human development but little or no growth« (UNDP 1996:1).

Despite data restrictions HDI is the best source available to correlate war-torn societies with human development because it builds on a set of five indicators: life expectancy at birth; adult literacy rate; combined first-, second-, and third-level gross enrolment ratio; real gross domestic product (GDP) per capita; and adjusted real GDP per capita in »purchasing power parity US dollars« measured and compiled by the International Comparison Programme (ICP) (*ibid.*).

3.1.2. INTERPRETATION OF THE TABLES AND FINDINGS

A total of 45 wars went on between January 1994 and December 1996 (*see: app. 2*). At the same time, 44 violent conflicts occurred (*see: app. 3*). Thus some 89 violent conflicts and wars have been going on throughout the world - most of them are still manifest. If one correlates these events with the country ranking drawn from HDI one gets some significant findings in terms of unequal distribution of violence among states (*see: app. 4*). Indeed, the global conflict geography is biased towards the South. Those states with high human development are far less conflict-prone than states with medium human development, let alone those ranking at the bottom of HDI.

Out of 57 high-ranking countries with a HDI between 0.799 and 0.999, only 11 are indeed conflict-prone; that corresponds to a conflict-proneness of 0.193 (*see: app. 5*). The high-ranking group encompasses the OECD countries, the Eastern European countries, the Community of Independent States (CIS) as well as Israel. Most violent conflicts or wars occurred in the latter three groups of countries, whereas the OECD segment was almost totally free of organized violence with the exception of Great Britain *vs.* Northern Ireland.

The segment of medium-developed countries encompasses 69 countries. In 31 of them there were violent conflicts and wars; this corresponds to a conflict-proneness of 0.449. The medium-developed countries' segment is the most heterogeneous one: we find countries with relatively high development and an established democracy but also a tendency to low human development, relatively poor state performance, and political instability. Some of the world's remaining tough military regimes are found in this country group. Thus some of the 38 medium-developed countries are highly at risk concerning sociopolitical instability and outbreak of violence. In general, conflict intensity measured in casualties and refugees tends to be higher in this country group than in the group with low human development (*ibid.*).

The group with low human development - between 0.1 and 0.499 - is most affected by violent conflicts and wars. In 44 countries out of a segment of 48 there was violence; thus the conflict-proneness is 0.854 (*see: app. 4, 5*).

The correlation is quite strong: present violent conflicts and wars chiefly occur in the segment of the least developed countries, secondly in medium-developed countries characterized by poor state performance and unstable political conditions (*see: app. 5*). If one compares regional aggregates of human development indices the trend of violence being characteristic of low human development is even more significant. Sub-Saharan Africa is at the peak regarding war-torn societies (*see: app. 10*). With the lowest regional average in human development (0.379), the regional aggregate of political freedom is also one of the lowest: 35.6 index points in comparison to 54.5 in Latin America (*ibid.*). Low HDI obviously coincides with little political freedom and poor state performance. This can be further illustrated if one takes other factors into consideration. With 5.4 million refugees Sub-Saharan Africa does not only account for almost one third of the world's refugees; with 9.4 million displaced people it also hosts about half of the world's internally displaced estimated at 20.8 million (*ibid.*). This is the highest migration rate per capita worldwide;

since the figures are based on conservative estimates, the rate is certainly higher. Especially the great number of internally displaced indicates that migration is not only a phenomenon of war-torn societies, although violence is the most important factor (Schmeidl 1995).

I suggest that a significant portion of the migrants stems from an *ecological* push or »man's inability to cope with natural forces« (Peterson 1958:259). An indicator supporting this argument can be drawn from figures of changes in total rural population. In 1960 about 85 percent of the total population in sub-Saharan Africa settled in rural areas. In 1993 only 70 percent of rural dwellers remained. As Susanne Schmeidl's comparative study has clearly shown, most migration to date is forced migration by either structural or physical violence or both (Schmeidl 1995:50). What applies for sub-Saharan Africa also holds for other regions affected by a relatively great number of violent conflicts and wars, such as the Arab states with the lowest political freedom index (PFI) of 24.0 points, South Asia, Southeast Asia/Pacific, and Latin America/Caribbean (*see: app. 10*). In all these regions the numbers of internally displaced exceed those of refugees, the only exception being South Asia. But here most of the 4.5 million refugees stem from one country: Afghanistan.

I compiled another data set to highlight what is behind these structural forces, which determine the regional aggregates on rural population dynamics, migration, low human development, little political freedom, poor state performance, and violence. For this purpose I compared 20 countries subdivided into four groups with five representative countries each. In accordance with the diverging conflict-proneness of country groups indicated in table 5 (*see: app. 5*) I selected five with high HDI and no ongoing violence; five with medium HDI and no ongoing violence; five with medium HDI and occurring violence, and five with low HDI and occurring violence. I then identified factors, which indicate access to natural resources being a factor of low HDI and violence. I compared social deprivation, demographic trends and labor force, trends in resource use, and political-institutional performance of states (*see: app. 6-9*). One major finding of this exercise is that the five countries with high HDI in all factors differ considerably from all the other 15 countries. The most significant difference is between countries with high and low HDI, whereas the difference between medium and low HDI is not as clear-cut as between those with medium and high HDI. Between the five medium-developed countries without violence and those five with violence no arbitrary line that can be drawn either.

To go more into the detail: the most striking difference between high, medium, and low HDI countries is the percentage of labor force in agriculture. In 1990 the labor force in agriculture amounted to 14.6 percent in the average of five high-ranking countries; in the ten medium-developed the average was 44.3 percent, whereas in the low-ranking countries 65.6 percent (it dropped from 78 percent in 1960; *see: app. 6*). The decreasing but still high percentage of rural labor force in the low HDI countries is an indicator for high dependence on natural capital, such as fertile land and water. At the same time, countries not highly developed in general and countries with very low HDI in particular have a common trait: urban arenas are generally better off than rural ones. Access to safe water in rural areas is far less guaranteed than in urban arenas. In some countries it is virtually at stake: in Haiti, for instance, only 23 percent of the rural population have access to safe water, in Nigeria 26 percent in rural compared to 63 percent in urban areas; in Algeria and in Ecuador 55 percent compared to 82 percent in urban areas.

Absolute poverty is definitely a rural problem. Heavy rural exodus to peri-urban arenas does not mean the poverty clusters have dissolved completely. There is still a high percentage of rural population living in absolute poverty with bad access to resources of any kind. In Rwanda, for instance, the labor force in agriculture dropped only slightly from 95 percent in 1960 to 92 percent in 1990, whereas rural poverty is 90 percent in comparison to 30 percent in the few urban areas. In Nigeria, the rural labor force dropped dramatically from 73 percent in 1960 to 43 percent in 1990. However, rural poverty remains high with 51 percent compared to 21 percent in cities. But also daily calorie supply per capita is not always higher in rural areas, as one might expect. In some cases it even dropped from 1965 to 1992 (Rwanda, Haiti, Nigeria, Guatemala, Zimbabwe, Mongolia). In some countries food production per capita was much lower in 1993 than between 1979-81: being fixed as 100 in 1980, Rwanda achieved only 70 percent in 1993, Haiti 67 percent (*see: app. 7*). Despite rural poverty and rural exodus, rural labor force is still growing in absolute terms due to population dynamics. The extremely high unemployment rate in rural arenas of low HDI countries indicates few off-farm opportunities (*see: app. 13a*). Relatively high annual population growth rates aggravate regionally high population density. In Haiti it increased from 211 people per km^2 between 1980-85 up to 247 people between 1989-94; in the Philippines from 182 up to 219. However, population density does not seem to be a good indicator for rural poverty and violence, since it differs tremendously regardless of the country ranking (*see: app. 7*).

Trends in natural resource use are diverse and therefore a more appropriate indicator than population density (*see: app. 8*). In Rwanda the area of cropland increased 8.5 percent in ten years from 1981-83 to 1991-93. During the same time, cropland available per capita decreased from 0.19 to 0.15 ha. During the same period of time forests disappeared at a rate of 4.8 percent in ten years. This indicates the clearing of marginal land in remote and hardly accessible mountain areas. Pasture - historically a domain of the Batutsi livestock breeders - shrank dramatically by 15.2 percent during these ten years. We will elaborate further down if and how this is an indicator for ethnopolitical conflict in Rwanda (*see: 5*). In Pakistan there was an overall cropland increase of 12.4 percent, whereas the hectares under cultivation per capita decreased from 0.21 to 0.16. However, pastures remained stable, whereas forests even increased. The annual rate of deforestation in most countries is relatively high, but there are no significant differences among the country groups (interestingly enough Canada, having repeatedly been criticized by INGOs for its deforestation policy, provides no data). In almost all countries the availability of cropland per capita decreased.

One may argue that this also applies for highly developed countries which, as a rule, have relatively low per capita croplands under cultivation. In Switzerland, for instance, it amounts to 0.07 ha per capita only (*see: app. 13a*). Even if this is generally true, there is the relative dependence on natural capital, which makes the difference. Highly industrialized countries with little cropland per capita depend on services, industries, import of food, and high agricultural productivity. The dependence on the natural capital is relatively low because technical and socioeconomic alternatives are available. In threshold or takeoff countries the dependence on the traditional as well as on the modern rural sector is still significant but decreasing, since there are substitutes available. Rural societies with few off-farm opportunities and little means to finance market alternatives depend highly on their natural capital. This concerns rural production schemes of low productivity, access to freshwater, patterns of land use and tenure systems, and also energy consumption. Most of the population depends on the land, whereas production of food and basic products depends heavily on noncommercial energy use. *Commercial energy use* per capita is therefore an outstanding indicator for economic development in general and for independence from renewables in particular.

The *»resource availability indicator« (RAI)* - drawn from the multiplication of cropland per capita in hectares and commercial energy use (CEU) in kilograms per capita (oil equivalent) - is supposed to adequately indicate both environmental stress in general as well as environmental discrimination against rural producers in particular (*see: app. 13a, b*). Rwanda, with 92 percent labor force in agriculture, has the lowest RAI of 4.0 (0.15 ha x 27 CEU). With 4.7 Bangladesh is extremely low as well. Niger, with 90 percent labor force in agriculture, has a RAI of 15.9 (0.42 ha x 38 CEU). The Philippines, with 60 percent labor force in agriculture, have a RAI of 45.9 (0.14 ha x 328 CEU). To compare: Germany, with 4 percent labor force in agriculture, has a RAI of 625.5 (0.15 ha x 4,170 CEU). Canada, with 12,357 and 5 percent labor force in agriculture (1.58 ha x 7,821 CEU), has the highest RAI worldwide because it has both high cropland availability per capita and high commercial energy use (*ibid.*).

The picture is more clear-cut than expected: countries with high percentage of labor force in agriculture, little cropland available per capita and low commercial energy consumption per capita belong to the group of countries with both high environmental stress and violent conflicts or wars going on. There are a few countries with low RAI and no violence. Here, a low political freedom index might indicate that a repressive regime is capable to suppress any organized violence. In China, for instance, the PFI with 8 is extremely low and the RAI 49.8. On the other hand, in India the PFI with 63 is significantly higher than in China, whereas the RAI with 45.9 is at a similar level but there is no major environmental conflict occurring either. Thus the PFI does not totally reflect the specifics of a political system in a country, nor does the RAI take cultural specifics in resource use, climatic differences, etc. fully into account. The constraints of statistical analysis provide once again a reason for carrying out in-depth case studies.

Military expenditure (MILEX) as percentage of combined education and health seems to be a good indicator for poor state performance and not necessarily for ongoing violence. In 1990-91 the average percentage in the five highly developed countries of our sample amounted to 17.6 percent. In the medium-developed countries without violence it was 22.8 percent and in those with violence the ratio was only slightly higher, namely 24.8 percent. In the five low-ranking countries the average with 87 percent was significantly higher than in the other groups (*see: app. 9*). These countries belong to those with low RAI and PFI, with a high degree of labor force in agriculture; they do not have democratically elected governments or a multiparty system; and the political participation of women in government at the ministerial level is weak (however, the latter is the case in most countries of the other groups as well; *see: app. 9*).

Concerning military expenditure and violence a further finding is noteworthy. Rwanda, which experienced the most intense conflicts in the 1990s, shows a very moderate MILEX *vs.* education and health ratio of 25 percent (to compare: Canada 15 percent, the Netherlands 22 percent, Poland 30 percent). Myanmar, on the other hand, claims the highest ratio worldwide of 222 percent. Thus a high ratio seems to express a highly stable repressive regime with poor state performance. An extremely oppressive military regime seems to be an inhibiting factor for major outbreaks of violence, even in societies with significant ethnic fragmentation and sociopolitical fault lines. High intensity violence - such as massacres and genocide - does not necessarily correlate with high military expenditures compared with education and health.[1]

---

1 As will be shown in chapter 5, Rwanda's MILEX sharply increased after 1990-91, that is, after the intervention of the Batutsi forces from Uganda. However, MILEX was still relatively low in 1994 (*see: 5.2.6.*).

Finally, the correlation with the type of regime and regime stability shows quite good results (*see: app. 9*). All countries without wars, including the medium-developed ones, have elected multiparty systems. The other ten countries of the sample are confronted with high political instability. The large group of medium-developed countries with violence shows extremely diverging tendencies: there is no clear trend towards democracy, multiparty systems, and high PFI nor to the opposite. Struggle for power seems to be one characteristic of this group of countries. In the low-developed countries there is neither democracy established nor are there multiparty systems. Changes of regime in 1990-95 did not occur through elections but through wars or coups d'état.

There is *ample evidence* for *correlations* between rural poverty, environmental stress in rural arenas, discriminated access to resources, and poor state performance. These factors coincide with political instability, *i.e.*, non-democratic change in regimes and/or violent conflicts and wars. These findings apply particularly for developing economies highly dependent on the natural capital of sensitive ecoregions. The results feed the supposition that the correlation between sensitive dryland and mountain transformation and violent domestic conflicts at a regional (subnational) level may be even stronger than it could be shown with the help of aggregated data. In the following rural poverty related dryland transformation will be examined in more detail. After that, historical and cultural-ecological aspects of the specific conflict-proneness of mountain economies will be highlighted.

3.2. Violent Conflicts and Wars in Arid Lowlands

3.2.1. ARID ZONES, POVERTY, AND CONFLICT

Throughout modern history there have been massive changes in use of fertile land worldwide. Forests, wetlands, grassland savannas, and arid lands have been altered profoundly in area and composition. Especially the area of human settlement and agriculture as well as the scale of timbering, mining, and other land transforming activities have grown enormously. A modeled estimate of world changes in land use over the past three centuries calculates that the world's forests and woodlands diminished by 1.2 billion ha or 19 percent of their total in 1700. Grasslands and pastures have declined by 560 million ha or 8 percent of the 1700 estimate, whereas croplands brought into cultivation show a net increase of 1.2 billion ha, that is a 466 percent increase in less than three centuries (Richards 1990:163). The reverse side of the transformation of landscape is that about 2 billion ha of formerly productive land were irreversibly lost, whether transformed into badlands or covered with water, buildings, or asphalt. About 1 billion ha of land suitable for agriculture remain as arable land reserve for human development, scattered in small patches throughout the continents and occupying more or less larger areas in savannas and forests in Africa and South America (Rozanov *et al.* 1990:205).

Transformation of Drylands
Major changes have occurred in the humusphere. In nature, transformation of humus proceeds rather slowly and marked changes require many decades or centuries. In comparison, transformation processes induce sharp changes in the ecological equilibrium. Within the past 300 years the total areas and quality of productive soils available for human and biotic use have decreased at an accelerating speed. There are five major causes of transformation processes, namely deforestation, overuse, overgrazing, croplands, and industries (Oldeman *et al.* 1991; *see: tab. 3.1.*). Anthropogenic degradation of the pedosphere occurs in several ways: 56 percent due to water erosion, 28 percent due to wind erosion, 12 percent due to chemical, and 4 percent due to physical degradation (*ibid.*).

degraded surface (billion ha)		*causes of transformation in percent* deforest.	overuse	overgrazing	croplands	industries
world total	1.96	30	07	35	28	01
Africa	0.49	14	13	49	24	<01
North/Central America	0.16	11	07	24	57	<01
South America	0.24	41	05	28	26	<01
Asia	0.75	40	06	26	27	<01
Europe	0.22	38	<01	23	29	09
Oceania	0.1	12	<01	80	08	<01

(Source: Oldeman et al. 1991)

Table 3.1. Causes of land transformation

Today *soil erosion* is one of the major environmental challenges worldwide. *Desertification* - as a specific type of soil erosion - is a process of sustained decline of biological productivity of arid and semi-arid land (Gorse/Steeds 1988:ix). It is the result of human-induced changes in soil organic matter, which transform cultivated soils into anthropogenic badlands. Desertification has been defined in the UN Plan of Action to Combat Desertification as «land degradation in arid, semi-arid and dry sub-humid areas resulting mainly from adverse human impacts« (UN 1977).

About 64 percent of the world's arid land are located in Africa and Asia. The present status of desertification is shown in an assessment done by UNEP (*see: tab. 3.2.*). At present, desertification affects 3.6 billion ha (Oldeman *et al.* 1991), that is about 70 percent of potentially productive drylands, or nearly a quarter of the total land area of the world (Tolba/El-Kohly 1992:134). Some 1.964 billion ha or 17 percent of such soils are severely degraded by human action (UNEP/ISRIC in: Oldeman *et al.* 1991).

	million ha	percent of total drylands
1. degraded irrigated lands	43	00.8
2. degraded rain-fed croplands	216	04.1
3. degraded rangelands (soil and vegetation)	757	14.6
4. drylands with human-induced soil degradation (1+2+3)	1016	19.5
5. degraded rangelands (vegetation degradation without recorded soil degradation	2576	50.0
6. total degraded drylands (4+5)	3592	69.5
7. non-degraded drylands	1580	30.5
8. total area of drylands excluding hyper-arid deserts (978 million ha) (6+7)	5172	100.0

(Source: Tolba/El-Kholy 1992:136)

Table 3.2. The extent of desertification in the world's drylands

The most obvious impact of desertification is degradation of 3.3 billion ha of the total area of rangeland, constituting 73 percent of the rangeland with a low potential for human and animal carrying capacity; decline in soil fertility and soil structure on about 47 percent of dryland areas constituting marginal rain-fed cropland; and degradation of irrigated cropland, amounting to 30 percent of the dryland areas with a high population density and agricultural potential. The question if and how soil erosion of millions of hectares of drylands is related to violent conflicts will be answered by focusing on dryland arenas in Africa, addressing transformation in the Sahel in particular.

Sahel (shore, coast) means the zone between the North African Sahara Desert and the grasslands. It encompasses the arid belt with 100-400 mm of precipitation per year and the semi-arid zone with 400-800 mm. The distance between Senegal and Eritrea is 6,000 km. With a latitude of 400 km from north to south, it covers an area of 2.32 million km^2. With a population of only about 44 million, the major traditional economic activity is nomadic and semi-nomadic pastoralism, which is highly adapted to the ecological conditions of the Sahel. Spatial and temporal variations in rainfall are accommodated by animal movements. The most important of these is *transhumance*, *i.e.*, the seasonal movement that occurs during the rainy season to fresh pasture in otherwise dry areas. Transhumance is followed by retreat to dry-season grazing near permanent water sources. But population growth in the more densely populated semi-arid areas has pushed cultivation into drier arid regions, transforming the environment more rapidly. At the same time, population pressure has increased demand for fuelwood, leading to increased degradation of vegetation.

Dryland transformation has its roots in poor state performance, which led to unmanageable changes of the socioeconomic and demographic structure. Many drylands could support population growth, rising demand for livestock and cash crops, and introduction of new technology. This was the case if and when traditional local authorities played a major role in establishing and protecting land-use and tenure systems in order to avoid both overuse and major conflicts. Natural resources were held in common rather than owned by individuals. One reason was that resources in arid and semi-arid areas are inadequate to support large populations with high density clusters. Especially in the arid belt, animals had to be moved over long distances and in unpredictable patterns. Subsidiary management of resources and conflicts changed completely during the colonial period and beyond, with centralization of political authority in the newly formed states. The process of centralization greatly reduced local land management by establishing and enforcing systems of individual land-use rights (Stryker 1989:90-93).

Centralization, privatization, and population dynamics have severely complicated the maintenance, enforcement, and adjudication of land use and tenure. Because of increasing demand animal prices have increased and thus encouraged expansion of individual economic activity within the livestock sector in the arid belt. Market economy has permitted ownership of animals to be transferred to people not subject to traditional forms of control. On the other hand, options for cash crops have increased pressure on cultivable soils in semi-arid areas. In addition, mechanization, new methods in water supply, and advancement in animal health have intensified pressure on traditional production systems. As a consequence, farmers have encroached on arid areas and herders have faced certain controls and sanctions to prevent them from causing crop damage in farmers' fields before the harvest. Rural unemployment has also been increasing because labor-intensive forms of land management - *e.g.*, terracing steep slopes or using herders rather than fences to guide animal movements - have been withdrawn simultaneously.

Thus the undermining of local authority, first by colonial regimes and later by centralizing authority in national governments, has weakened traditional systems of resource and conflict management (Beaumont 1993:57-96; Stryker 1989:94; Galaty/Johnson 1990:21-25).

Environmental Degradation in »Poverty Clusters«

Livelihood insecurity - having historically been a problem for traditional societies to cope with - has become worse under today's heterogeneous economic systems. One dimension is uncontrolled use of scarce resources and competition among farmers from semi-arid areas and (semi-)nomadic pastoralists from the arid belt. A second dimension is the marginalization of these producers altogether. This means central states with poor performance force them to settle in highly vulnerable ecozones. Impoverishment of the natural environment and social impoverishment go hand in hand once the producers find themselves prisoners of a vicious circle leading to their further discrimination as transformation goes on. Against the background of continuing retreat of poverty into low-potential agricultural areas and exploding peri-urban zones environmental discrimination has various aspects of hardship: poverty in female-headed households (Ardayfio-Schandorf 1995:3-15), landlessness of peasants, use of marginal and unproductive soils, lack of jobs in the modern agricultural sector, rural exodus to peri-urban areas, etc. (Leonard 1989).

The majority of the 780 million »poorest of the poor« live in sub-Saharan Africa and South Asia, and 370 million of the world's poorest live in rural areas where reduction of agricultural productivity, land-surface degradation, and water scarcity constitute a fundamental constraint on potential food production and therefore on rural self-reliance in tropical drylands (Falkenmark 1993:428; *see: app. 6*). In such landscapes the poorest people are more tightly clustered by regional location than they were several decades ago when poverty was pervasive throughout the developing countries: »Instead of being ubiquitous across the landscape, poverty is in many developing countries more and more concentrated into definable geographical areas. In many of these areas, poor people occupy marginal or ecologically vulnerable lands that lack appropriate infrastructure and technology« (Leonard 1989:22).

Socioeconomic Changes in Arid and Semi-Arid Zones

Major forces of socioeconomic change in arid and semi-arid regions are population growth, increased demand for marketed products (mechanized farming), and displacement of local political authorities by national governments. Malin Falkenmark et al. define the dilemma in the African »hunger crescent« as follows: short growing season; high risk of recurrent droughts due to inter-annual irregularities of the Intertropical Convergence Zone; strong soil degradation due to vulnerable soils subject to agricultural overexploitation; and very limited water surplus left after evapotranspiration to feed aquifers and rivers. She suggests that a crucial factor in driving rural poverty into the drylands is the recurrent drought reducing average crop yields and leading to land degradation, reducing rainwater infiltrability of soil (Falkenmark/Rockström 1993: 427-437).

There are basically three different land-use systems in drylands - pastoralism, rain-fed agriculture, and irrigated farming - that lead to erosion of the landscape. The following catalogue of major anthropogeographical factors of desertification, drawn and compiled from several sources dealing with dryland management, indicates the major obstacles transforming rural societies are confronted with (Leonard 1989; Stryker 1989; Sperling/Johnson 1990; Harborth 1992; UNEP 1992; Beaumont 1993; Falkenmark/Rockström 1993). The three systems are:

(a) *pastoralism*, with the following factors: overgrazing and uneven grazing; increasing number of animals; vaccination leading to higher herder productivity; burning off old vegetation to make tender; large concentration of cattle during the dry season denuding the landscape; construction of public wells and boreholes with unlimited access all over the Sahel, leading to extensive degradation around water points; governmental efforts to settle herders around water sources; increased orientation towards cattle and sheep (which are less adapted to the environment than goats and camels); growing demand of urban arenas for animal products from arid and semi-arid regions; rising prices causing a shift away from subsistence herding to raising animals for sale; increasing ownership of animals by farmers and traders expanding into areas of (semi-)nomadic stockbreeders;

(b) *irrigated farming*, with the following factors: badly managed irrigation; salinization and water logging; sterilization of huge areas of fertile land; reduction of available land for food crops because of increased cash-crop production; bad drainage systems: exploitation of scarce ground water or fossil water in motorized pumping stations; reduced infiltrability of the soil by rainwater.

(c) *rain-fed agriculture*, with the following factors: deforestation to gain cropland; complete land clearance exposing soils to erosion; dryland farmers encroaching on arid lands during a succession of good years with above-average rainfall; moving onto marginal lands and abandoning them in dry years; mechanized farming; intense farming without long enough fallow periods; short-term horizon because of immediate pressure to assure a subsistence level of food production.

Other dimensions of degrading dryland relationships are *deforestation* and its factors: growing demand for wood for cooking, heating, and construction; intense wood gathering leading to killing green trees and shrubs; exploitation of scarce and »sacred« trees (*e.g.*, Baobab); wood gathering for sale, especially in and around urban areas; increasing rainwater runoff, water erosion, and floods reducing the replenishment of groundwater reserves. Additional impacts on arid and semi-arid lands are urbanization and spread of squatter settlements caused by rural migration as well as large-scale mining and industrialization inducing local and sometimes interregional transformation, the latter due to pollution of waterways.

All these factors constitute just one side of transforming societal nature relationships; they provide the basis for potentially conflictive sociopolitical relations. Thus the reverse encompasses the socioeconomic fabric of dryland erosion. The following factors are seen as particularly conflict-prone: decimating livestock caused by competing use and overuse of pasture; reducing nomadic economy because nomads are forced to settle down; decline of crop yields caused by soil erosion in irrigated and rain-fed farmland; reducing available fuelwood and therefore long-distance trips in order to gather it; individual economic activities leading to increased degradation (Hardin's tragedy of the commons); migration and flight due to livelihood insecurity in arenas with heavy desertification; deterioration of life-support systems in general, unemployment, diseases, and starvation in particular.

Marginalization and Conflict Formations

Increasing poverty, densely populated poverty clusters, and transformation are consequences of conflict-prone social changes in the world's drylands. This is illustrated in an overview on desertification in the Sahel and on violent conflicts in the same arenas (*see: app. 1*). The overview shows empirical evidence for correlations between dryland erosion, socioeconomic change, and violence in most of the Sahel countries. In fact, the same factors that are said to lead to dryland erosion are also mentioned regarding causes of environmental conflicts; these are:

- loss of local regulatory mechanisms and traditional privileges concerning land-use and tenure systems;
- intertribal rivalries concerning water wells, pastures, and farmland;
- population pressure exacerbating disputes between farmers and herders, especially where permanent cultivation has been introduced in areas previously used by herders for seasonal migration;

- national governments trying to settle nomads;
- migration and flight-provoking violent conflicts between different actors in and between countries;
- short-term survival strategies of marginalized rural dwellers confronted with acute shortages due to persistent drought, etc.

The country list (*see: app. 1*) indicates that in *most* Sahel countries with major parts of their territory in either arid or semi-arid areas violent conflicts have been occurring during the last couple of years. Nevertheless, it would be short-sighted to blame dryland erosion alone for both socioeconomic change and violent conflicts or wars. The 1983-84 drought, for instance, was felt in all the Sahel countries, but serious consequences, such as starvation, decrease of cattle, crop failures, and flight were limited to a number of specific arenas within the countries. The socioeconomic effects of the drought were worst in those countries in which civil wars or deadly quarrels contributed to the disaster. Occurring violence aggravated the societal effects of the drought. In turn, environmental problems accelerated violent conflicts in which economic decline, political instability, and traditional conflict formations were already present. An example for this is Eritrea where 30 years of war had severe environmental and societal consequences. Large areas of forests and woodland have been decimated due to use for shelter, sale of fuelwood by the Ethiopian army, clearing risky areas of their tree cover, etc. Today Eritrea is left with less than 500,000 ha of forest covering only 4 percent of its territory. Its livestock wealth was reduced by 50 to 70 percent from 1977 to 1992. And one million refugees and displaced persons abandoned their villages. This meant a massive loss of both natural and human resources for a small country of 3.5 million people (Giorgis 1993).

The correlation between dryland erosion in the Sahel and violent conflicts or wars once more confirmed the concept of transformation introducing a syndrome of sociopolitical factors being at the same time responsible for both erosion of the landscape and violence. However, concentrating on degradation of individual resources would not lead to any valid results as it would exclude the most important factors transforming societal nature relationships: modernization *vs.* marginalization and poor state performance *vs.* centralization. Both elements are major forces for deepening structural heterogeneity with small dynamic centers and extended poverty clusters in ecologically sensitive arenas as its poles.

3.3. Violent Conflicts and Wars in Mountains

Mountains have been an arena of unrest, migration, and violence throughout history. After World War II mountains and highlands were the scene of some of the most devastating wars ever: the Caucasus, Northern Iraq, the Hindu Kush, Kashmir, Tibet, the Peruvian and Colombian Andes, the Vietnamese and Laotian mountains, and the Ethiopian highlands were affected by military activities of different intensity (Libiszewski/Bächler 1997). Some of the world's major hot spots with high escalation potentials - such as the Indian-Pakistani or Golan Heights conflict between Israel and Syria - concern mountains. In the 1990s, out of 45 wars and 44 violent conflicts 19 wars and 7 violent conflicts occurred in mountain areas (*see: app. 2, 3*). Compared with the mountains' share of the earth's landscape and population of about one fifth and one tenth respectively, the war-proneness of mountain arenas is quite significant. The question arises if a correlation exists between the mountains' geography and violence indicating a causal relationship.

Mountains and highlands are especially exposed as conflict-prone arenas because of ecological and socioeconomic vulnerability. On the one hand, mountain ecosystems are generally less likely than lowland environments to recuperate from substantial perturbations, such as widespread soil erosion or loss of vegetation. Due to relative impenetrability and difficult living conditions, mountains were almost everywhere excluded from national development and thus remained marginal both economically and politically (Omara-Ojungu 1990:143). For lack of off-farm alternatives, mountain communities in most parts of the developing world still rely heavily on agriculture for their subsistence. They are left with few means to cope with environmental challenges and they may, thus, precipitate rapidly into political crisis when the traditional socioecological balance is disturbed by internal population growth, external pressure, or global climatic change.

On the other hand, their inaccessibility and long-lasting marginalization have largely excluded many mountain regions from the market economy and therefore left them relatively unspoiled concerning their natural resource potential. But transportation means to access mountain areas have been greatly improved during recent decades. As natural resources are becoming increasingly short and degraded in plains, these natural reserves are more and more interesting for exploitation. Timber, fuelwood, water power, minerals, virgin soils, biodiversity, and the attractiveness of landscape as a tourism resource lead to the invasion of remote mountain arenas by both people and capital. The political implications of these processes - such as greater control by the central state, changes in the ethnic or national composition of the population, and often simply disruption of the indigenous peoples' way of life cause conflicts in many mountain regions of the world.

3.3.1. CULTURAL ASPECTS OF VIOLENCE IN MOUNTAINS AND HIGHLANDS

Is it correct to generalize that mountains imply scarcity, harshness, backwardness, and a scattered population?

Shortcomings of the Sanctuary Theory

Against the twin background marginalization vs. interests in mountains' resources, the assumptions of the so-called *sanctuary theory* must be questioned. Its protagonists maintain that mountains are generally less suitable and therefore less desirable for human settlement than plains. This owes to several environmental constraints, such as lower temperatures, scarcity of arable land, transportation difficulties, and the like. The crux of the matter is that, due to their assumed disadvantages, mountains should theoretically be abandoned ecozones. As a consequence, whenever a large population is found to settle in mountains, one might conclude that it has been forced to do so - displaced either by war, central governments, or destiny (Bencherifa 1990:370). But why then are there so many conflicts in deserted and unfavorable places?

It might be true that mountains belong to the poorest regions in the world, but there are extremely poor regions on the plains as well. Do they, in contrast to mountains, suggest abundance, ease, wealth, and good living in comfortable cities? As the French historian Fernand Braudel pointed out in his colorful and rich study of the Mediterranean during the age of Philip II, »the thickly populated plains«, which today are the image of prosperity, were »the culmination of centuries of painful collective effort«, originally a land of »stagnant waters and malaria, or zones through which the unstable river beds passed« (Braudel 1995:52). Some mountains are indeed quite deserted, mainly where little rainfall renders the land unfit to support even pastoral life. But there are also relatively rich mountain arenas. Equitable precipitation, a generally favorable climate, or the presence of abundant mineral resources ensure the attractiveness of some mountains. Historically, certain mountains have been densely populated and still are because of their agricultural and forest resources and not because lowland people have been driven from the plains by force.

There may be clear-cut geographical boundaries - 500 m, or so, as an arbitrary contour - between highland and lowland. Thus it can be inferred that mountains and highlands share a set of common characteristics determined by specific natural conditions - altitude, aridity, rainfall patterns, strong relief, soils, etc. - that distinguish them from lowlands. Producers, households, and mountain communities have problems comparable to those of producers in the world's plains, deserts, and deltas. However, the boundaries between highlands and lowlands may be blurred, indicating interdependence rather than complete separation. Therefore, the effects of distinct cultural geographical boundaries on mountain conflicts will be examined. These boundaries cannot easily be shown on a map because they shift over time depending on change in the human-nature relationship. This cultural-ecological change encompasses both a horizontal and a vertical dimension when applied to highland-lowland interactions.

The *horizontal dimension* encompasses the comparison of different interactions between environmental change and social behavior. The comparative approach highlights the rich variety of cultural and mainly agricultural systems throughout the world's mountain ranges; culture is defined here as the process of transforming societal relations with natural resources by production methods, resource use techniques, labor applications, and ways of life over a certain period of time. The Mediterranean mountains - even though marginalized in comparison with the plains and plateaus of the wider bicontinental region - are perhaps comparable to a certain degree to the Andes in Latin America. But in no way do they resemble the relatively impenetrable mountain areas of Central Asia, China, Indochina, or India. In the Atlas, for instance, agro-pastoralism has been and still is a vital resource-use system if not a dominant one, even though the major urban and economic activity is located along the Mediterranean littoral. Because it involves both farming and livestock raising, agro-pastoralism has survived as a risk-aversion strategy, which represents a response of traditional communities to the inherent environmental mountain hazards (Bencherifa 1990:373).

On the other hand, the Ethiopian highlands, in contrast to most mountains outside Africa, are not considered to have been originally hostile to human settlement. They not only hosted the famous ancient empire of Aksum but also provided an excellent human habitat, at least before the recent degradation of forests and soils (Hurni 1990:51). For centuries, the Ethiopian highlands were the most favorable areas of the country for both settlement and farming due to their temperature and rainfall patterns. The highland provinces comprise slightly more than 29 percent of the total area of the country, but they contain over 66 percent of the total cultivated land and nearly 45 percent of the grazing land. More than 63 percent of the urban and more than 69 percent of the rural population live in the highlands (Wolde-Mariam 1991:293). For the Horn of Africa alone, these figures disrupt the sanctuary hypothesis almost completely. Nevertheless, it is also evident that transformation of the highland's environment has placed enormous pressure on the landscape: settlement, crops, grazing, fuelwood consumption, and construction have all degraded the renewable resource base to such a degree that urban areas and lowlands are attracting more and more farmers and agro-pastoralists from the degraded highlands.

The *vertical dimension* focuses on highland-lowland interaction in single conflict regions. The crucial point can again be illustrated by reference to Braudel who suggests that the degree of contrast between plain and mountain is, first of all, a »question of historical period« (1995:53). He concentrated on cultural and socioecological disparities. In his sociohistorical analysis the disparities between old and new soils, that is, between exhausted and newly developed landscapes, are the key factors responsible for all existing forms of interaction such as vertical migration in both directions. In contrast to the premise of the sanctuary theory, it can be postulated: if a population is relatively dense in one arena, then the resources available necessarily make this concentration possible (Bencherifa 1990:370). If available resources do not support a certain level of population density, then alternative ways to sustain life must be found, or at least part of the population will be forced to evacuate in an attempt to locate more productive land.

The usual response to local overuse of old soils is migration. Yet the direction of such movement, whether uphill or downhill, will depend, at least in principle, on where the new lands are located. In reality, regular migration - *i.e.*, nomadism or transhumance, occasional migration between villages, for instance for marriage, and forced migration, such as flight or displacement - breaks eco-geographical determinism repeatedly and in various ways.

Another important factor is the means of transportation and travel, especially roads. Roads have a significant impact on the intensity and frequency of highland-lowland interaction; they serve as a kind of »extension of the plain and its power through the hill country« (Braudel 1995:41). Thus changes in the relationship between highland and lowland are mainly induced by modernization processes following the construction of roads or railways, however slow and imperfect these processes are. The pros and cons of opening up mountains by new transportation lines, *per se,* are nothing less than an expression of the collective will of a specific culture in a region. If politically and economically dominant groups concentrate on the lowlands for whatever reason, marginalization of the highlands becomes absolute. Nevertheless, drought and scarcity of resources in the lowlands can lead to a new understanding of the relatively favorable conditions of the mountains, as we will see with the example of Jebel Marra in Sudan (*see: 7.1.3.*).

Impenetrability as a Myth vs. Interdependence as a Reality

Conflicting highland-lowland migration patterns focus on two poles, described here by the notion of *impenetrability* for the mountain and *interdependence* for the lowland. Mountains have always been regarded as a world apart from civilization. At least for some mountain regions this may be not far from the truth. Because access has always been difficult, mountain people are deprived of important exchanges necessary for them to keep pace with urban civilizations developing in the foothills and plains. In »impenetrable« mountain areas, society, culture, and economy all bear the mark of backwardness and poverty. There is a lack of mutual exchange through movements while interdependence and functional solidarity are weak or absent. Thus social anachronisms persist - one of which is the Mediterranean vendetta and its system of honor and shame. Where vendetta flourishes there are mountains and they are usually not penetrated by feudal justice or by modern laws of the nation-state - be it communist or democratic.

Braudel gives us many examples on the historical origins of present conflict formations. In Europe the mountains were historically the lands of the free. This implies that people lived beyond reach of pressure and tyranny, having »no landed nobility with strong and powerful roots, no well-fed clergy to be envied and mocked« (Braudel 1995:39). The historical »mistake« of Syrian despotism, to name another example, was to stop at the first rocks of the mountains; and Turkish despots were the rulers of the roads, passes, and towns, but rather weak in Eastern Anatolia and in the Balkan highlands. Mountains were often considered zones of religious dissidence, sometimes real or perceived strongholds of foreign religious influence. When circumstances permitted, new religions were able to make substantial though unstable conquests in mountainous regions. In the Balkans in the 15th century, as Braudel describes, whole areas in the mountains converted to Islam, in Albania as in the Herzegovina around Sarajevo.

These evangelizing and modernization attempts, half-hearted invasions, and unstable conquests, or, in other words, these *imperfect penetrations* of »impenetrable« mountains turn out to be an underlying reason for ongoing struggles, social conflicts, and all-out wars. At present, Syria is struggling for water controlled by an upstream ruler (Turkey) over the headstreams of the Euphrates and Tigris. Simultaneously, the war between the central government of Turkey and the Kurds (the »Mountain Turks«) is still volatile, whereas the Balkans encompass various fragments of ethnopolitical groups accentuated by division between traditional rural mountain arenas and modern urban cultures that can be traced back to the early history of conflicting highland-lowland interactions: Albanians versus Serbs; Serbs versus Muslims in Bosnia-Herzegovina.

The concept of imperfect penetration can be supported by other examples. Where the population was so thinly distributed, widely dispersed, and separated by deep valleys that it prevented establishment of states, dominant languages, and important civilizations, the mountains were both the birthplace and battlefield of the struggle for liberty, democracy, and peasant 'republics'. The rather belligerent history of the Swiss cantons adds weight to this line of reasoning. In Rwanda, where the dispersed habitat on the »impenetrable« hills had been part of a decentralized culture for centuries, this very structure was perceived by the colonial power as »*un obstacle à la division du travail, à l'échange, à l'économie de marché*« (A.E.S.E.D:27. In: Guichaoua 1989:46). Thus the structure of the complex Rwandan society »*qui a résisté des siècles durant aux incursions des Etats voisins et des Arabes*« (Guichaoua 1989:44) was radically changed by the Belgians, a change important in explaining the deep roots of the present war and genocide (*see: 5*).

Finally, »impenetrability« is a one-way process; it almost always entails downhill migration (Jebel Marra mountains in Sudan being one of few exceptions). The harsh life, widespread poverty, hope of an easier existence, or attraction of good wages encourage mountain people to move downwards to more favorable ecozones. In fact, throughout the world large numbers of mountain dwellers have become indispensable to the life of towns, lowlands, and the sea coast. The disparities manifest themselves in social and cultural barriers that are raised to replace the imperfect geographical barriers in one way or another. This is the case in Central Asia, where strong regionalism among former mountain dwellers creates group identity conflicts in the plains and plateaus (Klötzli 1996:277ff). However, the migration issue relates more particularly to the second pole and embraces the concept of conflict emanating from *interdependence*, with mountains serving as regions of vertical human migration.

As a matter of fact, highland-lowland interdependence is founded upon a more or less dense network of routes along which both downhill and uphill movements occur. Transhumance is by far the most important of these migrations from the hills to the plains but, although it being a temporary as well as round-trip journey, it occurs on a larger scale and more regularly than other forms of movement. Mountain dwellers may be hired in the lowlands at harvest time; landless peasants, unemployed artisans, and casual agricultural workers may decide to seek cash income in the lowlands, be it in a formal or informal economy. Market places in the foothills or plains periodically attract producers from steep valleys and remote upland areas. In addition, resource scarcity, special occasions (*e.g.*, ceremonies in the plains), or military migration (mercenaries for regular armies or warlords) all induce downhill movements. Mountains have always provided a reservoir of labor for other people's use and have contributed to development of regions beyond their boundaries almost everywhere in the world and throughout history.

Intervention in Migration Patterns Causes Conflicts

Conflicts generated in the context of interdependence through various forms of migration are related to various kinds of interventions that induce a change in the pattern of movement, hinder or force migration artificially, create new forms of movement and interrupt or eliminate traditional forms, such as *nomadism* and *transhumance*. External intervention in highland-lowland formations changed interdependence and patterns of functional solidarity historically established between people in the mountains and on the plains. The imperfect penetration of mountains tends towards *conflicting dependency and dysfunctional partiality*.

To explain this in more detail: most mountain conflicts are associated with transhumance and nomadism - the regular movement of men and their flocks - which are the most distinctive characteristics of the mountain world. Changes in domestic consumption, household mobility, and the degree of commercialization are obviously important variables for all producer groups depending on mountain resources. However, today these changes seem to be less critical for more settled farmers and other mountain dwellers. It is the pastoralists, almost universally, who are challenged. They must compete with capital-intensive commercial ranching or dairying, on the one hand, and smallholding peasant husbandry, on the other hand. The latter indirectly reduces the role of herding and natural pastures by integration of animal husbandry and crop production. In addition, the pressure on pastoralists caused by both governmental attempts at resettlement and environmental transformation is accelerating (Galaty/Johnson 1990:1-5).

Today the phenomenon of transhumance is even more complex than in the past, and it involves many factors: physical, human, and historical (Galaty/Johnson 1990). It has become exceedingly difficult to classify the many different forms of transhumance.

There is *normal* transhumance, which is under pressure because of the scarcity of renewable resources. Cattle farmers, sheep farmers, and shepherds leave the lowlands, where they usually live in the early summer, because the season is unfavorable for stock raising. This vertical movement from the lowland winter pastures to the high-quality summer pastures in the mountains was once a sustainable way of combining two geographically and climatically distinct altitudinal belts. Today the mountain pasture may be either the property of other peasant farmers from the plains or rented out to the mountain dwellers; it may also belong to settlers and farmers living in the mountains. Problems of ownership, rental, and use of mountain grazing rights are by no means new. Braudel (1995:91) reports about a notorious conflict between the Grisons farmers in the southeastern Swiss Alps and Venetians; the former had to drive their stock to the southern Alps and towards Venice, to a mountain area rented by the

Venetians for their own herders who needed to pasture huge flocks. Over time, population pressure continued to grow, resource degradation accelerated, and productivity declined, so that the struggle for one and the same resource had become far more serious.

The problem has shifted from Europe to arenas where dependence on natural resources is much higher than in successfully modernized economies. The different phases of the highland-lowland conflict in the Sudanese Jebel Marra Massif clearly demonstrate the pattern of intensified potential for conflict between farmers and pastoralists arising from the last two decades of unusual drought in the Sahel (Suliman 1996:109-145; *see: 7.1.3.*).

Inverse transhumance, by contrast, involved shepherds and their flocks descending from the highlands, mainly for marketing purposes. Herders and animals rapidly left the high-altitude pastures at the beginning of the cold season to escape the harsh conditions and rushed into the lower regions »like an invading army«, according to reports on 17th century transhumance and related conflicts in the Mediterranean region: »All doors were padlocked against these unwelcome visitors, and every year saw a renewal of the eternal war between shepherd and peasant, first on the way down, until the flocks reached the open plains or the wide grazing lands ...« (Braudel 1995:86). When, because of land scarcity, the same grazing areas were used during the same season by practitioners of both inverse and normal transhumance, conflict became virtually inevitable. The traditional tradeoff according to the principle »our flocks may go to your land in summer, but your flocks come to ours in winter« fell by the wayside. Thus when the resource base became degraded, the different land-use interests determined the separation between peasants and (agro-)pastoralists.

All transhumance resulted from an agricultural situation that could not support the pastoral economy in its entirety; this induced therefore the offloading of its burden onto either the lowland or the mountain pastures. This has persisted until today. But governments tend to disregard the advantages of pastoralist resource use; furthermore, in recent years pastoralism has also been blamed for dryland degradation. Although this critical view of pastoralism has been challenged, climate change, population growth, advances in agriculture, and giant projects - such as mountain dams and land irrigation schemes on the plains - still threaten to break up the formerly pastoral ways of life (Galaty/Johnson 1990:21-30).

Transhumance is only one form of pastoralism, which concerns a specialized population: the shepherds; it implies a division of labor and also involves a combination of a settled form of agriculture with crops, fixed dwellings, and villages. Nomadism, in contrast, embraces the entire community, which moves over long distances: people, animals, and even dwellings. The policies of present-day governments, aimed at fostering sedentarization of nomads, have converted traditional nomadism into a modified pastoral way of life, a division of labor that increasingly resembles transhumance. Nomads have to defend themselves against the multiple challenges to their way of life, be it on the plains or in the mountains. Thus nomadism is confronted with similar problems to those faced by transhumance: the pressure on and interruption of seasonal vertical and horizontal movements through various kinds of interventions.

3.3.2. PATTERNS OF CONFLICTS IN MOUNTAINS

Conflicting patterns of migration in mountains are focused around the two poles impenetrability and interdependence. It is the *imperfect* and therefore *unstable penetration* of »impenetrable« mountains that has become an underlying reason for ongoing struggles, social conflict, and war at the one pole. The concept of an impenetrable mountain arena high above the plains with some rather wild-looking poor people is misleading in many respects, even in regions that have remained on the fringe of the great waves of civilization. On the contrary, it is the unsustainable, weak, and temporary penetration of the mountains that has caused and still causes conflict of interests among the different actors involved. Marginalization of mountain communities - which is a result of unsustainable penetration - forces, at best, separation of highland and lowland cultures through various interventions, such as establishment of clear-cut political or economic boundaries. In the worst case, the discrimination against mountain communities provokes formation of organized identity groups ready to defend their interests with military force.

At the other pole there is a trend towards conflict-prone dependence and dysfunctional partiality. Since the asymmetric structure favors the lowlands in most cases, all forms of vertical migration are challenged. Highland-lowland interactions which, in principal, have resulted from interdependent developments collapse. Change in essential environmental factors around which all agricultural and most pastoral economies are organized leads to painful and sometimes violent transformation processes: (re-)distribution of pasture, seasonal overuse of available land, struggle for control of water resources, the nature of animal and household mobility, division of labor, and interaction between highland and lowland economies.

Environmental conflicts in mountain regions almost always stem from their interaction with the lowlands. Although difficult to access, the mountain arenas of the world have always coexisted with the surrounding foothills and plains, *e.g.*, ecologically concerning the water cycle, socioeconomically concerning the use of pastures, and as a reservoir for the labor force or as a path for transportation. The various forms of highland-lowland interactions imply specific social and political organizations, which are severely challenged by the ecological consequences of population growth and the globalized modernization process. Social tensions and conflicts, which necessarily arise from these challenges, escalate into violence and war if and when environmental threats are perceived as existential, the traditional mechanisms of political regulation are overstrained, and the parties in question have the capacities to organize and arm themselves or eventually find allies for their claims (Bächler *et al.* 1996:308-309).

Stephan Libiszewski provides four patterns of conflict formations in mountain arenas (Libiszewski/Bächler 1997). These can be summarized as follows:

First formation: mountains as earth's water towers. Conflicts over the earth's freshwater supplies underscore the overriding significance of mountains due to control and distribution of sources of international river basins and by attracting much of their precipitation. At least half of humanity relies on water captured on mountains for drinking, domestic use, irrigation, hydropower, industries, transportation, etc. (Denniston 1995:41). The specific highland-lowland interaction given by the water cycle lies at the root of many acute conflicts, both international and internal. A first source of interstate conflict derives from the fact that most great rivers of the world are transboundary. Worldwide about 240 greater river systems cross national boundaries. Mainly international dam projects trigger conflict and military threats among neighboring states (Bächler *et al.* 1996:206-209).

Second formation: mountains as a niche of small peoples. A second factor favoring conflicts in mountain regions is their cultural and ethnic diversity. The heterogeneity of people divided through deep valleys and steep slopes is virtually due to a divided horizon. It is the direct outcome of local people learning to live in harmony with the mountains' extraordinary biological and morphological diversity. This built-in individuality or distinctiveness of groups contributes to the heterogeneity among the population often encountered in mountains. Regarding the number of remaining indigenous (tribal or native) peoples worldwide, estimates vary from 200 to 600 million. Although no precise figures exist on the total number of people living in mountains, the proportion is clearly high, since mountains account for a substantial portion of the landscape that has not been transformed by modern economies. Cultural diversity does not trigger conflict and violence *per se.* But such diversity provides dividing cleavages, which can easily be politicized if other factors of social, economic, and ethnopolitical stress appear on the scene.

A key factor that affects peoples in mountain arenas above all and often leads to conflicts between them and the central government are large development projects in their settlement arenas. Internal conflicts arising from the damming of rivers and the following inundation of settlement and farm land have already been mentioned. Of all the economic activities in the world's mountains, however, nothing rivals the destructive power of mining. For geological reasons, mountains are the *loci* of many of the world's mineral deposits needed for industrial production.

Third formation: mountains as poverty reserves. Resource depletion and degradation often affect economic productivity in poor countries, which tend to be highly dependent on their natural capital. Mountains share common physical attributes of steepness, instability, and ecological complexity that make them extremely vulnerable to ecological disturbances. Growing populations and inequitable distribution of and difficult access to land resources persuade farmers to abandon marginal lands, which are far more exposed to erosion. Disaggregated data indicate disparities in human-nature relationships that happen to result in violent conflicts. Economic decline caused by resource degradation disrupts key social and political institutions including the state itself. Poor state performance in general and in mountains in particular aggravates grievances between mountaineers and central governments, triggers rural protest, and increases rivalry among different factions belonging to the political elite.

Fourth formation: mountains as last frontiers. Environmentally exacerbated poverty not only causes social unrest and violent conflicts; it also induces movement of people. When available resources fail to support a certain population, alternative ways to sustain must be found or the people will be forced to leave a degraded homeland to find new soils. Movement of people and livestock have always been a characteristic of mountain regions (Libiszewski/Bächler 1997).

Due to vertical distribution of various environments and different seasonal conditions at each level, resources are stratified and need a staggered schedule for their exploitation. Nomadic pastoralism, transhumance, and mixed forms of grazing and farming cultures including different uses for various altitudes have been man's prevalent answer to these conditions on low, middle, and high altitudes. Moreover, migration into the plains has always been a means of coping with the population surplus, which the harsh mountainous environment could not support. Ethnic groups which have developed and practiced complementary ways of living may discover their distinct identity and become engaged in violent conflict when forced to compete with each other for shrinking resources in order to survive (Braudel 1995).

3.4. Conclusions

Some conclusions can be drawn against the background highlighted in the sections on arid lands and mountains: transformation and intractable poverty become increasingly intertwined in geographic arenas with fragile environmental conditions. This applies mainly for arid and semi-arid arenas in Africa (Asia, which was not mentioned here, would presumably differ little; Klötzli 1996:279-304) as well as for mountains worldwide. A digression with Fernand Braudel into the painful history of the mountains surrounding the Mediterranean has shown the overall dimension of the problem. The transformation conflicts in the Mediterranean mountains are the forerunners of similar events throughout the world's mountain systems of today.

Since both mountains and arid plains face comparable ecological problems, resource competition among highland and lowland dwellers has become even more conflict-prone than in the centuries Braudel focuses on. Moreover, geographical arenas with environmental degradation and poverty links are the very zones highly prone to social and political conflict. Neither poverty nor desertification affect countries as a whole. Degradation primarily undermines both geographical and political boundaries in sensitive ecoregions and not in nation-states or a politically defined territory: transformed drylands and mountains as endangered life-support systems comprise the structural pattern against the background of which conflicts occur in the political network where the confrontation between actors takes place.

The Janus-headed process of modernization *vs.* marginalization in mountain and dryland arenas promotes rapid transformation of all forms of the complex migration patterns, which have evolved over many centuries. Among the factors contributing to ecological conflicts structural change or even disruption of downhill-uphill traffic proved to be a major reason for conflict in the vertical dimension: penetration of the mountain peoples' resources control systems by the lowland centers became the main force in generating conflict. The real or perceived closure of geographical boundaries between highlands and lowlands led to clusters of partly isolated populations or to pockets of poverty in remote mountain arenas - cultural phenomena that were explained by the sanctuary theory *ex post facto*.

It is necessary to differentiate carefully between causes of hostility and what actually triggers people to fight. Ethnic, regional, and religious problems attributed to those conflicts described above may play a causal role, but in general less as a reason than as a channel that provokes exaggeration of group identity and triggers intergroup confrontation.

Today the marginalization of remote mountains and large arid plains is also the necessary outcome of poor performance of central states. The widening gap between the modern sector - on which state performance usually concentrates on - and the marginal sector - encompassing poverty clusters in marginal environments - constitutes a symbolic fault line, which in many places has become a real front line of environmentally caused violent conflicts.

CHAPTER 4

TYPOLOGY

Types of Conflicts and the Role of the Environment

Correlations between transformation of societal-nature relationships and violent conflicts are numerous, as has been pointed out in the previous chapters. In this and the following parts of the analysis the critical role of transformation regarding *causation* of environmental conflicts in certain arenas of developing countries will be examined. In order to highlight different patterns of causations, to select types of environmental conflicts in terms of different pathways leading to violence, as well as to stress the sociopolitical characteristics of environmental conflicts, one has to build on prepared empirical ground.

The empirical ground, on the basis of which the following typology is established, has been provided by the Environmental Conflicts Project (ENCOP). This international research project focused on the interrelationship between environmental degradation, maldevelopment, and violent conflict. 40 area studies were carried out by a permanent research team based in Bangladesh, Germany, Great Britain, Nigeria, and Switzerland. Additional regional specialists and inhabitants of the countries under consideration also contributed to the case studies.[1] The study group adopted a working definition of environmental conflicts in order to narrow the focus of the globally oriented study. According to this definition, »environmental conflicts manifest themselves as political, social, economic, ethnic, religious or territorial conflicts, or conflicts over resources or national interests, or any other type of conflict. They are traditional conflicts *induced by environmental degradation.* Environmental conflicts are characterized by the principal importance of degradation in one or more of the following fields:

- overuse of renewable resources;
- overstrain of the environment's sink capacity (pollution);
- impoverishment of the space of living«.[2]

The concept that led to this definition is twofold. First, the study group assumed that both structures and actors played roles if and when environmental problems

1 Most of the empirical studies this chapter refers to are to be found in Bächler *et al.* (1996) as well as in Bächler/Spillmann (1996 II, III). Some exist as draft papers only.

2 Working definition by ENCOP team (Libiszewski 1992:13).

contribute to conflict. Taken together the two determine the type of conflict that is triggered by environmental issues. Second, it focused on renewables resources and excluded minerals and nonrenewable resources. However, if mining, dam building, or industrial activities directly or indirectly lead to widespread disruption of landscapes, than these conflict-prone activities constituted cases, which match the definition.

After completing the ENCOP investigation - as well as regarding the previous chapters of this study - it is clear that neither apocalyptic scenarios of environmental catastrophes nor alarmist prognoses of world environmental wars are tenable. Environmentally caused conflicts escalate across the violence threshold only under certain conditions. Human-induced environmental change can be either a contributing or a necessary factor to both the emergence and/or the intensification of violent conflicts. On the one hand, violent conflicts, triggered by environmental disruption, are partly due to socioeconomic and political developments. On the other hand, social and political maldevelopment, partly due to degradation of natural resources, has become an international peace and security challenge. *Development and security dilemmas* are connected to a syndrome of problems, which produces environmental conflicts of varying intensity and nature.

The following discussion explores the specific triggering and inhibiting factors which determine conflict behavior. The framework for this discussion is a typology of conflict levels and parties to a conflict, and a generalizable examination of the environmental role in causing conflict and intensifying current conflicts. More details on the specific ENCOP cases can be found in the appendix (*see: app. 11*).

4.1. Environmentally Caused Violence: A Phenomenon of Developing and Transitional Societies (Hypothesis One)

Environmental conflicts are caused by certain ecological problems of particular intensity. Yet these conflicts remain social and political events. Therefore, several attributes of conflict must be considered together for a complete explanation of environmental conflicts: actor characteristics, interests, actions, and outcomes based on the results of actions (*see: 2.3.7.*). The first hypothesis holds:

Violent conflicts triggered by the environment due to degradation of renewable resources (water, land, forest, vegetation) generally manifest themselves in socioeconomic crisis regions of developing and transitional societies if and when social fault lines can be manipulated by actors in a way that social, ethnic, political, and international power struggles occur.

The ENCOP cases provide ample evidence for the assumption that developing and transitional societies or - to be more precise - discriminated groups within these societies are most affected by interactions between environmental degradation, social erosion, and endemic violence. Crisis areas susceptible to conflict are found in: (1) arid and semi-arid plains (drylands); (2) mountain arenas with highland-lowland interactions; (3) arenas with river basins subdivided by state boundaries; 4) zones degraded by mining and dams; (5) tropical forest belts; (6) poverty clusters of sprawling metropoles. In these sensitive arenas - to be found in Oceania, Africa, Latin America, Central and Southeast Asia - traditional society-nature relationships, regulated by cultural-specific approaches to the environmental problems, are acutely at risk.

All these conflicts have the phenomenon of marginalization of one or more actors in common. With one major exception: interstate conflict over shared river basins. Although there are cases where conflicts between upper and lower riparians occur in marginalized ecoregions of neighboring states, *e.g.*, Eastern Anatolia in Turkey versus Syria and Iraq, in most cases geopolitical and strategic security issues are in the foreground. Disputed water resources and rural development issues, *e.g.*, the farmers in Syria and Iraq, are shoved into the background. These priorities apply especially for regional water conflicts in the Middle East that transpire within the framework of historic territorial conflicts.

However, conflicts induced by marginalization of certain groups share the problem of discriminated access to natural resources. Thus the concept of environmental discrimination is crucial to all conflicts under consideration. *Environmental discrimination occurs when distinct actors - based on their international position and/or their social, ethnic, linguistic, religious, or regional identity - experience inequality through systematically restricted access to natural capital (productive renewable resources) relative to other actors.*

The conflict geography of environmental conflicts corresponds largely with that of regional conflicts. Conflicts tend to occur in the South, a pattern observed since World War II. In fact, the number of armed conflicts has risen sharply immediately after 1989 due to the collapse of the Soviet Union. In the early 1990s conflict occurrence in the Eastern transitional societies has ebbed slightly, and since 1994/95 there is a decline in the total number of violent conflicts and wars. This development notwithstanding, the number of ongoing violent conflicts in the South - particularly those of low intensity - is still high and probably increasing due to links with maldevelopment. This judgement results from various regional analyses which, in contrast to conventional war registers, include the assessment of unrest leading to bloodshed as precursors of environmental conflicts. For example, in Central Asia such incidents of unrest have already taken numerous human lives but are not registered in available data bases (*e.g.*, Fergana Valley). It is likely that relatively low-level conflicts of this kind will escalate either in the short or intermediate term. In these cases, environmental crisis serves as an indicator of likely state failure and thus expected major conflict (*e.g.*, Northern Ghana).

Most environmental conflicts are carried out between actors within a country (*see: fig. 4.1., A; app. 11*). In some cases, internal conflicts become internationalized (*see: fig. 4.1., B*). Most of those spreading across borders involve migrants or refugees coming from war-torn or marginal rural areas of a neighboring country. Seeking fertile land or jobs, they cause political, social, or ethnopolitical conflicts outside their region of origin. The internationalization of internal conflicts can also be the consequence of new states created after the collapse of an empire. With the dissolution of the Soviet Union, for instance, the new Central Asian republics face water distribution disputes, which suddenly have become international conflicts.

Finally, there are environmental conflicts between states, which have an international dimension from the very beginning (*see: fig. 4.1., C*). These conflicts may result from degradation of regional environments or the global commons. Contention surrounding ozone layer depletion and climate change (including sea-level rise) are political conflicts with no major military dimensions to date. But today international disputes arise especially between nations mutually dependent upon the cooperative use of international river basins. Although the cases examined in the ENCOP studies did not result in violent clashes, considerable potential for military actions persists among some upper and lower riparians (Bächler *et al.* 1996:158-165).

Distinguishing between the three levels A, B, and C only serves as a rough orientation to environmental conflict. The boundaries among the levels are literally fluid. Classifying a given conflict at just one level, especially over time, is often impossible because internal conflicts may be fueled by international events, whereas the latter may be the result of an escalating domestic war. A further fine-tuning is therefore needed. It is necessary to relate the type of environmental degradation to socioeconomic change and to parties to the conflict, herein referred to as *actors*. Seven ideal types of environmental conflict can be distinguished: ethnopolitical conflicts; center-periphery conflicts; regionalist migration/displacement conflicts; transboundary migration conflicts; demographically caused conflicts; international water/river basins conflicts; and international conflicts arising from distant sources due to neocolonialist exploitation of resources (the latter conflict is seen as a variant of center-periphery conflicts). Sometimes it is difficult to decide under which category a given conflict is best listed. Rwanda, for instance, may be type AII (ethnopolitical conflict) or BVI (demographically induced conflict). However, for analytical clarity, each case is listed in one category only.

There is no intention to provide empirical details concerning the cases here.

An encompassing table including the list of the cases related to each individual type of conflict, the indication of the environmental dimension of the conflict, the parties involved, and the conflict intensity can be found in the appendix (*see: app. 11*). The following synthesis of the case studies carried out in the framework of ENCOP refers to the cases included in this table. They provide the empirical background against which the *generalizations* provided in this chapter have to be seen.

	Conflict Levels:	*Adversaries:*
A:	internal	identity group* vs. identity group government vs. identity group government vs. migrants/refugees
B:	internal with interstate aspects/internationalized	government (and local population) vs. immigrants from third states
C:	between states (international)	government vs. government government vs. IOs/INGOs

Identity groups are groups within states that share common experience and cultural relationships. They organize themselves in order to defend their interests and they may be defined as: ethnic groups, indigenous people, religious groups, socially defined groups, regionally or geographically bound groups.

Table 4.1. Taxonomy of environmental conflicts according to the different levels in the international system

4.1.1. ETHNOPOLITICAL CONFLICTS (TYPE AI)

Environmental and ethnic discrimination coincide in many cases analyzed by ENCOP. Conflicts emerge because (1) two or more ethnic groups share one ecozone with degraded and thus unproductive resources, or (2) ethnic groups depend on neighboring ecozones with highly distinct degrees of productivity. In the first configuration, conflicts escalate because one or more ethnic groups have limited access to needed natural resources. In the second set of conditions, violence occurs if and when the environmentally discriminated group invades the territory of the neighbors who are environmentally better off.

Similar to the center-periphery conflict (*see: 4.1.2.*), ethnopolitical conflict represents a modernization conflict at its core. The difference is that the fault line does not run between a defined center and its periphery. Instead, the cleavages for ethnopolitical conflict run along group-specific traits within ethnically fragmented societies. In a few cases, population pressure on an ecologically sensitive region beset with scarce and degraded resources contributes to the hardening of interethnic relations (*e.g.*, Rwanda; Bangladesh versus Assam Province in India (*see: app. 11, BV. 25*).

Overuse of land, forest, and freshwater resources lead to ethnopolitically motivated conflicts if and when they combine with certain geographical and climatic factors. In many areas of rural societies, for instance, the traditional dualism between subsistence farming, on the one hand, and (semi-)nomadic livestock breeding and large-scale ranching, on the other hand, is at stake. Since the two different producer groups usually belong to distinct ethnic or indigenous communities, the competition over resources becomes the core of an ethnopolitical conflict. What colonialism was not able to do, the present struggle for fertile land definitely achieves: it opens up ethnically preserved environmental »niches« of groups that generally used to live apart. Owing to intensive use of both ecologically favorable arenas with high fertility and unfavorable arenas with low fertility by a growing number of rural producers, ethnically defined living space is giving way to ethnic mixing and social stratification. This process, which largely contributes to new clusters of ethnic groups facing environmental discrimination, is triggered by degradation of scarce resources. It may also be aggravated by modernization processes, protracted violence, or wars.

The competition over degraded lands has another dimension that overlaps with center-periphery conflicts. Huge areas of the most fertile land are under cultivation for monocultures (food and most often cash crops), or are serving commercial purposes of central governments (such as wildlife parks). As a result, increasingly less productive land remains exclusively for labor-intensive food crops. Rural populations put pressure on available land and further stress the landscape. However, identity groups - whether tribes, clans, or ethnic groups - are in general not willing to surrender land which they claim as the land of their ancestors. Historically land-use and land-tenure disputes have led to deliberate violent measures. But only recently such traditional conflicts have been aggravated by the negative ecological impacts of maldevelopment in the rural sector throughout the developing world. Both environmental discrimination against identity groups and the availability of modern small arms have made these conflicts even more brutal (*see: app. 11, AI.1-6.* Case studies by Bächler 1996II: 461-502; Klötzli 1996II:247-336; Lume 1996III:175-202; Suliman 1996II:109-180).[3]

3 The individual authors of the case studies are not listed separately in the bibliography attached to this study. All authors with either (1996II) or (1996III) indicated in sections 4.1.1. to 4.1.7. are included in Bächler/Spillmann (1996II, III).

4.1.2. CENTER-PERIPHERY CONFLICTS (TYPE AII)

Relationships between members of the center and the periphery in developing countries often assume precarious forms owing to environmental transformation. National elites and international investors in the modernizing centers have certain economic, environmental, and energy policy preferences as well as opportunities at their disposal because of their access to resources including power. At the same time, the socioeconomic opportunities for most levels of the rural population proved to be extremely limited. Arenas inhabited by poor rural dwellers are indeed often environmentally vulnerable and degraded. Environmental degradation also contributes to the development of rural and (peri-)urban poverty clusters by undercutting health conditions, constraining productivity, and shortening time horizons for adequate reactions. A lack of resources makes it extremely difficult for the poor to escape the environmental risks or to invest in risk-reducing and income-generating strategies (*see 1.1.4.*).

The catalysts to escalate center-periphery conflicts are primarily large cash-crop farming projects, dams, and mining. Globally oriented companies pursuing capital intensive, high-technology and high-energy projects are confronted with identity groups dependent on natural capital with low or no commercial energy input. This clash is heavily conflict-prone if and when the society-nature relationship for an arena, that has had little or no integration into the market economy, is transformed by third-party economic interventions that run counter to rural dwellers' interests and preferences. Mechanizing, collectivizing, or privatizing the rural economy marginalizes traditional cultivating methods, land-use patterns, and land-tenure systems. Landscape erosion - through soil and water degradation - from large-scale commercial farming, mining, and industries destroys the living space for peripheral populations.

Actors highly dependent on the disputed natural capital but facing environmental discrimination see themselves as prisoners of modernization or structural heterogeneity respectively. In poor countries modernization seldom provides valuable off-farm opportunities. Only relatively few among rural dwellers - which constitute up to 95 percent of the population of least developed countries - find regular incomes in the commercial sector. Still fewer receive financial compensation for the loss of natural capital or for being forced to use marginal land. Companies and central governments often break promises on the positive impacts of huge projects. Marginalized groups endure more discrimination through project-driven transformation of the environment. Modernization without participation makes them feel more disadvantaged materially, socially, culturally, and spiritually than ever before. Intensified rivalry for scarce water, shrinking settlement areas, and productive soils belong to the inevitable consequences of this maldevelopment.

Center-periphery conflicts differ greatly depending on the degree of periphery dependence on the center and the level of underlying power asymmetry. In contrast to dam or mining projects, the interdependence between center and periphery concerning agricultural and irrigation projects is usually high. If necessary, mines or dams can be protected militarily; large areas of irrigated land cannot.

There is an international variant of center-periphery conflicts (*see: C.VII*). Whereas the center-periphery conflicts discussed above pit national elites (governments/economic sector) and international firms against less powerful populations, the international type sets outside powers against developing country governments. Severe conflicts surround *neocolonial exploitation* of the natural environment against the will of the governments of an affected region. Since the power asymmetry among the two actors is significant, the escalation potential of this kind of global conflict is minor in comparison to the environmental problem it creates. Concerning the only such example we analyzed, namely the French nuclear tests in the Pacific, the worldwide protests were relatively moderate. Bloody violence and direct confrontation between the French army and protesting groups (including Greenpeace), however, occurred only in Polynesia, *i.e.*, in the immediate neighborhood of the nuclear explosions. The conflict was contained in the local arena far from the center of the globally acting player France. Although this kind of globalized center-periphery conflict is an exception (Bophal in India and other »accidents« did not have the same level of global response), it is indeed not inconceivable that in the near future asymmetric socioeconomic impacts of climate change, ozone layer depletion, and sea-level rise will lead to similar constellations (*see: app. 11, A.II.7-19*. Case studies by Böge 1996II:503-720; Claus 1996III:269-264; Okoh 1996II:181-246; Schönenberg 1996III:315-358; Schwark 1996III:359-408; Wegemund 1996III:285-314; König 1996III:149-174; see also: Böge 1993; Carino 1993 (manus.); Quimpo 1993 (manus.)).

4.1.3. INTERNAL MIGRATION CONFLICTS (TYPE AIII)

Internal migration conflicts are triggered by either voluntary migration or forced displacement of inhabitants from one region to another within one country. The geographic origin of migrants or displaced persons is the primary criterion for conflicting social and political relationships between the actors. Migration is induced by structural changes, such as persistent drought, flood, and soil erosion (desertification). Its direction leads from depressed areas to more favorable zones such as fertile rural or (peri-)urban arenas. Although both push and pull factors work together, the push factors are stronger here. Forced displacement and expulsion, on the other hand, are almost entirely due to push factors often in connection with large (agro-)industrial, mining, and dam projects.

Interregional migration and displacement - as a special type of internal dislocation - pit people of the same ethnicity from different regions against each other. The most important fault lines are those between highlanders and lowlanders, pastoralists and farmers, rural and urban population. Mountaineers, for instance, drawn downwards by the quest for jobs, income, and land get caught in competitive situations with indigenous populations. The distinct society-nature relationship of newcomers and settled populations triggers tensions, clashes, and in some cases violent conflicts.

Thus a myriad of social interactions emerges. In locally overpopulated and degraded mountain regions with nomadic cultures and few off-farm opportunities, environmental degradation and stress prompt major migration waves into irrigated areas and the urban fringe with resident farming cultures. Integration of former livestock breeders is difficult in large irrigated areas with monocultures (*e.g.*, Himalayan pastoralists in the plains of Central Asia). On the other hand, farmers also migrate from eroded highlands into fertile valleys settled by semi-nomads (*e.g.*, in the Horn of Africa). Thirdly, conflicts emerge if semi-nomadic pastoralists flee from persistent drought and soil erosion to semi-arid and subtropical mountain regions settled by farmers (e.g., in Sudan).

The basic patterns of the three interaction systems between highland and lowland residents are comparable. Interregional migration conflicts are in part determined - as are the other conflict types discussed here - by environmental discrimination against actors who are heavily dependent on scarce natural resources. A second factor of significant influence is *poor state performance* in marginal areas. High dependence on natural capital combined with poor state performance are two main reasons why discriminated groups are attracted to rich rural arenas and country capitals. These two factors are critical in countries with great regional disparities not offset by the rule of law and democratic mechanisms. However, poor state performance is also an inhibiting factor for large-scale migratory movements, namely in cases when poor state performance coincides with illegitimate and oppressive regimes, which have been able to put vast territories under military control (e.g., the Kurds in various countries).

Interregional conflicts are commonly confined to local arenas and cover neither the entire country nor the core of state power. Violent conflicts (skirmishes, clashes, riots) occur in disputed rural zones. Some conflicts spread to (peri-)urban arenas and blend in with violence and criminal gang behavior, at times involving former soldiers or mercenaries. Interregional migration or resettlement leads to political struggles for state power if and when groups that had been discriminated against succeed in penetrating the ruling elite or driving it out of power (*see: app. 11, AIII.20-23*. Case studies by Faath 1996III:203-268;

Schönenberg 1996III:315-358; Smil 1996 III:127-148; Melber 1996III: 409-440).

4.1.4. CROSS-BORDER MIGRATION CONFLICTS (TYPE BIV)

When migrants or refugees cross national borders voluntarily, resettle in rural border arenas or resettle in cities of a third country, they represent socially and at times politically orchestratable conflict potential. Even though the term »environmental refugee« is rejected by the United Nations High Commissioner for Refugees (UNHCR), the linkages between migration and environmental discrimination have been acknowledged by UNHCR head Sadako Ogata (Ogata 1994:41-47). UNHCR is concerned with environmental disruption as a serious consequence of large refugee movements (mainly in large camps depending on fuelwood). However, transformation of the environment is itself a reason for migration and flight. Migrations channeled by environmental discrimination intensify conflicts where the economic situation is eroding and political instability deepens lines of conflict. In some cases the use of violence opens pre-colonial divisions between rival identity groups.

Environmentally induced migration normally takes the form of slow infiltration over a long period of time. People move into arenas, which either permit survival or provide favorable living conditions. Only in exceptional situations, such as acute drought, do massive flights occur spontaneously. The escape routes are diversified. In many regions it pays to cross the national border because more favorable foreign destinations are geographically nearer than the remote capital of one's native country. Frustration and despair can explode into violence in host countries or transboundary regions populated by hostile identity groups or by earlier migrants from common identity groups who show hostile behavior towards the newcomers. Occasionally the routes also lead to the northern industrial countries.

The following key factors hold for both internal migration and displacement as well as cross-border migration and flight.

(1) *Problems arising from poverty and poor state performance:* as mentioned above, the largest proportion of populations in developing countries settle in rural areas. Some poverty clusters suffer not only from environmental discrimination but also from insufficient infrastructure, unclear or competing land ownership, subdivision of already small plots, and lack of credits. Phenomena as varied as soil erosion, heavy rains and flooding, drought, salinization, deforestation and woodland clearing, and overgrazing of savannas accelerate the dissolution of traditional living orders. These include specific ensembles of economy, culture, neighborhood, and kinship groups (families, lineages, clans). Reaching a point of no return, people have no choice but to

give up their homestead. At the same time, market economies absorb only a few rural dwellers. The market induces a highly selective dissolution of traditional structures. Thus landscape degradation belongs to the very transformation that has produced most of the migrants and refugees leaving their degraded environment to date.

(2) *Problems arising from the modernization*: problems of modernization include mechanized farming, mining, and urbanization. The various side effects of these activities - such as total loss of land, use of fertilizers, salinization, and pollution - urge rural dwellers to withdraw. They have in fact only two alternatives: either to move to more marginal lands and clear them, or to join the marginalized in (peri-)urban areas. Shrinking lakes (Aral Sea, Lake Chad), flooding, irrigation, loss of biodiversity, and the spread of epidemic diseases force resettlement, expulsion, and escape.

(3) *Problems arising from the location of population growth:* population growth occurring in marginal arenas creates more potential victims of natural events (such as landslides, earthquakes, and volcanic eruptions). These perils are incorrectly perceived as »natural catastrophes« and not as »social catastrophes«, which, as a matter of fact, they are.

A conflict-prone situation worth mentioning is intercontinental migration from southern to northern continents. Along the North-South fault lines, industrial countries are trying to stop the entry of illegal immigrants and to facilitate their return. Some of the migrants come from environmentally degraded arenas and can be viewed as victims of a globalized resource distribution conflict. As a rule, however, refugees from environmental degradation lack the necessary resources and health for long and costly trips. For this reason, the destinations of most migrants and refugees lie close to home. If one pursues the route of a migrant from his homestead to his possible destination in an industrial country, conflicts occur at various stops along the way: in neighboring ecological regions, in the national capital, in the arena just across the national border, in third countries, and only in few cases in other continents. Various obstacles represent significant inhibiting factors for large-scale and long-distance migration (*see: app. 11, AB.IV. 24*. Case studies by Faath 1996III: 203-268; Bächler 1994; Gallagher 1994:65-72; Ogata 1994:41-47; Suhrke 1994:93-100).

4.1.5. DEMOGRAPHICALLY CAUSED MIGRATION CONFLICTS (TYPE BV)

High population pressure in ecozones of low productivity causes either local conflicts or migration, which can lead to conflicts in the area of destination.

Demographic developments matter for environmental migration conflicts in three different ways: population scale in relation to resources available (density), population growth rate, and resource redistribution through migration and displacement. It is difficult to highlight the causal linkage between population pressure, environmental degradation, and violence. Yet in a few cases only (Rwanda, Bangladesh/Assam, Indonesia/Java) ample evidence suggests that such connections do exist.

The repeated subdivision of land in smaller and smaller inheritance shares is an indicator for these connections. Fragmented arable land, decreasing yields per hectare, and a lack of off-farm alternatives coerce large parts of the rural population to migrate towards urban arenas, where a conflict potential accumulates both concerning land use and distribution in growing peri-urban arenas as well as concerning the environmental decline in mega-cities (Girardet 1996:67-115).

Another indicator of demographically induced migration is the clearing and cultivation of new land in remote mountains, deltas, and ecologically sensitive coastal areas. Landless people and semi-nomads gradually move into protected zones in urban areas or into national parks. Social unrest can recur as these movements provoke clashes with governmental troops and contribute to politically unstable situations (*e.g.*, the Maasai in Kenya and in Tanzania). Acute conflicts occur if discrimination is perceived as tremendous by the actors affected. The threshold of the beginning of discrimination depends greatly upon the perception and varies from case to case. Generally speaking, discrimination is particularly perceived to be unacceptable when social and/or ethnopolitical factors accumulate facilitating group identity building (*e.g.*, between Bengali immigrants and residents of the Assam Province in India).

Population dynamics accelerate the impact of other key factors, such as poverty, inadequate land-use and land-tenure systems, environmental transformation, and poor state performance. This constellation of factors encourages cross-border migration, which - in the context of violent coups and civil wars - assumes the form of mass flight, *e.g.*, in the Great Lakes region in Africa (*see: app. 11, B.V. 25-27*. Case studies by Hafiz/Islam 1996II:1-108; Ehrensperger 1993 (manus.)).

4.1.6. INTERNATIONAL WATER CONFLICTS (TYPE CVI)

International river basins are the most obvious example for the general contradiction between ecoregional boundaries and state borders. The asymmetric dependence of upper and lower riparians on an international river basin triggers political tensions, international bargaining, and military threats. Since lower riparians are more vulnerable than upper riparians, they can easily receive discriminatory access to freshwater resources.

River pollution and water distribution conflicts are distinct problems. The former refers to the substantial degradation of resources, whereas the latter refers to economic scarcity. Pollution conflicts are represented as strife over an indivisible public good, which affects the degree of pollution, the political responsibilities, and the economic costs. Since neighboring riparians have a vested interest in solving pollution problems cooperatively - in win-win solutions - such conflicts are easier to resolve than those over access to the resource *per se*. Distribution conflicts turn out to be conflicts over divisible public goods. They are perceived as zero-sum games. Discriminatory access to scarce water resources affects national sovereignty and integrity more directly than pollution aspects. Both pollution and distribution can obviously appear in combined forms, which complicates the search for cooperative solutions.

International conflicts over water use develop in the context of strong riparian interest in securing access to the shared water resources, of asymmetric power distribution among riparians, and of the quality of the multilateral relations in general. Conflict dynamics also depend on climatic and geographical conditions, population growth, the economic structure, and the state ability to cope with vulnerability. Therefore, in addition to given hydrologic conditions, the political and socioeconomic *milieu* is of central importance for settling international water conflicts. There is no direct linkage between water pollution and distribution, on the one hand, and the intensity of conflicts, on the other hand; it is the political context which matters.

In regions that suffer from seasonal drought if not permanent water crises (*e.g.*, the Middle East), distribution and discrimination are highly sensitive issues, which are treated as threats to national security. Because water flow is easy to manipulate by riparians of a shared basin, scarcity conflicts in crisis-prone regions get inevitably mingled with other contributing factors. But the example of the Arab-Israeli peace process shows that negotiations about water management are possible even under conditions of acute scarcity and protracted conflict. This process is possible because all actors perceive water issues to contribute to no-win solutions. On the other hand, water talks can easily be canceled if and when the political situation changes.

There is no automatic spiral towards violence. So far there have been no open wars caused by water distribution issues alone. Even in arid zones, where states are extremely dependent on external water resources (Egypt), there has been a balance, albeit a precarious one, between threat and cooperation. The geographic course of a river is a power factor worthy of attention. If a country is the upper riparian and well equipped militarily, it holds all the trump cards. It can discriminate against lower riparians by regulating the cross-border water flow. If superiority is overwhelming, cost-benefit analysis will keep lower riparians from engaging in a warlike dispute despite its discrimination. The presence of a hegemon controlling the sources of a basin has a thoroughly stabilizing effect if the power demonstrates some flexibility and competence to enable mutually satisfying technical solutions (*e.g.*, to some extent US-Mexico concerning the Rio Grande and Colorado River Basins).

Within the context of institutionalized and cooperative relations power relationships are mediated by legal barriers and rules of behavior derived from custom. The best case scenario for avoiding the escalation of water distribution and pollution conflicts are regimes which focus on realities. Therefore, water conflicts in and between industrial countries are settled with peaceful means because of the parties' high degree of negotiating competence and existing regulatory mechanisms at the policy level; see for instance the dispute settlement capacities of the Rhine Commission in Western Europe. Environmental conflicts become a catalyst for cooperation if political compromises are seen as desirable and technical solutions as feasible. Successful compromises or even institutionalized mechanisms of dispute settlement reduce the danger of water-use conflicts getting out of control.

Only if water issues coincide with extremely unfavorable political conditions will they become a trigger of warlike actions (*e.g.*, between Israel and Syria in the prelude to the Six Day War). The asymmetrical geographic positions in the basin come into play when the upper riparian puts pressure on highly vulnerable neighbors. Some water-related conflicts coincide with center-periphery conflicts. If dominant riparians turn out to be authoritarian regimes with poor state performance, water issues further delegitimize central governments in the eyes of discriminated groups, which highly depend on water for agriculture (*e.g.*, Kurds in Anatolia, Syrian and Iraqi farmers below Turkish dams on the Euphrates and Tigris rivers). On the Indian subcontinent river basin conflicts are imbedded in a context marked by extreme poverty, ethnopolitical schisms, and the hegemonic demands of a regional power. These factors induced sociopolitical conflicts within the lower riparian Bangladesh or between Bengali migrants and inhabitants of the Indian Province Assam (*see: app. 11, C.VI. 28-39.* Case studies by Bächler *et al.* 1996: 117-166; Gleick 1996III:1-26; Hafiz/Islam 1996II:1-108; Klötzli 1996 II:247-336; Libiszewski 1996II:337-460; Okoh 1996II:222-232; Rogers 1996III:25-64; Thomas 1996III:65-126; Wegemund 1996III:285-314; Bächler 1997; Durth 1993; Klötzli 1993).

4.1.7. GLOBAL ENVIRONMENTAL CONFLICTS (TYPE CVII)

Climate change and stratospheric ozone depletion demonstrate the globalization of environmental transformation. Globalization is considered to be a rather conflict-prone process by various authors (Myers 1993; Renner 1996). Predicted developments towards a global environmental conflict formation notwithstanding, specific statements about the socioeconomic and ecological effects of climate change *cannot* be made based on the case studies carried out in the framework of ENCOP (Bächler *et al.* 1996:329-332). Sociopolitically significant sea-level rise, for instance, will be a phenomenon of the intermediate and long-term future. Continuing drought in the arid and semi-arid zones is not clearly attributable to the anthropogenic climate change. Nonetheless, international concerns (*e.g.*, the global campaign of the Small Island States) indicate that there is potential for future conflict. Due to the development dilemma, the victims of global transformation will be found where environmental discrimination has already provoked a precarious situation. If current conflicts can be traced back to global environmental phenomena at all, they presumably concern mainly domestic conflicts of types AI (center-periphery conflicts), AII (ethnopolitical conflicts), and AIII (internal migration conflicts). In other words, acute conflicts do not arise along the great fault line between North and South, but rather where climate change contributes to the collapse of local rural structures and regional political authorities.[4]

4.1.8. CONFLICT TYPES: CONCLUSIONS

Discrimination against actors in sensitive ecological arenas and a high level of dependence on natural capital are two key factors determining the conflict potential of transforming society-nature relationships. Conflicts about degraded renewable resources manifest themselves as international, center-periphery, interregional, and group identity struggles exacerbated by migration and displacement, and in some cases accelerated by population dynamics. Actors with access to state power have typically access to the most productive arenas, whereas identity groups facing environmental discrimination are forced to use and degrade marginal arenas with low productivity, thereby perpetuating impoverishment. In addition, groups against which environmental discrimination works are confronted with environmental deterioration *beyond* their control: deforestation by loggers destroys the livelihood of indigenous forest dwellers, dam building degrades land both upstream and downstream, mining leads to widespread contamination of the landscape, industrial water pollution leads to the depletion of inland and coastal fisheries.

The hypotheses of ECACP (*see: 2.4.*) are confirmed by the cases evaluated above. There is indeed little empirical support for the first hypothesis that environmental scarcity causes violent conflicts or wars between states. Thus alarming

---

4 ENCOP conducted a case study dealing with the globalized conflict in French Polynesia concerning the nuclear tests carried out by the French government. The study addressed the environmental disruption through testing as well as the protests of indigenous population, liberation movement, and INGOs against the policy of the French and the intervention of the navy (Danielsson 1993 (manus.)).

statements such as »water wars« or »green wars« are definitely to be questioned. Environmental conflicts tend to be persistent, diffuse, and subnational.

Simultaneously, there is substantial evidence to support the second hypothesis that environmental scarcity causes large population movements, which in turn cause conflicts. There is only one finding where ENCOP results differ from ECACP. While Homer-Dixon suggests a linearity between large population movements and group identity conflicts, ENCOP suggests that migration is linked to different kinds of conflicts: socioeconomic conflicts between highland and lowland producers, conflicts between rural and urban dwellers, and conflicts between rural producers and central states' forces. Migration also causes conflicts within one and the same ethnic group, which may be divided by geographical or national boundaries.

Empirical evidence partly supports the hypothesis that environmental scarcity simultaneously increases »economic deprivation« and »disrupts key social institutions« despite the fact that ENCOP refers to different theoretical concepts than ECACP. »Deprivation conflict« as one general type introduced by Homer-Dixon has comparable connotations as the ENCOP types: center-periphery, ethnopolitical, internal migration, and global environmental conflicts. The concept of »disruption of key social institutions« is grasped in the context of this study with the concept of *marginalization* induced by discrimination against certain actors, on the one hand, and by poor state performance in certain arenas, on the other hand.

Moreover, in most cases environmental conflicts involve rural populations in developing countries struggling for survival. Modernization and high dependence on degrading resources challenge the livelihood security of rural dwellers. There is a high probability that conflicts will escalate when

(a) a major contradiction exists between economic expectations and/or a larger demand for resources, on the one hand, and limited development perspectives, degraded resources, and poor state performance on the other hand (*e.g.*, few off-farm alternatives, lack of technical skills and financial means);
(b) at least one of the actors involved perceives the resort to violence as the best alternative to other solutions.

It is necessary to include many »if-then« clauses when examining violent outcomes of environmental conflicts. Environmental degradation may be an underlying reason of a certain conflict, it may be a factor leading to channeling along lines between distinct groups, and it may even be a triggering factor to conflict dynamics. However, passing the threshold of violence definitely depends on *sociopolitical* factors and not on the degree of environmental degradation as such. Among them are the lack of institutional capacities for peaceful conflict settlement, the readiness and/or capacity of authorities and leaders to organize and mobilize collective actors, the (mis-)perception of alternatives to resorting to violence, the preferences and opportunities of actors, and their limitations. These topics have to be examined in more detail to better understand when and at what point environmental conflicts turn violent.

4.2. Inevitable Situations and the Lack of Regulatory Mechanisms (Hypothesis Two)

When considering the interests and the behavior of actors, action can be seen as the result of two consecutive filtering processes of decision-making (*see: 2.3.7.*). Concerning the first filter the question arose: how does transformation influence the opportunity sets of individual and collective actors? Related to the second filter the question was: how does transformation shape actors' preferences so that violent conflict is considered the mechanism for solving environmental conflict?

In all 40 ENCOP case studies transformation of society-nature relationships was perceived as *serious* in terms of both degradation of renewables and discrimination against actors highly dependent on their shrinking natural capital. Yet only *18* of these cases crossed the threshold of violence. In *eight* cases out of them there were wars, whereas in *ten* cases there were violent conflicts below the threshold of war. In *22* ENCOP cases - which now serve as control cases - there was neither war nor violent conflict going on. In *11* cases out of them minor incidents of violent action occurred that were below the threshold of violent conflict. *Nine* cases experienced either military threat or political tension only. And in *two* cases the disputed projects were dropped or postponed (*see: app. 13a, b*).

Against the empirical background the conclusion is reached that the resort to violence only occurs if and when some of the following *five key situations* coincide:

(1) *Inevitable environmental conditions*: group survival is dependent on degraded resources for which no substitutes are apparent and eventually faces an inevitable and therefore desperate environmental situation. Inevitability does not stand for a deterministic or functional approach to human behavior. *Inevitable circumstances* are those upon which an individual or a collective actor cannot rely rationally or deliberately.

(2) *Scarcity of regulatory mechanisms and poor state performance:* when a political system is incapable of producing certain social and political conditions, goals, such as sustainable resource use, become unattainable. Scarcity of regulatory mechanisms happens either due to lack of state output regarding resource management and livelihood security or due to disruption of (traditional) social institutions designed to regulate access to resources. Migration, for instance, can be a result of the first type of scarcity (state

output) and thus provoke the second type of scarcity (disruption of institutions).

(3) *Instrumentalizing the environment:* the environment is instrumentalized or manipulated by dominating actors to pursue specific group interests, so that environmental discrimination becomes an (ideological) issue of group identity. Instrumentalization can reach from upstream manipulation of the seasonal water flow for political reasons to the deliberate poisoning of water resources.

(4) *Opportunities to build up organizations and find allies*: actors organize and arm themselves in political settings - often behind a strong leader - and gain allies either from groups affected by similar problems, from certain (fraternizing) factions of the elite, or from foreign groups such as I(N)GOs.

(5) *Spill-over from a historic conflict*: environmental discrimination occurs within the context of an existing (historic) conflict structure and, as a result, the conflict receives new impetus. This push can come either from the perceived consequences of environmental transformation or from intensified resource competition between polarized actors.

These five contexts constitute the arena for actors who choose violence as their best alternative to nonviolent activities.

4.2.1. INEVITABLE SITUATIONS

Rural dwellers experiencing environmental transformation depend more and more on less and less productive or available natural resources. Time horizons are very short as rapid resource extraction meets present needs - despite the resistance of third parties. If survival is at stake on a day-to-day basis, even minor incidents can force a decision about whether to stay and probably die or to flee. Violence may be viewed as a way to escape when no substitutes are available for degraded resources and/or the relationships between parties are stressed with no alternatives.

An inevitable situation has two sides: a lack of off-farm alternatives and poor state performance. High socioeconomic pressure on rural populations is paired with the powerlessness of policy makers to achieve desirable goals, such as the wider distribution of more productive land, the creation of jobs, and the alleviation of poverty. The crisis of subsistence economies is therefore caused by both external and intrinsic factors. Both are closely intertwined. External factors are the commercialization of agriculture and related issues. The major intrinsic factor is the relatively low and even declining productivity of subsistence economies due to poverty-driven population dynamics in marginal ecozones.

Inevitable situations tend to lead actors in a certain arena to hold mutually exclusive goals. This, in turn, provokes rigid polarization and - as a possible consequence - the resort to violence. Most environmental conflicts have to be seen as part of a new wave of rebellions, where rural producers find themselves in inevitable situations. The arena's groups encompass peasants, smallholders, subsistence farmers, landless people, rural workers, lifestock breeders, and other rural groups. The opposite party in this conflict formation encompasses large-scale farmers, agro-business, *latifundistas*, rural entrepreneurs, international companies (*e.g.*, mining), urban dwellers, and certain political elites.

Inevitability is a *relational* rather than an absolute term. Many situations perceived as inevitable - such as »natural catastrophes« in a densely populated arena - are in fact avoidable social or *humanitarian* catastrophes. Inevitable situations are determined by major social stress induced by an overwhelming density of conflictual interactions. A few of these *stressors* include: for almost exhausted land regulatory interventions to preserve productivity must occur faster and more intensely than for fertile land (shorter fallow periods, if at all; multifunctional land use; more frequent change of crops, more frequent movements of herds, sharper competition, etc.). Simultaneously, a growing number of activities takes much more time than before (hauling water, collecting fuelwood, finding pastures and water wells, cooking with solar energy, etc.). Time is literally running whereas dependency on scarce resources urges a larger time frame for many productive and reproductive activities.

4.2.2. LACK OF REGULATORY MECHANISMS

Inevitable situations cannot be avoided because there is a lack of regulatory mechanisms, be they traditional conflict and resource management, modern law, or international regimes. If instruments for managing resources and regulating conflicts become ineffective over time, actors may come to view violence as a thoroughly rational means of pursuing their own interests. Instruments become blunt because traditional means are inadequate for new challenges of environmental discrimination. And new institutions based on modern law are not yet available to meet these challenges. In marginalized regions, the central state has not succeeded as an administrative and law-enforcing apparatus or as an institution founded on the rule of law, legitimized and accepted by the local actors.

With respect to environmental conflicts, the role of legal and civil institutions cannot be overemphasized. Civil society as a subsidiary conflict-regulating corrective is (largely) lacking in the countries analyzed, although big differences between some medium-developed and least developed countries exist. Nonetheless, political pluralism or the ability of opposition movements to make their claims is weakly developed in case studies, where violent conflicts and wars occur. The less stable and developed regulatory mechanisms are in a given society, the more susceptible the society will be to violence. The more established the rule of law and civil society, the lower the level of violence and the more meaningful the use of force. In participatory societies countless and sometimes serious environmental conflicts are resolved by legal and political means. Negotiations, compromise, and mediation play a central role. The organized use of force is not a central part of political strategy.

A state's authority to act consistently vis-a-vis environmental transformation should encompass a large array of economic, social, and institutional instruments: to assess suitability and to support crop choices, to enhance the workability of land, to provide access to markets, to make credit and cash available, to introduce land property rights, etc. Most of such instruments are hardly available in the ENCOP case studies, where an *appropriate choice* would sometimes have made the difference between degradation and sustainability. Local and regional arenas if not the state as a whole are subordinated to the interests of the center, which is often more concerned about adhering to international standards on commercial and investment law than about internal developments outside the capital district. Parts of the marginalized population see the state as a bureaucratic apparatus, a hostile agent for foreign interests or both that plunders national resources without redistributing the revenues to provinces and communities.

The establishment of subsidiary conflict and resource management mechanisms would presume more than a mere economic distribution logic. Yet precisely the lack of conflict-resolving mechanisms prevents innovative practice. In many places the ruling political culture allows little latitude to manage resources subsidiarily, that is, on the lowest level possible (except on marginal and degraded lands of minor value). As a consequence, there is widespread insecurity concerning property rights. Property rights disputes have rarely been solved satisfactorily, depriving arenas of a prerequisite for effective local self-government and sustainable resource management. Property rights enhance livelihood security and thus contribute to labor-intensive improvement of the productivity of sensitive soils.

States with poor performance are unwilling to adapt existing international regimes to new challenges (*e.g.*, Nile riparians). Nor are they committed to delegate substantial authority to supranational regional organizations which aim at acquiring dispute settlement capacities (*e.g.*, International Governmental Authority on Development, IGAD, in the Horn of Africa). Existing environmental agreements often express goodwill, but they show a considerable lack of binding legal power and strict implementation. The search for the least common denominator, weak enforcement mechanisms, and »free riders« characterize regional agreements.

On the one hand, weak states are not committed to assuming political responsibility for the ecological crisis. Instead, governments tend to count on internationalizing responsibility for the crisis and waiting for foreign assistance. Due to the weakness of civil society, on the other hand, (re-)privatization of state power occurs through relatively small and inaccessible cliques usurping the state's monopoly on the use of force and changing its function into a spearhead against the population experiencing environmental discrimination. Only in relatively few cases is the disadvantaged group capable to respond with organized violence to the poor performance of their state and/or to the robbery of local natural capital by national elites.

4.2.3. INSTRUMENTALIZING THE ENVIRONMENTAL PROBLEM

Because of the great importance of safe water supply for vulnerable states, international river basins are easily instrumentalized as political means of pressure or blackmail. As discussed earlier, a strong upper-riparian state can carry through geopolitical interests against lower-riparian neighbors. The lower riparian has clearly fewer means of pressure at its disposal. However, it can seize the water issue in exemplary fashion in order to denounce the upper-riparian state's unethical behavior. This strategy helps to create international awareness and to mitigate the asymmetry between the actors.

In cases where heavy environmental damage is caused by third parties (*e.g.*, mining companies), the protection of nature proves to be a good mobilizing factor for local groups. This strategy can also be beneficial because environmental consciousness is rewarded on a global level (by the UN, INGOs, etc.). Using the ecological vocabulary, though previously little concerned with nature protection, is often the only way for marginalized groups to get attention concerning their worsening living conditions (*e.g.*, Ken Saro-Wiwa and the Ogoni in the Delta State of Nigeria).

Opposition groups tend to instrumentalize ecological crises in their criticism of the state. Organized actors in opposition sometimes use segments of groups facing environmental discrimination for ulterior political purposes. Remnants of communist guerillas with recruiting problems side with the demands of protesting farmers against deforestation and export business. Indigenous peoples with close and mythical nature relationships see a political advantage in making environmental disruption central in their criticism of the modernizing state. Thus - feeling uncomfortable with the infiltration of the modern world - they emphasize the cultural and spiritual dimensions of deep human-ecological relationships. The destruction of sacred »mother earth« by foreigners is rejected as extremely immoral and as a threat to humanity.

4.2.4. OPPORTUNITIES TO BUILD UP ORGANIZATIONS AND FIND ALLIES

Instrumentalizing or manipulating environmental transformation is not necessarily explosive. A military conflict only emerges if specific fault lines accentuate it, if polarizing actions drive it forward, and if groups are organized and mobilized. In order to do so leadership - and arms - are required.

It is noteworthy that the threshold of organized use of force in environmental conflicts is relatively high in comparison to the large amount of environmental discrimination occurring throughout the world. Given the dimension of environmental discrimination with at least 500 million of the poorest people living in ecologically sensitive zones (about 400 million in rural and about 100 million in urban zones), how can one explain the fact that force has been used in so few cases to regress these grievances? There is certainly no linearity between environmental discrimination and the use of violence. Many inhibiting factors are involved. The same cause has, as a matter of fact, different effects: individuals prove to be unwilling or unable to join a group with strategic goals; various communities and identity groups have virtually no capacity to go to war. Different reasons account for these outcomes: marginalization, lack of means and organizational skill, lack of leadership, state repression, poor health conditions, wide-ranging apathy, fatalism, defeatism, and religious mystification at one's own situation.

Only groups which actually organize and arm themselves seize the means for collective resistance against other parties in the arena. The ability and opportunity to ally and build coalitions with other actors are important prerequisites for organized violence. Mainly rural actors capable of waging conflict need powerful coalition partners from different social levels to support their goals, for instance, part of the *intelligencia,* members of the middle class, or a charismatic leader of an ethnic minority at risk. Many scholars of studies in peasants' revolts come to the same conclusions as Barrington Moore, namely that »(p)oor peasants and landless laborers (...) are unlikely to pursue the course of rebellion, unless they are able to rely on some external power to challenge the power which constrains them« (Wolf 1969:290).

It is the weaponry that makes a difference. Through proliferation of cheap weapons, especially widely available small arms, individual conflicts - *e.g.*, between farmers and nomads - assume a more dramatic turn of events than was probably intended by parties to the dispute. Parties that have historically fought each other with traditional weapons often underestimate the lethality of modern arms. The capacity to wage an armed conflict is especially reinforced where marauding armed gangs, militant youth groups, and demobilized soldiers or mercenaries are entrusted with the supply of weapons and/or take part in military actions.

Yet coalition possibilities are not always available to groups attempting to redress environmental discrimination. Communist guerillas with greater fighting experience have dropped out as potential allies with few exceptions (*e.g.*, Malaysia, Philippines). The UN has recently demobilized the last armies that fought against the colonial powers (*e.g.*, Mozambique). As a result, it has become more difficult than in the 1970s and 1980s for rural populations under discrimination to find combat-ready allies.

Furthermore, national and international environmental organizations (I/NGOs) are committed as a rule to nonviolent forms of resistance. Their solidarity and expertise are aimed at environmental concerns and they are prepared to oppose government positions (*e.g.*, IUCN concerning the Okavango Basin in Botswana). But these groups do not resort to using armed force. In the international context of the United Nations, increasing emphasis is put on peaceful resolution of disputes within nations. Therefore, groups prepared to use force cannot count on support as they could for colonial and post-colonial liberation movements.

4.2.5. CONTEXT OF AN ONGOING ARMED CONFLICT

The remarks on the limits of using force apply if environmental conflicts are not drawn into the maelstrom of historical armed conflicts or even provoke them anew. Environmental problems, such as the scarcity of river water, can have massive consequences within the framework of a comprehensive historical conflict, *e.g.*, water issues in the Middle East.

A more frequent phenomenon is the twofold interaction between environmental destruction and war. Many domestic and international Third World wars waged within the context of the Cold War had a devastating impact upon renewable resources. Deforestation, destruction of vegetation, expulsion or killing of farmers and livestock caused massive degradation in regions, such as the Horn of Africa where highland areas were already threatened by heavy erosion. In the aftermath of protracted civil wars, rural development is more at risk than before the war. The environmental destruction - previously a result of protracted conflict - may itself become a contributing or triggering factor of a future confrontation.

4.3. The Role of the Environment as a Cause of Conflict

Which role does environment play if and when it causes a violent conflict? Is it a deep-rooted reason for violent conflict? Does it shape group identity? Or is it an intervening variable contributing to the escalation of violence?

As most of the area studies show, transformation of society-nature relationship does play an interesting role in the causation of violence. In the cases under consideration environmental degradation is supposed to be an exogenous or necessary factor to the conflict; this means the conflict would not have occurred in the same way - or not at all - without environmental degradation being an important variable.[5] The evaluation also indicates that the role of environmental discrimination - depending on the individual case - can be quite different. Its role ranges from discrimination being an underlying reason to the point where it is a proximate trigger to a violent conflict. In the causal analysis, therefore, it is critical to distinguish clearly between the various impacts environmental degradation has on the conflict behavior of actors (*see: 2.3.8.*).

4.3.1. REASON

First, transformation of society-nature relationships plays a role as a *reason* for a conflict. It is perceived as almost predestined and, from the viewpoint of the groups affected, hardly within their power to ward off. This is the great background role of the environment being the permanent »noise« in the system. Transformation of landscapes in a historical dimension and its effects »act« as either hidden or clearly visible »system powers« by touching the opportunities and preferences of affected actors in many ways.

Due to complex interactions it is often difficult to distinguish between the role of transformation of the landscape and the role of economic decline. However, there is an important difference between *economic* scarcity and *environmental* scarcity, which has always been neglected in classic economics. Economic scarcity addresses the distribution problem of man-made goods between those interested in access to these goods. Environmental scarcity, on the other hand, highlights the input side of a third (external) factor, namely of natural resources provided by the landscape as a life-support system (land, water, minerals, coal, oil, gas). Economic conflicts are political conflicts, which deal with the production and (re-)allocation of human and physical capital; environmental conflicts are political conflicts, which are concerned with the availability of natural capital. The latter is a necessary prerequisite for any economic activity. The subject of environmental conflicts are degraded sources and overstrained sinks.

---

5 From a social science perspective those cases are interesting only where environmental scarcity is a necessary factor. With the configurative case studies approach the distinction between contributing and necessary factors depends highly on subjective judgements. Therefore, biases were diminished by periodical discussions among the researchers, which contributed different case studies to ENCOP. On the other hand, the development of a testable model will be necessary if one aims at empirically falsifying the distinction between contributing and necessary variables (*see: 2.3.6.*).

For example, eroded and marginal land triggers conflicts over access to productive land. Polluted water resources trigger conflict over access to rich fishery resources. Hence, the reason or the *casus belli* is the increasing availability of »common bads« and the discriminated access to scarce »common goods«.

4.3.2. TRIGGER

Second, transformation of the landscape plays a role as a *trigger* if actors perceive discrimination against themselves as inevitable. A trigger causes an actor, who was previously averse to violence, to prefer violent action. Sudden events, such as crop failures, trigger migration and flight which lead to violence. Violent action cannot be excluded as a possible outcome if livelihood security is at stake and organized actors face environmental discrimination (*e.g.*, ethnopolitical conflicts).

The transformation of the landscape triggers violence if it is obviously caused by projects of third parties (*e.g.*, mining company/central government on Bougainville and Dutch Shell/central government in Ogoni land in Nigeria). Generally speaking, transformation triggers violence if discrimination against actors is immediately linked with specific events leading to the destabilization if not dissolution of the social order. The latter may be caused by the use of marginal land or by specific project-related activities, which create »national sacrifice areas«. Therefore, conflict analysis has to focus on how political institutions operate, how socioeconomic structures fall apart, and how traditional ways of life are at stake.

4.3.3. TARGET

Third, environmental concerns become a *target* of discriminated actors if transformation of the landscape is what the conflict is all about - at least in the eyes of one of the actors. In many conflicts sustainable resource use may be an ultimate goal of actors highly dependent on their natural capital. However, in the foreground is most often the protection of one's own resources against the intervention of third parties.

A target usually encompasses different subgoals not always internally consistent. Resistance against foreign intrusion into one's own environment is also the expression of a dilemma discriminated actors are confronted with. On the one hand, it is a matter of safeguarding the natural and cultural environment against invasion of modernity; on the other hand, it turns back the threat of marginalization by the interest to participate in modernization and development. Thus environmental concerns - first having been a target - become a reason for the targeting of new options. As we have seen in some cases, the struggle for self-determination, autonomy, and secession becomes the main target putting the environmental concerns on a sidetrack. Central governments tend to react to the call for self-determination by upholding national sovereignty and territorial integrity, if necessary by use of military force. Since the latter is perceived by discriminated actors as a further proof of centralization and delegitimization the targeting on self-determination is justified once more.

4.3.4. CHANNEL

Fourth, environmental concerns serve only indirectly as a *channel*. A channel is a line of political, social, economic, or national cleavage. Channels are thus designed to shape the group identity by manipulating existing sociopolitical fault lines.

Even though a high level of environmental degradation in a certain arena shapes threat perceptions, channeling moves the environment to the background as ongoing conflicts proceed. Once a conflict escalates to war, it will hardly be waged primarily over the original reason or the trigger of the conflict itself. In the hot conflict phase hostile parties tend to resort to fundamental legitimization patterns and ultimate goals. Slogans such as »to be or not to be« or »they« destroy »our« resources are mobilizing channels more than »land scarcity« as such.

Nonetheless, group leaders fighting for autonomy or secession may promise at the same time a solution to environmental problems. If self-determination will be achieved - so the assumption goes - »we« will not act as irresponsibly as »they« did. War is therefore not waged directly to solve ecological problems, even though they may be a reason or a trigger. Similarly, war is not waged in order to defend the traditional way of life against the »attacks« of modernization. War is often about self-determination and national sovereignty. Once this goal is achieved self-determination is supposed to contribute almost automatically to the realization of previously formulated ecological goals. This, however, almost always turns out to be a miscalculation.

In politicized identity conflicts and center-periphery disputes environmental damage is used as a means to realize larger goals. Marginalized groups may conclude that they can only find coalition partners and international recognition if the environmental damage caused by them can be used for solidarity to realize a further goal: *e.g.*, independence from a corrupt or nepotistic central government. Indeed, this mobilizing strategy forms the basis of clearly perceptible and perhaps even dramatic environmental destruction. However, the environmental problem is overemphasized or it is taken selectively as a factor from the other context and reinforces the attempts to shape identity (*e.g.*, Ogoni in Nigeria or the Bougainvillean Revolutionary Army against the central government).

4.3.5. CATALYST

Fifth, there are only few cases where transformation of the landscape becomes a *catalyst.* However, sudden events such as floods or cyclones may unexpectedly contribute to the further deterioration of renewables and - as a result - to acute problems in supply of food and therefore intensify ongoing resource conflicts. The damming of water leading to acute downstream scarcity or the severe pollution of freshwater resources also suddenly enhance tensions between conflicting parties. As a rule one can state: all actions which are carried out intentionally in order to deny someone else's access to resources lead to the environment being a catalyst. Moreover, if environment is designed to be a catalyst it may also be a valid instrument for channeling (*e.g.,* Delta Region in Nigeria).

4.4. The Intensity of Environmental Conflicts

Actors alone shoulder responsibility for triggering and supplying the motive for violent conflict. A distinction must be made between the structural causes of a conflict and its dynamics or intensity. While environmental discrimination plays different roles in the causation of a conflict, its intensity depends not on the degree of the physical and chemical degradation of the landscape. As pointed out earlier, there is no linear correlation between the quality or quantity of natural resources available and the intensity of violence; there exist a lot of accelerating and inhibiting factors.

In disputes between the *center and periphery,* all-out wars are rather unlikely. This generalization applies especially for mining and dams. In such settings escalation to war occurs only in exceptional cases (*e.g.*, Bougainville and Chico). Violence is prevalent at a relatively low level with only few casualties. Center-periphery conflicts engender almost everyday endemic and diffuse violence by groups facing discrimination. But these groups hardly display organization towards developing »war parties« with clearly defined strategic goals. Conflicts often escalate in a spiral of violence if acts of sabotage prompt government troops to take punitive actions directed arbitrarily against communities and settlements. If escalation to violent conflict actually occurs in connection with large projects and accompanying ecological degradation, most of them remain below the war threshold. The conflict is often contained within the especially sensitive arena, such as a national sacrifice area, by the militarily superior center.

The greatest conflict potential lies in *ethnopoliticized* conflict settings and in *interregional* or *demographically driven migration conflicts* in countries with poor state performance. The actors are as numerous as they are diverse: minorities versus majorities, tribes versus tribes, clans versus clans, native people versus immigrants, settlers versus nomads, nomads versus governments, subsistence farmers versus multinational corporations and central governments, unemployed versus the financially better-off, and rural classes versus the central government and nomenclatures. The diversity of the actors shows that two well-equipped armies with heavy weaponry seldom face each other. Often more or less motivated government troops see themselves confronted with lightly armed groups. Despite these trends the risk of arming marginalized groups should not be underestimated. Struggles for resources have historically been relatively confined and partly ritualized by tradition between indigenous groups. But modern weaponry often brings about a more lethal level of dispute between opposing troops.

In certain ethnopolitical wars of medium and high intensity resource degradation, competing land-use rights and tenure systems, population growth, ethnosocial stratification, regionalism, and maldevelopment accumulate into an insoluble problem syndrome causing and/or triggering violent responses. A high intensity of violence with all its excesses follows, such as war crimes, rape, massacres, and crimes against humanity including genocide (*e.g.*, Rwanda, Sudan).

In the foreseeable future environmental conflict will not be a »world war« with a global front. A war between the United States and China to preserve the ozone layer, for example, would be absurd. Even classical interstate wars - for instance between riparians of the same river basin - may remain an exceptional phenomenon due to intensified efforts concerning international agreements. However, a certain threat potential in some cases warrants careful monitoring (Middle East, Central Asia, Nile Basin, Mekong Basin).

The growing problems of supplying agriculture, industry, and households with freshwater will become domestic problems. They will be linked to conflicts due to marginalization of rural poor or to the creation of national sacrifice areas. Either way they are two sides of the same coin, namely environmental discrimination. Conflicts in marginalized ecoregions and national sacrifice areas are by definition related to some clusters within states. Thus they fail to induce an overall conflict pattern affecting countries as a whole. Often central governments try hard to contain violence within the arena at stake. These attempts, if successful, lead to protracted low-intensity conflicts in focal areas. As a result, heterogeneity increases between highly productive, rural farming arenas and efficient urban centers, on the one hand, and ecologically sensitive rural arenas with low human development, on the other hand. The front line between the two sectors becomes the fault line of ongoing conflicts: the same key factors lead to further transformation of society-nature relationships and violent conflicts: environmental discrimination, overuse of renewable resources by actors highly dependent on natural capital, unclear and competitive tenure systems and property rights, and political mobilization against poor state performance in marginal arenas.

CHAPTER 5

CASE STUDY

Why Environmental Discrimination Caused Violence on the »Mille Collines«

The 1996 humanitarian disaster occurring in numerous Rwandan refugee camps on Zaïre's territory made commentators forget about the deep roots of the internal conflict on the *»mille collines«* of Rwanda. But the behavior of the international actors also indicates that the reasons for violence, massacres, and genocide have always been ill-understood and ever more so. When we go back to the starting point of the present refugee crisis - 6 April 1994 - we get to some ready-made explanations for escalation of the conflict in Rwanda. At that time the presidential plane was hit by a rocket and crashed on its approach to Kigali, killing the presidents of both Rwanda and Burundi. There were reports about tribal feuds between Bahutu and Batutsi, bloody chaos, and a Hobbesian battle of all against all.

The viewpoint focusing on interethnic clashes has to be broadened. What, at first sight, appeared to be an interethnic bloodbath is at a closer look much more complicated. The disaster consisted of three major front lines: firstly, genocide against the Batutsi and massacres among the moderate Bahutu, which had been plotted before the plane crash by a relatively small number of conspirators in the presidential guard, the army, the police, and the militias[1]; secondly, the war between the *Front Patriotique Rwandais* (Rwanda Patriotic Front, FPR) and the *Forces Armées Rwandaises* (Rwanda Armed Forces, FAR); and thirdly, interregional conflict which - roughly speaking - led the northern provinces with Bahutu strongholds (and where the influential tribe of President Habyarimana's wife stems from) against the southern provinces of the country, where the Bahutu were considered to be more Batutsi-like than »real« Bahutu.

These were the crossing conflict lines. Now the question about the underlying causes for the violent outbreak of conflict must be asked. Given the ethnic configuration in the Great Lakes region in general (including Burundi) and in

1 In its interim report the UN named 55 persons supposedly responsible for the massacre and genocide (Security Council 1994:para 29). See also: The final report of the Commission of Experts (1994), The Report of African Rights (1994) as well as Abdulai (1994).

Rwanda in particular: which key factors were a necessary prerequisite for worsening of political, interethnic, and interregional relations? Which intervening factors contributed to accelerating the dynamics of the conflicts? My preliminary answer encompasses three key factors to the conflicts: environmental discrimination against most rural producers by a tiny ruling elite, population dynamics, and historically grown social stratification, which was ethnopoliticized intentionally several times during recent decades. These factors include components of the historical and the more recent past.

In order to explain the syndrome of factors running together in the Rwandan arena, I refer to the concept of causal roles introduced in chapter 2 of this study. To briefly repeat the five *causal roles*: there are *reasons (I)* or underlying causes (sources), which lie far back in time but still have an impact up to the present, although they may no longer be in the consciousness of the actors; *triggers (II)* provoke an actor to prefer violence instead of other means used before. They often assume the role of »occasions«. Indirect and direct occasions can be distinguished; while the latter may have triggered the current conflict, the former are decisive for earlier conflicts without which the present conflict is inconceivable; *targets (III)* are what the conflict is all about in terms of interests and objectives of the actors; there are *channels (IV)* created by the actors. Channeling aggravates distinct existing fault lines in order to harness them in light of the targets. It builds group identity with social, ethnic, or other attributes; there are finally *catalysts (V)* or factors, which control the duration of the conflict and its intensity. Catalysts are amplifiers, which accelerate the conflict once it has been triggered and the channels have been built up (*see: fig. 5.1.*).

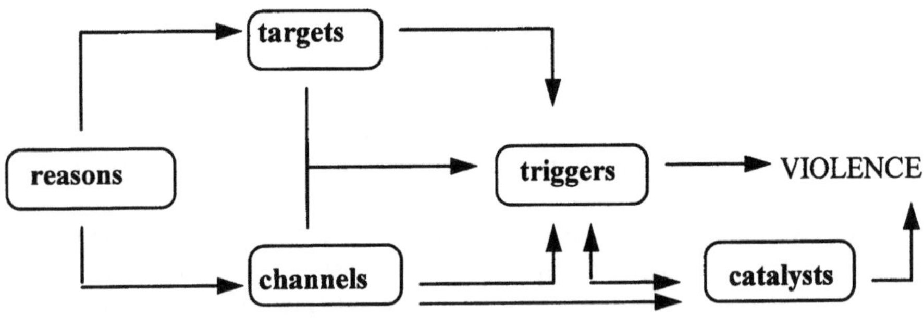

Figure 5.1. Multiple causal roles concept

This chapter explores one particular case more in depth. It is aimed at leading to a more standardized conflict explanation, which must be seen as the necessary preparation for modeling environmental conflicts. If the concept of causal roles holds and if key factors - such as discrimination - can be integrated in such an explanation, we will be one step closer to a model of transformation conflicts (*see: 6*).

5.1. Propositions

If we define *ethnic groups* as communities tied together by blood, language, and common history, then Bahutu and Batutsi belong to the same ethnicity. Both speak the same language (*kinyarwanda*). The members of both groups belong predominantly to the Catholic Church, both used to live in the same settlement areas, at least before April 1994. Assignment of the individual to one of the two groups is, after generations of mixed marriages, only possible because the paternal line alone defines genealogy. It is widely seen as the paradox of Rwanda and Burundi that far-reaching divisions by the law of heredity on the basis of an ethnic mythology have been able to survive until today and have done so *despite the remarkable cultural unity* of the population in the Lake District of Africa (Chrétien 1992:121-140). Yet, this does not properly answer the question how in the ancient »ethnic nation« of Rwanda ethnic identities and with that »subjectively perceived« contrasts were formed between two groups whose different origins got lost in the clouds of the distant past.[2]

My proposition is as follows: the conflict in Rwanda cannot be reduced to a question of genealogy in accordance with birth and blood.[3] Instead, the distinct ethnic groups must be defined as »psychological communities« whose members share a distinguishable and continuous collective identity based on cultural features and way of life (Gurr 1993:3). Understanding how distinct identities have emerged or, in other words, how ethnogenesis worked means tracking down the reasons for grave violations of human rights, crimes against

---

2 Schildkrout follows the definition of an »ethnic group« given by Max Weber when she writes: »The minimal definition of an ethnic unit is (...) the idea of common provenance, recruitment primarily through kinship, and a notion of distinctiveness whether or not this consists of a unique inventory of cultural traits« (Schildkrout 1979:184).

3 This is suggested in the social science study by Maquet, which set the tone for a long time. Maquet (1954) sees no fundamental changes between pre- and post-colonial ethnic relations in Rwanda (Maquet 1961). See the criticism of this position by Newbury (1988). She emphasizes transformation of interethnic relations by colonialism. At the same time, she investigates the role of the sociopolitical Batutsi and Bahutu categories in forming the state based on the »patronage-clientele paradigm«.

humanity, genocide, and torture.[4] In determining the key factors, one has to examine whether or not relationships between the two actors are characterized by *rank order* and how a hierarchy has been formed. Moreover, it matters how groups are divided in the available *territory* concerning their distribution on good and bad lands, what access they have to resources, the latter also in terms of what *positions* they occupy in public administration and politics (Horowitz 1985:24-26). In other words, one has to focus on various aspects of discrimination in general and environmental discrimination in particular; one has to explore concisely who has discriminated against whom, when, and why.

In the context of the argument about ethnic groups defined as psychological communities with a certain rank among them two simplifications should be avoided. Both the assignment of Batutsi and Bahutu to two different groups related by blood and the definition of the two ethnic groups based on purely stratum-specific categories are rejected. For instance, the concentration on genealogies cannot explain the formation of *multiethnic* clans or how it was possible to *convert* from a Muhutu to a Mututsi[5] and vice versa; the latter just on the basis of social criteria. The argument on social stratification suggests Batutsi always to be rich rulers and Bahutu always poor oppressed people. Then, by definition, neither rich Bahutu nor a Bahutu government could exist, since everyone at the top would automatically become Mututsi and each poor Mututsi would become Muhutu. Although there were societal tendencies that would support this very argument, it is contradicted by several facts. Among them are the formation of a Bahutu elite serving several Batutsi dynasties, *e.g.*, as »hill chiefs«, the social revolution of 1959-60 which led to the first Bahutu republic, and the existence of subordinate Batutsi who have always been farmers and not typical Batutsi stockbreeders. *Ethnic intermingling, social mobility,* and *political opportunism* have led to complex relationships within Rwandan society, which cannot be reduced to a common denominator - be it ethnicity or social stratification.

I further propose that the conflict was caused and escalated by significant discrimination, which grew over the years, and unfulfilled needs at both levels: the material level encompassing territorial claims and resources including the resource *power*, and the psychological level encompassing group identity, individual and social recognition, local and regional autonomy, etc. As Herbert Kelman recently pointed out, »psychological needs as identity and security - in contrast to more material interests like territory and resources - are not inherently zero-sum in nature«, although in protracted conflicts they are often perceived in zero-sum terms (Kelman 1996:105). In Rwanda - and Burundi - both

4 Thus the charges of the Independent Commission of Experts appointed by the UN in their reports (UN Security Council 1994).

5 There is one Muhutu (sing.), but several Bahutu (pl.), one Mututsi and several Batutsi.

ethnic groups or so-called psychological communities are supposed to have seen the material as well as the psychological level as zero-sum in nature. There are four factors that contributed in different ways to the discrimination against Bahutu: ethnopoliticized hierarchy, distribution of natural resources, territory in the political sense, and discriminatory access to power in terms of administration, politics, and armed forces.

Since this study deals mainly with environment-related causation processes, I will emphasize the interrelatedness of ethnogenesis and environmental discrimination. Other factors will be dealt with in a cursory and thus more superficial way.

5.2. Causes and Their Roles

In the following section I shall investigate the four factors mentioned above in the light of their roles in the process of causing the recent violence in Rwanda (and, to a similar extent, in Burundi).

5.2.1. REASON I: ETHNOGENESIS AS HIERARCHY

When assessing the role of ethnic hierarchy in relation to the present situation, several observations have to be taken into account. The hierarchy, which from a very early stage included elements of caste formation in the Weberian sense, was neither an anthropological constant nor a historical premise. The hierarchy was socioeconomically and politically *constructed*. Ethnic hierarchy developed between early Batutsi clans and dynasties at the top and Bahutu clans and marginalized kingdoms at the bottom. It was not until very recently that the centralized Batutsi kingdom - urged on by the colonial powers (first Germany, then mainly Belgium) - transformed the complex intraclan and interclan relationships. How did this ethnic hierarchy lead to a situation that was well addressed by Catharine Newbury (1988) as *»dual colonialism«*? In order to answer this question one must explore how material and psychological needs historically remained unfulfilled and how these two facts reinforced each other. Thus after a word on the quality of historical sources, I will first address political and psychological ethnization and secondly the socioecological side.

Historical Sources

The only sources available on pre-colonial events in the Great Lakes region are oral traditions. The extremely rich and accurate documents of history, social life, and warfare are divided into so-called codes, such as the military code, herders' code, and administrative code. Precise information, rules, and ceremonial customs have been transmitted from one generation to the next by *rapporteurs* who inherited this role from their ancestors. Each rapporteur had to learn only a relatively brief part of the codices by heart to make sure that they did not forget important parts. Thus the individual rapporteurs had no general knowledge of the history. To listen to the traditions as a whole was a privilege of the kings alone.

In the 1950s the codes were written down for the first time ever by Abbé Alex Kagame, himself a Mututsi, with the help of the Belgian Jean-Jacques Maquet. Concerning the methods involved Kagame noted: »*Voulant m'aider en cette tâche, il (*Maquet, G.B.) *entreprit une enquête qui lui fit visiter toutes les provinces du Rwanda, où, grâce au concours de l'administration européenne et du roi Mutara III, l'attendaient les notables de ces diverses régions, chefs, sous-chefs et particuliers, Batutsi, Bahutu et Batwa, qui avaient vécu sous le régime décrit par le Code*« (Kagame 1952:11). After a meeting held by chiefs coming from all regions of the country, they completed the list of elders who had total or partial knowledge of it. All of them were interviewed on the basis of a questionnaire with one hundred questions. Kagame published the codes (most of them in French) in a series of the Belgian Colonial Museum; the latter being the best place for the study of Rwandan history.[6]

The military code, for instance, has 197 paragraphs, which regulated the defense of the territory - most often through establishing camps near the border and *vis-a-vis* the camps of the hostile kingdom across the border. The army consisted of several armies founded by different kings. They were called »social armies« because their major task was not warfare - »expeditions« into foreign territory lasted only two days - but to monitor the complex system of distributing goods and gifts between the king and the population.

The herder's code has 133 paragraphs. It describes an organization parallel to the armies, the king himself being the herder-in-chief, whereas an army chief had to protect the herds. The next lower level in the hierarchy was the chief of the longhorn cattle (*nyambo*). This inherited role requested special skills in breeding as this cattle was regarded as being sacred. Then the chief herder followed with a number of families who had to take care of the herds of longhorns and ordinary cattle - part of them belonging to successful soldiers (Kagame 1952:84).

Besides codes on political institutions there were various other codes, which dealt with life at court, with poetry, etc. Kagames lists »*le code*

---

6 It took me at least three weeks to go through them. Although their value for the present work may not be very obvious, I can only recommend them because they reopen a world whose richness may have been forgotten against the background of the present tragedy. In fact, if one reads about the differentiated societal order established over centuries, one would never expect such a slaughter as happened in the 1990s.

cérémonial ésotérique de la dynastie, la poème généalogique de la dynastie, la poésie dynastique, la poésie guerrière épique, les morceaux de harpes, les chants guerriers, la poésie guerrière lyrique, les armées sociales, les armées bovins, la poésie pastorale, et l'histoire des familles (Kagame 1972).

These are the sources which help us to highlight the background of today's situation. However important the early work of Kagame and Maquet has been for understanding the Rwandan past, their approach was also criticized as biased. Catharine Newbury pointed out that »the questions posed in the standardized questionnaire used in his study generated normative responses about how the Rwandan political system was 'supposed' to have functioned in the eyes of the elite and beneficiaries of the system. Moreover, Maquet based his analysis on the responses of 300 Tuutsi informants (...). It is clear that the primary norms represented were those of central Rwanda, and particularly those of the central court« (Newbury 1988:6). They barely considered the regional autonomies and the missing state organization. Thus a history from below - that is outside the central province and perceived by the majority of poor farmers - was supposed to lead to quite different results.

Peaceful Penetration and Emergence of Inequality

The first contacts between Bahutu and Batutsi go back at least to the 15th century.[7] However, there is no scientific clarity on this point. Various myths of origin are being discussed. There is some scientific evidence for the Hamitic origin of the Batutsi coming from the wider Horn. This suggests in no way that the racist associations of the »Hamitic hypothesis« are also accepted; in fact, it has to be refused because it turned out to be a dangerous channel for all actors, although at different times. On the one hand, authors such as De Lacger and Kagame but above all the colonial powers wanted to prove the natural superiority of the Ethiopians or the »Caucasoids« as quasi-Europeans in comparison to black Africans. They constructed for the first time the »Hamitic hypothesis« virtually on instructions of the royal dynasty of the *munyiginya* lineage. On the other hand, in the 1990s this very hypothesis served the Bahutu-owned Radio-TV *Milles Collines* for its exterministic propaganda against the invaders of Rwanda, that is the Batutsi. To be clear on this point, the propaganda not only condemned the Batutsi FPR invaders of 1990 coming from Uganda and settling in the northeastern part of Rwanda. The media accused them of having invaded

7 Since the history of the Rwandan kingdoms has been handed down orally, anthropologists argue how far the genealogy of the royal dynasties reaches back and at what time the »empire« of the mythical ancestors begins. In the genealogy of Kagame the first king, Gihanga, ruled between 1091 and 1124 (Kagame 1972: 90), whereas according to Vansina the first king ruled from 1386 on (Vansina 1962:56). See the different chronologies provided by Vansina (1962:18), Kagame (1959:86), and d'Hertefelt/Coupez (1964:510).

the Great Lakes region in the 14th century or so and therefore to still be »foreigners«.

There is little conclusive information on migration and their directions in pre-colonial Africa. Based on available sources it is probable that both Bahutu and Batutsi came from different directions at some time and migrated into the region, where they continued to display high mobility up to the 1930s. Noteworthy here - and crucial for further discussion - is that Abbé Alexis Kagame in his work on Inganji Karinga I, Kabayi (Kagame 1943:26), partly translated into French by Nahimana (1993:198-206), maintained that all Bahutu kingdoms go back to one Batutsi kingdom in mythical prehistoric times, that is the 14th century or so. Interestingly enough, he revised this hypothesis and admitted in 1972 that there have been various Bahutu kingdoms whose principles of organization were taken over by the Batutsi (Kagame 1972:23-24).

According to Mairieu's theory, which I basically follow here, the Bahutu, belonging to the Bantu culture, stem from the sub-Saharan region near Lake Chad. Migration of Bantu groups in diverse directions was triggered by expansion of the Sahara at that time, by population growth, and by the need for new fertile land. They were said to have brought their agriculture, their cattle, and their political organization into the Great Lakes region (Mairieu 1972:26; see also Davidson 1962). Concerning the Batutsi immigration, Mairieu follows the proposition of Davidson that over centuries thereafter invasions of Galla, Somali, Maasai, Bahima, and Louoo from the wider Horn destroyed the relatively highly developed Bantu civilization. Above all a remarkable increase in cattle had caused the Batutsi to look for new pastures in the Great Lakes region. This persuaded the Bahutu to settle in the more mountainous and forest-covered hills in the western part (Mairieu 1972:46; de Lacger 1961:103; d'Hertefelt *et al.* 1962:17).

The thesis of »peaceful penetration« - although unable to answer all questions (for instance, why the Batutsi language disappeared) - is widely accepted among scholars. During a long period of time the Bahutu peasants and the Batutsi pastoralists were supposed to have lived side by side in relative harmony. The Bahutu used fertile areas on the hills, whereas the Batutsi used relatively thinly populated grassland arenas in the eastern part, practicing transhumance.

Mairieu also supports the view of Vansina that Bahutu kingdoms with clan structures and patrilinear lineages existed long before the first Hamitic invasion. The clans acquired sociocultural unifying functions; they were determined economically by the *stockbreeding, agricultural,* and *craft* sectors, clan boundaries being open in relation to professional specialization. A group of clans was often headed by a king *(umwami*), who had the highest supraclan authority, though the latter had no political function and was not a formal leader.

Mairieu and others suggest that a clan with the name *banyiginya-basindi* moved further to western parts to evade the pressure of Hamitic neighbors in the east. It was this very clan that counted on perpetuation of its power transforming itself into a sacred kingdom according to the example of structures found in the invaded areas. They especially adapted the social and military role of the king, the important role of the king's mother, the genealogy, and the tambour as a symbol of power. Being nomads in a bigger sedentarized population of farmers, they even imitated rules of public life, social structures, religion, and philosophy. The only way to explain this phenomenon is that they acknowledged the Bantu peasants' cultural superiority - which, after all, had produced iron tools not known to the Batutsi before. Pierre Pagès could not but state that *»les Tutsi se sont adaptés aux mêmes façons de vivre, de manger, de se vêtir et de se loger que les Hutu (...). Même langage, même nourriture, mêmes costumes, mêmes habitations«* (Pagès 1933:15).

The Batutsi adapted the Bahutu culture while transforming the political structure at the same time. The *munyiginya* dynasty became politically dominant and started to compete with the Bahutu concerning land use and above all the land-tenure system. The different use of the existing ecoregions triggered the departure from relatively peaceful coexistence of earlier Bahutu and Batutsi communities, which were based on equal rights concerning access to natural resources. Instead, the *political ecology* of the Batutsi livestock breeders became dominant because of the relative military strength of the nomadic culture and the degree of central organization (Mairieu 1972:44-45; Guichaoua 1989).[8]

With population growth and the evolution of small kingdoms it happened that both groups used the same favorable ecoregions more and more (Nkurikiyimfura 1994:37; Moser 1994:54). This led to dissolution if not destruction of socioecological niches with distinct producer groups, on the one hand, and to emergence of political structures dominated by Batutsi, on the other hand. As Nkurikiyimfura (1994:47) writes: »It is therefore from the reign of Kigeli I Mukobanya (1520 -1543, G.B.) that the remark of de Lacger (1959) can be justified: 'The munyiginya monarchy is as much martial as pastoral. It has imposed itself by the spear and the raid as much as by the cow and pasture'«. Clans and little kingdoms developed in which original ethnic differences slowly

---

8 Political ecology deals with the social embedding of man in his environment and thereby the power, distribution, development, and peace policy aspects of man-nature relationships (*see: 1.1.*).

disappeared and developed into greater forms of social differentiation, from family-related associations to caste-like gradations.[9]

Cultural adaptation and takeover of political power and struggles over land form a syndrome of factors in the *mille collines* arena, which has caused violent struggles to this day. In the causation process for violent clashes this syndrome plays the role of a reason. A reason is a theoretical concept, an abstract relationship of factors present in a certain arena over a certain period of time. The abstract relationship may be specified in terms of inequality, hierarchy, or discrimination against the sedentarized Bahutu farmers. As we will argue later on, the state-building process brought the Batutsi inevitably into conflict with resettlement and - being subsistence farmers - territorially oriented Bahutu. The conflicts manifested themselves in competing land rights within certain territories claimed by both groups. In this context enforcement of the emerging Batutsi law, which mainly concerned land use and tenure, intensified social mobility and led to fundamental displacement of Bahutu and Batwa, the latter being only a very small part of the population.

Most scholars agree on the later historical development, although there are different standpoints about the timing of the genealogies. In the 16th century or so, the Batutsi nomadic pastoralists began to build a state and later, following the example of the Bahutu, a kingdom with sacred features, the king (*umwami*) being the supreme judge, chief of the army, and ruler over the entire land. Well into the 20th century the arena controlled by Batutsi encompassed only Central Rwanda, most of the south, and parts of the northeast. Before the arrival of the first Europeans in 1895, Rwanda was neither a nation-state nor a united kingdom. However, there have been strong unifying forces at work for centuries.

The extraordinary expansion of Rwanda began in the 16th century under King Yuhi Gahima, followed by the successful King Ruganzu Ndori in the 17th century and led to military reform under Cyirima Rujugira in the 18th century. The strength of Rwanda was founded on an internal hierarchy with a clear tendency towards despotism during the 19th century. The weakness of the neighboring kingdoms (bushis), on the other hand, was mainly due to internal struggles between dynasties and clans. Two dynasties were chiefly responsible for establishing central despotism: the *banyiginya* or *nyiginy* dynasty producing

9 I use the concept *caste* as a primordial or religiously determined category in the Indian sense as does Maquet, for instance, but rather in the sense of a principle of social stratification, which ran across the *clan* structures (Weber 1958:189). Interestingly enough, in the Great Lakes region the clan is a social category including various lineages (genealogy associations) and is of multiethnic nature (Newbury (1988:96, d'Hertefelt 1971).

the kings and the *bega* clan producing wives for the king.[10] As from the 18th century the *nyiginy* dynasty came to the fore. It exercised increasing control over a territorial unit in which both small Batutsi and Bahutu chiefdoms or kingdoms coexisted. An alliance was formed between the predominant dynasty and various lineages and clans. As a consequence, a strong aristocracy developed.

The basic goal of the Rwandan kingdom was to unify the whole region under one central kingdom. Therefore, mainly during the *nyiginy* era, a »definitive peace« with the neighbors was not possible any longer (Kagame 1952:54, para. 121). There were several non-aggression pacts signed by the four Hamitic kingdoms existing between the 16th and 19th century. It was the Burundian King 'Ntare III Kivimira who broke with the era of armistice and declared permanent war. At the end of his reign in 1885, Batutsi King Kigeli IV Rwabugiri tried to come to an agreement with Burundian King 'Mwezi IV Gisabo only to be broken again by the border chief (Kagame 1952:55, para. 121d, fn. 29). There were three main reasons for warfare during this period: the so-called *razzia du bétail* (or theft of cattle), territorial expansion of the kingdoms, and defense of one's own territory.

Despite expansionist and unifying forces originating in Central Rwanda until the mid-19th century the situation in the country was characterized by enormous flexibility in relation to ethnic identification, by regional heterogeneity in relation to imposition of royal power, and finally by political differentiation in relation to the individual chiefs' or kings' range of power. In precolonial Rwanda - not coinciding with the *pre-political* era which ended before the arrival of the European powers - family, lineage, subclan, and clan - combined with the different codes mentioned above offered major opportunities for identification and guidelines for public life. As long as the influence of the central administration was negligible in some peripheral arenas considerable regional autonomy based on kinship relations prevailed (Czekanowski 1917; History and Chronology of Rwanda (anonymous); Nahimana 1993; Newbury 1988; Vansina 1962). For instance, in the southwestern province of *Kinyaga*, located on the shore of Lake Kivu, neither ethnic Batutsi-Bahutu formation nor royal central power played any part. On the contrary, the ethnic and social variety was remarkable due to immigration from present-day Zaïre. At the same time, refugees from the kingdom in Central Rwanda strengthened the opposition to Batutsi dynastic power (Newbury 1988:97). Even stronger was the resistance of the Bahutu kingdoms in the north and northwest which, despite occa-

10 During Musinga's rule 1931-1940 the main provinces were ruled by the *bega* clan. There were about 24 hill chiefs directly responsible to the king who controlled about 80% of the country.

sional military defeats, lost their autonomy only during Belgian efforts to build up a nation-state.[11]

Autonomy in peripheral regions significantly decreased only during the second half of 19th century. At that time fundamental changes in the power structure occurred shortly before colonization of the country during the rule of King *Rwabugiri* of the *nyiginy* dynasty from 1860 to 1895. He was described as an absolute monarch, as »warlike« conjuring Batutsi supremacy similar to a military dictatorship out of nothing, just as Shaka invented the »Nation of the Zulus« in South Africa at the beginning of the 19th century or, to give an European example, as Friedrich Wilhelm I had done in Prussia almost 100 years earlier.[12] In order to impose central power in peripheral provinces, which had been conquered at the end of the 19th century, population was classified more rigidly by ethnic and social-strata criteria. Wealthy lineages began to be regarded as Batutsi lineages *per se*, while Bahutu were defined simply by the fact that they were not Batutsi or rich. As a consequence of this rather voluntaristic approach traditional chiefs - whether Bahutu or Batutsi - were replaced. Under pressure from the growing central power more and more hill chiefs were appointed by the central administration. This act resembled establishing a political control system by the central government introducing a new structure, which definitely undermined the social order of subsidiarity on the *mille collines*.

Due to related political discrimination against non-Batutsi outside the central province, the shaping of group identity took place rapidly - especially in remote rural arenas; a process described by Newbury strikingly as the »cohesion of oppression«. Identification with oppressed Bahutu was the starting point of both the social and psychological unfulfillment of non-Batutsi needs. It reaches back to the pre-colonial 19th century; it was only accelerated by foreign rulers or during the period of »dual colonialism«.

To summarize: social stratification combined with power struggles between different clans as well as a strong hierarchic order and centralization by absolutist dynasties in Central Rwanda, already during the 18th and 19th centuries, led to armed clashes and small wars. However, only with evolution of the political sphere in the modern sense and with unification of the kingdoms into one »nation« ethnically motivated riots and massacres began to take place.[13]

11 The resisting kingdoms were the following: Bushiru, Cyingogo, Bugamba-Kiganda, Bwanamwari, Buhoma, Rwankeri, Ruhengeri, Bukonya, and Kibari. In addition, some enclaves in the midst of the kingdoms were ruled by the chiefs of certain lineages (Nahimana 1993: 23).

12 Thus the somewhat sloppy but vivid comparison of Ansprenger (1994:574).

13 The years 1959, 1963, 1966, 1973, 1990, 1991, 1992, and 1993 were especially marked by massacres. In October 1990, in January and February 1991, and in March 1992 mass kill-

The Cohesion of Oppression

To grasp the key factors for the conflict model I refer to explanations of the ethnogenesis phenomenon that are consistent and widely accepted among scholars as well. One such explanation was given by Catharine Newbury, who did her field study in a remote province in the south of Lake Kivu (Kinyaga). Her findings on regional diversity and the differentiation between the center and the provinces provide a better ground for understanding the 1994 events than most other approaches. She aimed at analyzing »how the growing power of the state undermined kin groups, transformed patron-client ties, and created new forms of ethnic identities. These changes occurred in the context of and interacted with alterations in the social relations of production. Transformations in state power, clientship relations, and labor control mechanisms helped to create the social preconditions to revolution, by contributing to a heightened political consciousness and solidarity among Hutu« (Newbury 1988:208).

As stated above, the late 19th and early 20th century saw an erosion of traditional kin group status, while the extension of politically motivated clientship ties brought together men of various kin groups in a common position of political subordination. These two developments - linked to the expansion of central state power - served to broaden corporate identity beyond kin criteria in order to embrace ethnic criteria. Such fundamental social changes reflected the increased power of the Batutsi dynasty before and during the colonial period as well as »the transformation of clientship linkages from ties of alliance to instruments of exploitation« (Newbury 1988:209). In Kinyaga province - as in other provinces - the already existing hierarchical patterns of stratification and subordination in Central Rwanda came relatively late to the peripheral region: »The growth of self-consciousness identity among groups of Tuutsi had emerged in the central core of the state before the onset of European colonialism. Group awareness among Hutu came later. This was partly because Hutu relationships to the central structures of political power were both diffuse and distant: as Hutu were increasingly included in a common structural role within a single state apparatus, they also developed an 'ethnic' identity, which was no less powerful than that of Tuutsi« (*ibid.*:208). State building under Batutsi central power failed to promote integration and amalgamation of a socially, ethnically, and geographically fragmented population in the Great Lakes arena. As a consequence of the spread of central power patron-client ties became highly exploitative, fostering cohesion among subordinates. For all non-Batutsi the grievances that basically shaped the consciousness and provided the motivating

---

ings occurred. According to the fact-finding mission of the International Federation of Human Rights about 2,500 people were murdered between December 1992 and February 1993 alone (UN Security Council 1994:12).

force for political action were products of the socioeconomic challenges of everyday life. These grievances were felt more deeply by peasants in the hills than by the relatively small Bahutu elite, because the nominated Batutsi chiefs on the hills created Bahutu mass consciousness or »the cohesion of oppression« among a large part of the population (*ibid.*:209).

There were both class and ethnic elements in the emerging conflicts in the late 19th century, which got mixed with anti-colonial attitudes during the first half of the 20th century. Since economic problems were a critical factor in rural unrest, as was the struggle over access to the means of production, ethnic status and class overlapped for most of the population: most of the poor and exploited people were categorized as Bahutu. Although a few Bahutu were better off than others (teachers, traders, etc.), they were viewed as subordinate. The ethnization of social conflict and power struggle was due to the fact that virtually all who controlled the state - the chiefs and subchiefs - were perceived as Batutsi. In the late 1950s - thus shortly before the social revolution of 1959-60 against the monarchy and Batutsi oppression - it was clear for the first time in history that the appeal to Bahutu solidarity had become the most effective rallying point for revolutionary activity by the opposition leadership. Since the »feudal« oppression machine was perceived as Batutsi-like, mobilization of ethnic solidarity was much more potent than an appeal to all poor people would have been.

One cannot overemphasize the fact that the long-lasting processes from a parallel order of ethnic coexistence to a hierarchical class order with *caste-like* features and finally even to an absolute monarchy never evolved along the single track one may call *ethnic determinism.* There has always been at least a second track: strong identification with one of the two ethnic units always had the dimension of socioeconomic and environmental discrimination. One may even add a third track separate from the second one. It could be characterized as opportunistic cost-benefit evaluations. It happened that Bahutu associations allied themselves with the ruling groups and became Batutsi by definition. Simultaneously, in some state affairs, *i.e.*, »community affairs« on the hill, some Batutsi were even subordinate to a Mutwa. The opportunistic interchange of members from one ethnic group to another was strongly related to the *political* role lineages got.

With decreasing political and geographical distance to the epicenter of power, the question of genealogy became less significant. Ethnicity was hardly relevant in the countryside, far away from the lion's den. On the contrary: the kings tried to safeguard their power by an ingenious interethnic system of »*divide et impera*« in order to bind potential challengers of the king, *e.g.*, ambitious chiefs of rising lineages, into firm structures and dependent relationships in remote arenas. However, ethnic affiliation or a personal decision to assign oneself to a group became more important the closer one was to the monarch. Therefore, the ancestral line was most important in the vicinity of royal power. The king of Central Rwanda, whose power was greatest at the end of the 19th century, ruled in an increasingly arbitrary manner. His main objective was control over or, if necessary, breaking the power of the aristocracy, which represented a constant threat. The higher an official rose in the hierarchy, the stronger he was disciplined; if he became too powerful, the slightest reason was sufficient to disgrace him and his family. The struggle for power went on within the elite and scarcely affected the populace. Thus the ethnic boundary at the top of the hierarchy turned out to be razor-sharp. And should a rift occur in the central circle, social and cultural relationships could be severed all the more easily far downwards and outwards into peripheral circles. This rule even applied to the presidency of the first republic: the »little house« *(akazu)*, as the circle of in-laws of President Habyarimana and his powerful wife's clan was called. It also applied to the free-trade elite, misleadingly described by some observers as the *fourth ethnic group*, which had great influence on the government due to its self-defined ethnic proximity to the center.

Caste or Clan?

During the 100 years between the beginning of the Rwabugiri rule in 1860 and the revolution in 1959-60 the *caste-like* structure of Rwandan society became ever more pronounced, though not in the sense of the quite fruitless anthropological dispute on the question of »clan or caste?« (d'Hertefelt 1971). Like many others, Newbury rejects the concept of »caste«. I suggest she does so because she starts from too narrow a definition encompassing the following three points: rigid and impermeable hierarchy, religious ideology, and pronounced economic specialization. Indeed, none of these three points is strictly accurate for the situation in Rwanda. Such a rigid standpoint would not even apply to India, being the classic case for discussion of castes. I agree with Newbury that the term caste is not adequate at all if only used to distinguish between Batutsi and Bahutu communities (Newbury 1988:13). As a matter of fact, the caste-like structure lies right across the primary identity-building institutions, such as families, lineages, subclans, clans, and small independent kingdoms. On the contrary: the shape of the specific type of caste-like societal entities goes together with political transformation of these very structures. How to explain this rather difficult cross-cutting relationship?

To strengthen the central power in the peripheral provinces, the power of the chiefs (whose regional status was heavily drawn from their position at the central court and *vice versa*) was to be built up. Centralization and political degeneration of the patron-client relationships promoted a rigid classification of society according to complex social strata and ethnopolitical criteria. The predominant royal *munyiginya dynasty* used to be at the top, followed very closely by other influential dynasties, such as the *bega* clans producing the king's wives, the clans producing the royal consorts, the clans producing the bewares of the various codes, and thereafter a number of selected and, hence, privileged chiefs and subchiefs designed to rule the distant hills. Most of these clans were ethnically mixed. This is the main reason why it was not ethnic distinction as such that shaped the caste-like structure but primarily the opportunistic »Tutsiisation« of the ruling group, which also permitted downgrading, namely »Hutuisation« or marginalization of the rural population. In accordance with the Weberian concept of castes as societal status groups, the following description of the term caste exactly reflects the interethnic situation in Rwanda:

> »... the caste structure transforms the horizontal and unconnected coexistences of ethnically segregated groups into a vertical social system of super- and subordination. Correctly formulated: a comprehensive socialization integrates the ethnically divided communities into specific political and communal action. In their consequences they differ precisely in this way: ethnic coexistences condition a mutual repulsion and disdain but allow each ethnic community to consider its own honor as the highest one; the caste structure brings about a social subordination and an acknowledgment of 'more honour' in favor of the privileged caste and status groups« (Gerth and Mills 1958:189, cited in: Horowitz 1985:23).

Finding I

The socioeconomic system of super- and subordination created fault lines, which could easily be ethnopoliticized, first in order to establish a kind of absolute monarchy and later on to mobilize against this very system.

Presumably the ethnosocial system of Rwanda was modeled by a polydominial structure over a long period of time. This would have meant that the position of a group within the hierarchy could change. In Rwanda collective mobility between ethnic groups had been possible since the mythical beginnings in the 16th century without destroying the nascent hierarchy. Flexibility, promotion, and relegation in the hierarchy were altogether possible - even »ennoblement« of the underdog Batwa was not impossible. Only with the emergence of absolutism, further exacerbated and strengthened by the colonial powers, did a steep hierarchy develop with firm role assignments in which a participant was never supposed to be subordinate in one area and dominant in another. The integration of the two groups in the Rwanda arena already failed in the 19th century.

Dissolution of traditional mechanisms and kinship relations and transfer of the *social hierarchy* to *politics* led the expanding central state to become an arena, where ethnicity was the predominant channel to either hold power or compete with it. The central state itself - created only after 1860 - became an important tool in this struggle: the winner could easily turn the state into an agent of his own interest, which was supposed to be ethnic interest - the latter following the logic of a winner-takes-all system in a fragmented society.

In this perspective, the social revolution of 1959-60 represented seizure of power by a subordinate group which, in the course of collective identity building, began to establish a coherent leadership attacking the caste-like structures.

5.2.2. REASON II: SOCIAL AND ENVIRONMENTAL DISCRIMINATION

Reasons I and II are closely interrelated; as a matter of fact, they form a strong unity. Each reason can only be explained and understood in the light of the other. In this study they are only separated for heuristic purposes to gain scientific clarification about both intergroup and societal-nature relationships. Whereas reason I dealt with ethnogenesis and ethnopolitical hierarchy, reason II consists essentially of the socioeconomic, ecological, and demographic complex in the Great Lakes region. It represents the materialistic aspect of ethnopolitical conflict formation. As became clear in the section above, ethnopoliticization of power is ill-understood unless related change in the socioeconomic fabric is taken into consideration.

Historical Development of Patron-Client Ties

During the 18th and 19th century territorial and political-administrative claims, which derived firstly from the extensive livestock economy and secondly from subsistence farming, were no longer compatible with establishing geographically expansive dynasties and kingdoms, which encompassed several Bahutu and Batutsi clans. Territorial separation of the two groups was gradually replaced by ethnic mingling and social differentiation. Social stratification and ethnopolitical hierarchy - combined with power struggles between clans and the evolution of a centralized Batutsi dynasty in Central Rwanda - led to numerous ethnopolitically motivated riots and small wars (Guichaoua 1989, Newbury 1988).

The background of conflict patterns along ethnic fault lines is that between 1860 and 1959-60 the Batutsi-Bahutu relationship was characterized by patron-client ties of an increasingly discriminatory nature. The majority of the Batutsi constituted the stratum of wealthy landowners and livestock breeders. Most of the Bahutu were small subsistence farmers, agricultural workers, tenants, or herdsmen. Although some had the chance to rise to the status of a lord of the manor or a patron, the rural majority was discriminated by the small elite in its access to the natural capital of Rwanda - and increasingly so during the 20th century.

The extensive and expansionist Batutsi livestock economy proved to be more effective a motor of social change than the Bahutu subsistence economy.

Take the case of livestock 'leasing' (*ubuhake*) as an example. Traditionally, *ubuhake* was a well-regulated patron-client relationship of »functional solidarity« (Guichaoua 1989) or »unequal friendship« (Newbury 1988) between livestock owners and those who 'leased' part of the stock to take care of it and gain some benefits from this job (animal products). Both sides were supposed to profit from this relationship in terms of a »win-win game«.

During the late 19th century, but above all on the eve of the 1959-60 revolution, the relationship had degenerated into glaring discrimination: the win-win solution was transformed into a win-lose game, obviously favoring the Batutsi patrons. The patron could enhance his power and wealth by increasing the tax burden or 'leasing rate', whereas many clients had to cultivate inferior soils or smaller plots because of both expansionist pastoralism and high population dynamics. Together with the colonial »work hard, work better« imperative this led to enormous individual and collective stress of the rural producers.[14] This fact applied particularly to the Lake »Kivu Complex«, which is synonymous with colonial exploitation in the form of high taxes, forced labor on public works, obligatory cultivation, and recruitment of workers for European enterprises or the Catholic Church. These severe conditions served to heighten the antagonism between the Bahutu population and Batutsi intermediaries. At the same time, one should not overlook the fact that, despite the heavy burden imposed by the colonial powers, they opened for the first time off-farm opportunities and thus offered the possibility for Bahutu to escape degenerating clientship ties and to shape elements of group cohesion against Batutsi oppression (Newbury 1988:210).

However, denial of access to resources was not only an interethnic problem. This is proven by the fact that, despite the social Bahutu revolution of 1959-60, the situation for most rural dwellers - who belonged to the same ethnicity as the government after 1960 - did not change fundamentally. On the contrary: because of unchanged economic parameters during the first republic of President Grégoire Kayibanda from 1962 to 1973 and the second republic of President Juvénal Habyarimana from 1973 to 1994 deterioration of living conditions seemed inevitable. However, dissatisfaction did not turn against the

---

14 With regard to the comparability of the pre-colonial system in Rwanda and European feudalism, Kagame, after detailed studies of European feudalism and its determinants concluded that »forms of feudalism had existed in the social institutions of Rwanda«, though the feudal relationship related solely to the pastoral system of the *buhake*. This system consisted of a feudal contract between families aimed at strengthening the power of the leading family. Undoubtedly, according to Kagame, this system and its tribute obligations had a great influence on the political system of Rwanda. Nevertheless, it is not a specific political element. Yet it is a fact that Rwanda - long before the arrival of the Europeans - had reached the stage of an »absolute monarchy«. Extensive hierarchization, which undoubtedly took place, and feudalism should not be confused with one another (Kagame 1952:6).

government. Instead the new Bahutu elite under President Habyarimana *(akazu)* was able to channel collective resentment against the Batutsi population in general and exiled Batutsi groups in particular. Social concerns and transformation of the environment may serve as a factor explaining why the massive return of exiled Batutsi imposed a major threat to most rural dwellers, the latter with post-revolutionary expectations that had not been fulfilled by their »own« Bahutu government.

Some facts may serve to highlight the discriminatory policy against the rural population.

Land Law as an Indicator of Competing Interests

The social system established by the Batutsi is based economically on their position as extensive livestock owners and culturally on the »cow cult«, regulated by means of a compact network of laws and customs (Buschmann 1986:53). In Rwandan land law it was established that »the pasture belongs to the cow« (Mairieu 1972:56). Accordingly, the cow is regarded as a subject of law and is both holder and owner of its own pasture rights. The pastures - like air and water - were a *res nullius* (no man's land) and accessible to all in accordance with the needs of the herds of cattle (Buschmann 1986:58). The nomad livestock owners had no need for territorial rights. Nevertheless, the pastoral law system became a pillar of political support for the ethnosocial hierarchy: while the cow held pasture rights, the animals of the Bahutu (sheep and goats) were locked out of the pastures.

In sparsely populated regions with large pasture areas or in infertile zones not used for agriculture or dense forests, the land law of the Batutsi was hardly a cause for conflict. In densely populated agricultural areas it had to be pushed through against the traditional »clan« land law of the Bahutu *(ubukonde)*. In the course of extending Batutsi rule, the decline of the Bahutu clan organization, and the increase in population three types of land law developed: *igikingi* in the formerly freely accessible pasture regions, *ingobyi/isambu* in the agricultural regions, and *buhake* as a pastoral feudal contract (Buschmann 1986:72).

With the institution of *igikingi* individual rights to pasture domains were created in intensively exploited areas with high population density. The holder of *igikingi* had the right as a patron to allow his subordinates, clients, or farmers to take their livestock to the pasture. In return he reserved the right for himself to graze his livestock on the cultivated plots after the harvest. This right was claimed more and more because of the increase in population and livestock. First the fallow areas for sorghum were annexed, then the fallow areas for peas and beans, and finally all fallow land suitable for pasture (Buschmann 1986:62). The concessions were heritable in principle and passed to the male successors. The holder could also lease pastures, but he was not allowed to divide the *igikingi*. Even though the land law on pastures was applied throughout the country, it never had the same influence in all districts.

In agricultural regions the tenure system was subdivided in that of the extended family line *(ingobyi)* and that of the head of a nuclear family *(isambu)*. In the *ingobyi* system the head of the lineage held a common land right. For the purpose of common land use a parcel of land without firm boundaries was allotted to individual members of the lineage. *Isambu* was based on the political authority of Batutsi, who in case of quarrels in the clans could intervene as mediators. They allocated a piece of land from the clan property or previously uninhabited land to individual families. In this way, a politically decreed individual law developed beside the clan law (Buschmann 1986:65). This gave the Batutsi considerable advantages. They could increase the tribute income through special taxation and at the same time break the power of the clans through their political disintegration. In addition, parcels of land without a heir no longer went to the community but to the Batutsi chiefs, who gradually degraded their subordinates to the status of land clients. A further disadvantage of this law was that the land was divided between the sons of a deceased head of a family household. In this way land in densely populated areas was divided up continuously. At the same time, ever more farmers were compelled to enter into sublease contracts with the Mututsi chief and became even more dependent.

In the cleared forest regions the clan system of the *ubukonde* was in force at first. But in connection with occupying free land and clearing forest, the systems of *igikingi* and *isambu* (one family holding) spread to the former forest areas, and use of the forest was more and more jeopardized - despite the king having granted the right to use the forest to the Batwa (who, in return, had to hand over a share of their hunting bag: ivory, skins of leopards, lions, otters, etc.).

The *buhake* goes back to the 18th century. This is a voluntary contract between a person of lower rank *(client or umu-garagu)* and one of higher rank *(shebuja)*. It constituted a certain social relationship *(ubuhake)*, a form of pastoral servitude, or a client relationship by means of a livestock leasing contract. The *ubuhake* functioned under conditions of sufficient land and sparse settlement in the second half of the 19th century and was not abolished until 1931 (Buschmann 1986:60). The increasing scarcity of land due to expansion of cultivated land by ever more settlers made it indispensable to strictly regulate rights of access to pasture land.

Cultivation Caused Land Transformation

Rwanda can be divided into favorable and unfavorable ecogeographical arenas. The central highlands are an underdeveloped mountainous region with steep slopes and deeply weathered acid soil of limited fertility. The fertile volcanic soil in the northwest and the previously almost unusable swamp and savanna region in the south and east are being used to the limits of their capacity. On the western boundary up to the Central African Rift Valley, even the most extreme slopes are cultivated. Today the once fertile soil is degraded, while geographical alternatives are rapidly diminishing.

Total land area is 2,634 million ha. Some 20.9 percent are forests and woodlands, 32.5 percent arable land, and 0.5 percent irrigated land. The loss of humus is high, amounting to 10.1 MT/ha on average. While 50 percent of the soil in relatively favorable areas shows a comparably moderate loss of 3.7 MT/ha, it is above all marginal soil in the relatively unfavorable areas which suffers from very high erosion; more than 36 MT/ha on 5 percent of the soils and more than 68 MT/ha on 1 percent (*see: table 5.1.*).

	percent of area			
	Mean	**50%**	**5%**	**1%**
1st season	04.9	1.4	18	68
2nd season	05.2	2.3	18	32
1984	10.1	3.7	36	68

Source: SESA, December 1986.

Table 5.1. Estimated humus loss 1984 (in metric tons per hectare)

Over the centuries large areas of Rwanda were deforested and transformed into cultivated land. Of the primary forest, which covered around 80 percent of the country, there is only 5 to 8 percent left. During the 1980s the deforestation rate was 2.3 percent or 2,000 ha per year (HDR 1996:181). The 56 percent expansion of acreage between 1970 and 1986 pushed cultivators onto poor soils in marginal regions of the country. Altitude gradation of vegetation was greatly influenced by this. Whereas the lower limit of the mountain rain forests was previously at 1,800 m, today land is being cultivated at altitudes of up to 3,000 m (for example at Bisoke Mountain); be it by state-sponsored large-scale agriculture, countless land-seeking farmers, or returning refugees.

Mountain rain forests and mountain mist forests predominate in only two forest areas. Between 1958 and 1979, the mountain rain forest of the Virunga-Chain range was reduced from 34,000 to 15,000 ha, above all at the expense of the bamboo forest. All forest areas are extremely endangered by the high degree of pressure from land and wood use.

Natural vegetation was driven back by dense and permanent settlement even in the previous main pasture areas, the shrub and tree savannas with their high grass fields, thorn bushes, acacias, and last but not least the characteristic candelabrum euphorbia. In the savanna of the Bugesera arena there were still more than one million hectares of dry forest in 1958. Some 20 years later two thirds of the forest had been destroyed. This was in particular the effect of the peasantry beginning to migrate to this region; in 1962 only about 16,000 people lived here; in 1980 there were already 200,000 (Werle/Weichert 1987:36).[15]

Persistent Rural Poverty and Human Deprivation

In the country-ranking system of the Human Development Index (HDI) Rwanda with 0.332 points ranks 152nd of 174 countries (*see: app. 2*). 51 percent of the population are at the upper poverty line which, according to criteria of the World Bank, means absolute poverty (World Bank 1996:288). Also in the *capability poverty index* recently developed by UNDP Rwanda belongs to the low-ranking countries. The index is composed of three indicators that reflect the percentage of the population with capability shortfalls in three basic dimensions of human development: living a healthy, well-nourished life, having the capability of safe and healthy reproduction, and being literate and knowledgeable. Out of 101 countries Rwanda is at rank 77 (HDR 1996:112). Although absolute poverty and the low rank indicate a correlation between maldevelopment and organized violence, this coincidence *per se* provides no satisfying explanation of the extent, which ethnopolitical violence reached during the 1990s. Other countries are in a similar position without this spectacular extent of violence. In addition, there must be some underlying socioeconomic tendencies, which support the proposition of discriminatory character in the inter-group conflict within Rwanda.

The World Bank *development diamond* is more revealing concerning the extraordinary resource scarcity Rwanda faces than the capability poverty index. The diamond, which is based on four indicators, shows the average level of development in Rwanda compared with the average of the low-income group of countries it belongs to (*see: fig. 5.3.*).[16]

---

15 Intact areas, though small and scattered, still exist in extreme locations in the east (back-water marshes with an impenetrable papyrus jungle in the border region with Tanzania) and in the high mountains. In the terraces of the mist forest and the Afro-Alpine zone only areas above 3,500 m remain more or less unspoiled.

16 The figures were ascertained before the outbreak of the war in April 1994 and thus represent only harbingers of the dramatically worsened situation as it presents itself today.

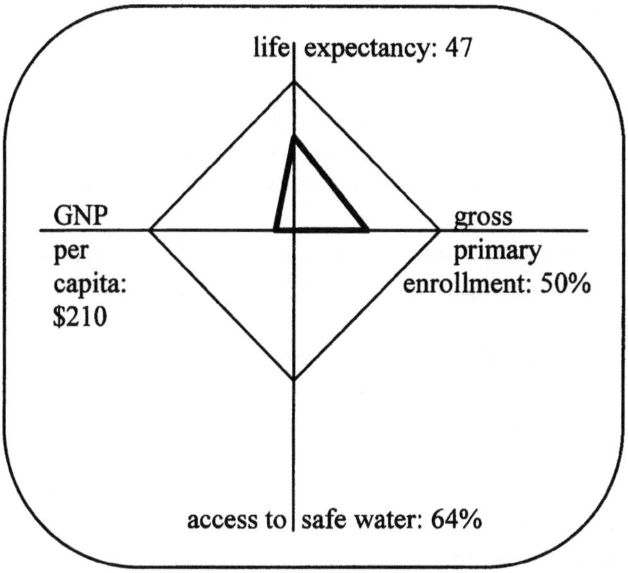

Source: World Bank 1996:288.

Figure 5.2. Development diamond

The diamond indicates significant deprivation in human development. Whereas Rwanda's social institutions (health care, school) are not far below the average of the low-income group countries, access to economic and environmental resources deviates significantly from the low-income country group average. While some *social* indicators even showed a positive trend in recent decades (declining infant mortality, increasing life expectancy, rising literacy levels, growing income), above all the *socioecological* indicators worsened in every respect. There are several reasons for the extremely poor state performance in this regard.

Rural-Urban Poverty Gap

At first sight the rural-urban poverty gap is accentuated enough to provoke the question if it is a crucial part of the causation process leading to endemic violence: in 1993 94 percent of the population lived in rural areas. Out of a total population of 7.6 million people 1.5 million had no access to health services, 2.6 million no access to safe water, and 3.2 million had no access to sanitation (HDR 1996:179). Because of the relatively low urbanization rate and the high population growth rate of 2.6 percent annually, the figures on human deprivation are expected to turn even worse in the countryside: compared to 75 percent with safe access to fresh water in urban areas - chiefly in the capital Kigali - in rural arenas only 62 percent have been safely provided with freshwater; the tendency has even been decreasing over the last two decades (*ibid.*:149, 153). Between 1975 and 1980 68 percent of the population had access to drinking water, between 1988 and 1990 the figure was only 64 percent (Planning Ministry 1992:265-267).

According to the »Falkenmark indicator« (1995:19), comprised of the total number of people existing from one flow unit of water (or one million m^3 freshwater per year), countries with more than 600 people per flow unit belong to the category of water-poor countries. More than 2,000 people per flow unit would mean absolute water scarcity (*see: 1.2.; 6*). With roughly 1,200 people per flow unit, Rwanda belongs to the heavily water-stressed countries of the world (HDR 1996:181). Since only 2 percent of the available amount of freshwater is used effectively per year, the deprivation of the rural population is extremely accentuated or, in other words, Rwanda performed very poorly in developing water resources.

Rural poverty is also expressed in low rural income compared to the average GNP. The annual rural income per capita was estimated at $150 compared to $220 nationally (World Bank 1996:288). Only about one third of rural income is derived from monetary income. Simultaneously, agricultural loans have hardly been available to smallholders. The well-organized *Banques Populaires*, founded in 1975 with the aim of mobilizing rural savings and providing loans to farmers, soon found themselves in a dilemma; they succeeded in tapping small farmers' savings, yet the savings were lent mostly outside the productive rural sector. In 1981 about 53 percent of total deposits came from farmers, but the bulk of lending was to finance housing, construction, and commercial activities. The lack of demand for credit for directly productive purposes owed to the practical nonexistence of innovations, limited improved inputs in the rural areas, and the high risk of providing credit to small farmers with low capacities (Studiengruppe 1996:77-84).

Lack of rural income in cash terms is a major reason for the very limited possibilities for economic restructuring. A shift from subsistence production (basically destroyed by political interventions of the colonial powers and later by the modern central state) towards trade and industry, producing on the basis of division of labor, is immensely restricted by the lack of the population's purchasing power but also due to the precarious resource situation. Ore deposits are nearly exhausted. There is not enough wood available to produce bricks, tiles, and charcoal burning. The extensive state-owned or cooperative plantations made farmers closely dependent if they abandoned their subsistence economies, exposing themselves to the risks of the world market. Yet, those farmers who remained in the countryside had to escape the consequences. At the beginning of the 1980s the government invested massively in production plants for quinine and thus encouraged farmers to produce the raw material, the cinchona tree. The collapse of the world market led to a drop in production from 565 MT in 1980 to 300 MT in 1990 (Planning Ministry 1992:226). There are still enormous peat deposits. Their commercial mining would not only cause enormous ecological problems but also necessitate large-scale resettlement operations and thereby aggravate social problems.

Demographic Dynamics and Livelihood Insecurity

One aspect of rural poverty in a society with a 94 percent workforce in agriculture is *demographic dynamics.* In Rwanda the precarious balance between population size and food supply has historically been of immediate concern. Because of epidemic diseases, drought, locust plagues - and repeated military expeditions by colonial rulers (so-called punitive actions) - livelihood security has never been very high in the Great Lakes area. Starvation was a major problem in 1890, 1895, 1887-98, 1989, 1900-1903, 1904-08, 1909, 1910, 1911, 1912, 1916-18, 1921-22, 1924-26, 1927, 1928-29, 1943 (Nahimana 1993:268-78). The last severe famine, which reduced the population by more than 20 percent, occurred in 1943. Even though food insecurity has historically been a major problem of rural producers, it has again increased significantly during recent decades. Annual food production dropped on a scale of 100 points in 1979-81 to 70 points in 1993 (HDR 1996:163). Rwanda is still on the FAO list of countries at risk (see: FAO website).

A reason for the widening gap between socioeconomic and politico-ecological development can be found in the combination of a high population growth rate and the preservation of precarious agricultural relationships, which developed in the course of modernization from removal of the traditional subsistence economy. With 2,947 inhabitants per 1,000 ha in 1991, Rwanda displays the highest population density in Africa; at the same time, it is one of the most densely populated countries in the world. In the mid-1980s density related to cultivated land was estimated at about 390 capita per square kilometer. This figure is about four times higher than, for instance, in neighboring Zaïre. In the intensively exploited heartland of the highlands 400 to 550 people used to live on one square kilometer of productive land. The population increase of 3.3 percent per year has to be compared with the annual increase in total agricultural production during the same period of time. From 1980 to 1988 the increase in agricultural output amounted to a poor 0.3 percent annually (Diessenbacher 1994:14-16).[17]

The reverse side of the coin is that population growth has been restrained by high mortality owing primarily to famine and diseases. High infant mortality is supposed to be a strong indicator for both maldevelopment and state failure: 46,000 children under five died in 1992 (HDR 1996:147).[18] Migration has had a mitigating effect on clusters of high-population density in favorable ecozones. In Rwanda the migration to less heavily populated arenas has generally occurred from the northwest to the northeast. At the same time, migration to urban arenas has been far less important than migration within or between rural arenas. An analysis of movements shows that mainly the poorest farmers (*i.e.*, those with little or no land) migrate to other rural areas. This indicates that the high-population density stress chiefly affects smallholders discriminated against in overused areas with few off-farm opportunities. According to the World Bank (1983) rural unemployment already reached 30 percent in the early 1980s.

Since it is difficult to connect the isolated farms and atomized settlements on the hills - in Rwanda there are no closed, scattered, or linear villages - to traffic and energy systems, problems of infrastructure enhanced the inclination to migrate to the few towns of the country.[19] Mainly the younger generation was abandoning the exhausted land. However, only a small number of educated young people left for a paid job to urban arenas (World Bank 1983:102-103).

There are further aspects to be considered, which affect livelihood security in one way or another. One aspect concerns marketing food crops. Since traditional smallholders are structurally predominant, rural marketing was not well developed. Only 35 percent of the food crops produced was marketed. And this amount was mainly sold locally or was derived from an exchange of goods.

Another aspect is the declining value of the cattle, which once played a crucial role in the sociopolitical system of Rwanda. Chiefly the increasing population pressure on the land is forcing a steady decline in the national herd.

---

17 Under pre-war conditions about 5 million people had a secure nutrition basis. In 1993 population had already exceeded the 7 million mark and was expected to reach 10 million in the year 2000.

18 In the « state failure project« carried out by Ted Gurr *et al.* from the University of Maryland infant mortality strongly correlated with the phenomenon of failed states. Since the project (results not yet published) dealt with 600 indicators, the significance of this indicator was somehow quite surprising. Infant mortality obviously indicates a syndrome of socioeconomic, demographic, ecological, and political stress factors.

19 The urbanization rate is comparatively low, but rapidly increasing: from 2 percent in 1960 to 8 percent in 1991 and an estimated 11 percent in 2000. The annual increase in urban population amounts to 7.5 percent. In 1970 Kigali had about 50,000 inhabitants; in 1981 there were a little over 150,000 and in 1987 about 267,000.

In the mid-1980s less than a third of all farms owned cattle. Small stock was replacing cattle as a source of animal protein, the supply of which is generally very low (*ibid.*:111).

Overuse of Fertile Land through Shrinking Farm Sizes

During the 1980s population growth did not only generate greater demand for jobs and social services. It also exacerbated pressure on rapidly shrinking land resources. By the end of the 1980s the available land was more or less completely exploited. The World Bank estimated that by 1990 a further reduction in farm size would be practically unavoidable (World Bank 1983:110).

In addition to the negative effects of population dynamics, the tradition of dividing real-estate inheritance resulted in ever smaller farms. In densely populated areas the size is already below 0.5 ha. This, among other factors, means the fallow cycles can no longer be maintained (*see: tab. 5.2.*).

classification	**enterprises in percent**	**in percent of cultivated land**
< 0.5 ha	26.4	06.9
0.5 - 1.0 ha	30.3	18.4
1.0 - 1.5 ha	15.6	15.7
1.5 - 2.0 ha	11.1	16.1
> 2.0 ha	16.4	42.9
total	99.8	100.0

Source: Ministry of Agriculture, National Survey 1984 (In: Guichaoua 1989:59).

Table 5.2. Distribution of farms by size in hectares

In the light of the exceptional demographic data of Rwanda, the Malthusian causal chain *overpopulation - land shortage - lack of resources - catastrophe* is often applied to the Rwandan crisis. Indeed, the system of rural self-sufficiency, which in the recent past provided insecure but nevertheless basic supplies for a large part of the rural population, could not withstand the rapidly growing population pressure. However, it is not only population dynamics, which have become an obstacle to necessary economic and social restructuring. Yet, if one considers the figures in the context of access denied to natural capital, unequal distribution seems even more significant than parceling out agricultural land: 16.4 percent of those farms larger than 2 ha own more than 42.9 percent of agriculturally productive land. In other words, of a total number of 1,112,000 farms in 1984, 182,000 farms owned about half of the agriculturally productive land.[20]

The distribution problem contrasts with the World Bank's observation that physical constraints were becoming particularly acute. Analysts predicted that by the end of the 1980s remaining arable land would have been fully exploited (World Bank 1983: 122). About ten years later, the Rwandan Planning Ministry pointed out: »Degradation of soils due to overexploitation has resulted in the collapse of the agricultural sector«. The ministerial report concluded that »(m)ore than 500,000 people are presently in a state of declared famine« (1992:266).

The fact that the World Bank and international agencies have focused on diminishing population pressure and not on restructuring agriculture indicates part of the Rwanda syndrome as described above. The reason was that »land reform, through legal regrouping of land holdings and restructuring of traditional inheritance rights, is probably not a viable option given the complex legal, social and technical issues which it raises« (World Bank 1983:122). This observation is confirmed by frequent violation of local land law during the last few decades, which has increasingly led to violent conflicts between the indigenous population and land-hungry migrants (Sirven/Gotanegre 1974:93).

It is no exaggeration to maintain that the land-tenure system and the scarcity of available natural capital in relation to the users are at the core of the Rwandan problem. Since socioecological fault lines and ethnopolitical cleavages coincide and mutually enforce each other, land use and tenure is highly conflict-prone and has failed to be addressed adequately by the Rwandan government or the international community.

Low Commercial Energy Use

Another significant deprivation factor is commercial energy use or the mere absence of any available resource. Most Rwandans are forced by their low income to use traditional fuels, wood and charcoal when available, and agricultural residues when necessary for cooking and other basic energy needs. The high demand combined with high population density has made fuelwood increasingly scarce and has put reforestation efforts in competition for land with food and cash crops. Almost 95 percent of energy requirements are met by wood. Scarcity of fuelwood has led to increased use of agricultural residues as energy source rather than organic fertilizer, which has exacerbated the adverse impact on soil fertility of the long-term decline in farm size. Since Rwanda is landlocked, importation of oil through Kenya and Uganda is expensive, and

20 In more detail and from a critical Malthusian point of view regarding ecological-demographic structure: Guichaoua (1989:47, 59).

since borders follow the major waterways, major hydrological resources require at least bilateral agreements. The same holds for hydrocarbon resources. Since large quantities of methane dissolved in the lower depths of Lake Kivu, which is shared with Zaïre, development of this resource also needs an agreement - as well as expensive technical solutions.

If one compares Rwandan energy consumption with other low-income economies, once more there is deprivation. Commercial energy consumption per capita and per unit of GNP remains far lower than in some other countries with similar size and income level, except Burundi which is no better off (World Bank 1983:135). Considering the index of available resources developed to indicate the degree of dependence on natural capital (*see: app. 13a; 6*), we find that Rwanda is indeed at the bottom of all countries. With an average of 0.15 ha per capita and a commercial energy use of 27 MT per capita per year in 1993, Rwanda has a *resource availability index* of 4.0. A comparison with other countries belonging to different income groups shows this to be a very low figure: Niger, for instance, amounts to 15.9, Ghana to 24.9, Sudan to 31.8, India to 45.9, and an industrialized country, Germany, to 625.5. Rwanda, in other words, has the lowest resource availability index whatsoever and - as we have seen above - suffers by having few off-farm opportunities, which could provide a way to escape high dependence on natural capital.

Coffee Crisis Enhances Rural Vulnerability

Coffee and to some extent tea were supposed to be a promising commercial agricultural sector. Especially after the total collapse of the mining sector in 1985, coffee became the most important cash crop. During the early 1980s income generated from coffee increased tremendously, with a significant peak in 1986 (*see: tab. 5.3.*). This was mainly because coffee market predictions about economic development in Rwanda tended to be quite optimistic at the time. However, prognoses about success stories were rather shortsighted. They basically ignored that high profits were mainly due to effects of drought in Brazil. This was the main reason why world coffee prices were projected »to rise in 1986 to levels in excess of their previous peak levels of the late 1970s, representing an increase of more than one third in a single year« (World Bank 1986:x). Because of windfall profits, Rwanda's overall balance of payments registered a large overall surplus that turned out to be unsustainable. The World Bank thus recommended: »While providing some room for 'maneuver', it will be essential for the Government to bear in mind that the coffee 'windfall' of 1986-87 will be only temporary. This should allow it to avoid repeating its serious mistake during the previous coffee boom of the late 1970s« (*ibid.*:xi).

	1966-70	71-75	76-80	81-85	1986	1987	1988	1989
coffee	939	1952	7236	7839	**13903**	8350	7125	5164
tea		242	879	1573	493	1202	1365	1905
mining	620	929	2282	1453	257	78	30	541

Source: Studiengruppe 1996:56.

Table 5.3. Coffee export in million FRW (annual average)

Thus the world market outlook for coffee - both for prices and growth in demand in the countries which participated in the International Coffee Agreement (ICA) - would have required further justification in economic and financial terms (World Bank 1983:120). Yet, concerning the third Rwandan five-year development plan, the government failed to diversify the structure of exports. The collapse of coffee prices and markets in 1989-90 was no big surprise, although the government was not at all prepared to face this very fact. The coffee boom of the mid-1980s - similar to that of the early 1970s - left the economy extremely vulnerable (World Bank 1986:iii). Government expenditures - although dependent on a cash crop highly unpredictable in terms of prices and quantities - increased significantly during the 1980s. The public sector was more and more financed by income drawn from the rural sector. This state of affairs did not induce a dynamic interdependence between rural and urban arenas. Instead, the small industrial sector was heavily subsidized by state income drawn from coffee, whereas the agriculture situation did not change or even worsened increasingly, having been divided in an income-generating cash-crop sector and a marginalized non-monetary sector of the rural poor (Studiengruppe 1996:60-61).

Finding II

In Rwanda - as in other agrarian societies - fertile land was traditionally scarce and led to distribution conflicts. Historically, land distribution conflicts led to clashes over fertile soils in relatively favorable ecoregions between actors in search of a better and above all richer life. Land scarcity was therefore a question of supply and demand. Today, the case of Rwanda indicates a qualitative difference between simple distribution conflicts and environmental transformation conflicts. The latter manifest themselves as struggles for survival based on exploited and heavily degraded natural capital. When favorable arenas transform into unfavorable arenas due to human intervention, demand-side economic scarcity changes fundamentally into supply-side environmental scarcity.

In the Rwanda arena two factors cumulated and thus accentuated the conflict potential of environmental scarcity. Firstly, demographic dynamics led to high density clusters of small rural producers, which - due to lack of financial and human capital - were unable to escape the deteriorating society-nature relationships. Secondly, social stratification led to discrimination against the majority of small rural producers. This fact alone may not explain the escalation of violence well as the same phenomenon holds for many if not most societies throughout the world. The crucial point in Rwanda is that the rural-urban gap was overarched by clear-cut ethnopolitical segregation used by the ruling elite to exclude the other group almost completely from any resources whether power, education, natural capital, or the entire country (the latter referring to Batutsi exiled in 1960 who were not allowed to return to their country).

Demographic dynamics and transformation deepened the fault lines between Bahutu and Batutsi and contributed to formation of coherent strategic groups, which perceived that they were fighting each other for ethnic reasons and not because of economic, demographic, or ecological ones.

5.2.3. CHANNELING: INCREASING CONFLICT POTENTIAL BEFORE THE REVOLUTION OF 1959/60

Towards the end of the monarchy land-use and land-tenure conflicts intensified between the Bahutu majority and Batutsi feudal elite, on the one hand, and the colonial power, on the other hand.

There were major conflicts over cultivated single-family land because the individual right *(isambu)* was cast on traditional or clan law. This law of the Bahutu peasantry, which was based on communal land ownership reinforced by customary right *(ubukonde)*, was removed forcibly by submission through the Batutsi chiefs. Political law - to which both *isambu* and *igikingi* belong - was encoded by colonial legislation. It pushed holders of the clan right - chiefly Bahutu - little by little into a lawless state, followed by expulsion from ground and soil (Buschmann 1986:46). *Isambu* was applied in the whole country, but the colonial power was unable to protect this right from infringement or to prevent confiscation of land by chiefs. In the 1940s, as a consequence, many landowners sold their individual pieces of land and left for the sparsely settled Congo (Guichaoua 1989). The colonial power allowed selling the *isambu*, at least in principle. This caused a proliferation of land-selling by poor smallholders throughout the country. This development notwithstanding - the Rwandan government failed to counter wealthy purchasers from acquiring land, which they did not cultivate with a law until 1976 - *isambu* represented the basis to maintain the livelihood of a large part of Rwanda's population and provided the holders with a hitherto unknown extent of legal security (Buschmann 1986:86).

The Belgian colonial power reacted to the increasingly urgent problems of Rwanda's agrarian constitution with various, partly radical measures. However, they failed to tackle fundamental questions of legal equality and legal security. Country-wide extension of *igikingi*, abolition of *ubuhake,* and promotion of individual private property contributed to the emergence of three competing types of land-use and tenure systems. These conflicting systems molded the social structure immediately before the revolution of 1959-60. This can be shown by highlighting the microcosms of local and subregional power relations on the hills, the latter constituting the highly fragmented societal arena of Rwanda:

(a) Abakonde vs. Batutsi chiefs. Mainly in the northeastern regions of the country a specific conflict formation had been caused by incompatibility of traditional clan law and the political land law. In fact, the formation has two dimensions. The first encompasses holders of *ubukonde* land rights *(abakonde)* and the Batutsi chiefs and their clients. If *abakonde* decided to sue, they experienced that the courts were mostly composed of Batutsi who protected their own clientele against legal claims of the *abakonde*, the latter defending their clan right. The second one escalated between *abakonde* and individual land clients. The main reason for this emerging conflict pattern was the incorporation of plots, which had gone to chiefs where there had been no heir. This development resulted in a gradually increasing shortage of *ubukonde* land. It also worked to the advantage of the chiefs. In addition, tribute payments, which had to be paid by atomized families for their individual land holdings, became ruinous. Very often the clients could not pay their tribute and therefore lost the land - which again resulted in further loss of *ubukonde* areas in favor of the chiefs.

A further issue creating conflict with holders of the clan law were the ever more extensive community fields *(amashiku)*, which had been set up by the Belgians in 1932 for the increasing group of landless farmers. The condition was an agreement to carry out seasonal cultivation. Because of this system 4,000 ha of new arable land had been obtained in marginal areas by the end of 1944 (Buschmann 1986:80-82; Mairieu 1983:52, 174).

b) Abakonde vs. subclans (abagererwa). The *abakonde* generally refused to make any concessions beyond the *usufruct* principle towards individual land rights to immigrant subclans or atomized individual land clients. However, under the double influence of the colonial administration and the political law of the Batutsi chiefs, individual land rights became increasingly attractive, especially since the clients often had to use their own hands to cultivate the land they were farming. The growing politically motivated shortage of *ubukonde* land and the demographically induced relative shortage of productive soils forced the *abakonde* to claim more land for their own use. Clan members were placed on better soils and gentle slopes, while (poor) land clients lost their land or had to move to more marginal, unfavorable arenas. Despite lower productivity, tribute payments were not reduced in return. Thus the poor landless Bahutu moved more and more into the front line drawn between Batutsi chiefs and powerful *abakonde* representing the traditional right.

(c) Abakonde vs. clients of the Batutsi. The fact that clients of the chiefs settling on former *ubukonde* land were forced to pay taxes only to the chiefs directly and not to the *abakonde* intensified conflicts between representatives of clan law and modern law. In November 1959, for instance, serious clashes took place in the Gisenyi and Ruhengeri provinces and, as a consequence, the Batutsi chiefs were expelled with their clients. Their land was again added to the *ubukonde* (Buschmann 1986:82). Such clashes constituted a harbinger of coming events. Land-related conflicts and economic as well as environmental scarcity of land, on the one hand, and a confrontation of traditional and modern land-tenure systems, on the other hand, were crucial preconditions for the sociopolitical revolution of the Bahutu.

The importance of the land question, among others, is shown by the fact that - right after the takeover of power by the victorious Bahutu - a hastily appointed land-law commission presented a proposal to the »Provisional Special Commission« as early as March 1960, advocating that preference should be given to old clan land law over political land law. The incompatibility of the two laws was justified by the original occupation and cultivation of the land by the *abakonde*.

On 26 May 1961 the legislature passed an edict reforming the system of land clientele.[21] The *ubukonde* estates not under cultivation at that time could be put under the control of local authorities. The tribute payments of the clientele were replaced by a lease. The arbitrary nature of expropriation was abolished. Clients of former political chiefs had to pay rent to the clan of the *abakonde* or to purchase the land. If a plot of land had constantly been in the hands of one family for 100 years, it acquired all legal rights. Two hectares of cultivated land was established as the minimum size for a family with five children. For every further child a family could get another 30 acres, up to a maximum size of 3 hectares (Mairieu 1983:232).

Although this arrangement originally met the demands of most of the Bahutu population for a stable land law with egalitarian and social norms (according to the European model of private ownership), some problems have remained to this day, for example, that of the countless impoverished land clients on the *ubukonde* estates. No right of ownership was granted to them, while former clients of the old chiefs could purchase their land.

---

21 Edit No. 530/1 du 26.5.1961 relatif au régime ubukonde dans les territoires de Gisenyi et Ruhengeri, J.O. du 15.1.1962, 3-7-.

The revolution of 1959-60 merely turned the distribution of resources between the Batutsi and Bahutu upside down without leading to a system more just for all. The emerging ruling Bahutu elite - with strongholds in northern provinces of the country belonging to clans with huge land resources - aimed at maximizing benefits for their clients as the Batutsi elite had done before 1960.

Even today the land law is still complicated, diverse, and hardly takes either parceling out or environmental transformation into account. This leads, for instance, to a situation in which farmers have to cultivate land around the *rugo* (little farm protected by a fence), in remote mountain areas, and in swamp-land; this often occurred simultaneously as owner, tenant, lessor, or holder of usufruct rights. The legal diversity and differences in legal position hardly served to motivate people to obtain loans or to preserve the land's fertility for future generations.

Apart from that, traditional dualism between farming and pastoralism remains prevalent today. Cattle herding is done mainly for prestige; the ultimate goal of breeding is a beautiful set of longhorns. But the necessary reduction of »luxury cattle« led to a situation in which fewer and fewer of the more than one million farms had a cow of their own. As a consequence, the supply of animal proteins is insufficient throughout the whole country (Werle/Weichert 1987: 121).

The modern Rwandan state has adopted the land law enacted by the colonial power more or less unchanged. The code of civil law, for example, makes the same distinction between traditional rights and those entered in the land register. And due to the growth of cities, the government designated so-called city boroughs, which relate to the ten prefecture capitals and two other cities. Within the city boroughs the system of written law determines the issue of land ownership and private property. Land is a commodity, registered in a land register, divided into plots, and alienable to individuals. The growth of cities forced the resident population, which could not pay the purchase price of their plot of land, to migrate away (Buschmann 1986:96).

Concerning land issues, the Roman Catholic Church in Rwanda came also under pressure. The church had acquired enormous estates, which led to conflicts with the *abakonde*, clients, and adjoining farmers. The church insisted vigorously that it had acquired private property rights through purchase - as early as the German protectorate under King Musinga - and this stood in contradiction to clan land law. Yet, this did not prevent the church from using the system of *socage* of pre-colonial Rwanda to the best of its advantage. The missions behaved virtually like traditional Batutsi patrons.[22] Against the back-

22 In 1903 a curate said to his priest; »A large landed property is necessary in areas where there is *socage*. Without it the mission would have no workers, and people could not be brought in to work if they were needed« (In: Buschmann 1986:99).

ground of population growth and restricted *ubukonde* ownership, conflicts flared up with the church, which often enough had not even cultivated the land (Ruhashyankiko 1977:113). The revolutionary government of the first republic aimed at changing this situation. In an edict of 1962 it was emphasized that »a society can possess only such landed property as is necessary in order to attain the goal for which it was founded«.[23] The size of a plot was fixed at 10 ha per unit; as the church owned more than 250 ha, this was a deep cut into the church's property.

Finally, some severe conflicts were induced by decrees of the colonial power to combat soil erosion. In 1947 the colonial administration began laying out terraces, planting hedges, and starting reforestation work throughout the territory of Rwanda. Since many of these measures were implemented by military force, anti-erosion terraces were perceived as a symbol of colonialism and were therefore destroyed by angry farmers on the eve of independence (Uwimana 1982:80).

Finding III

The power struggle of the lineages and clans was entwined with a conflict between semi-nomadic livestock herders and settlers, which manifested itself periodically in land-use and land-distribution conflicts. The distribution conflicts over scarce resources - due to agricultural conditions and combined with demographic dynamics - led to agricultural overuse of land, forests, and available water resources and thus resulted in ethnopolitically motivated transformation conflicts.

Because of the ethnopolitical hierarchy and the environmental discrimination against most rural producers, it did not prove possible to dissolve the socioeconomically and environmentally harmful agrarian structures, to overcome the spatial fragmentation and political weakness of the peasants and landless, and to induce sustainable development on a broad basis by means of a solid agrarian constitution. The environmental discrimination under the quasi-feudal Batutsi agrarian constitution and the related socio- and ethnopolitical conditions became a modernization trap, which could have been avoided only by means of innovative social strategies to overcome ethnization of the agrarian institutions. At the same time, one must be aware of the fact that agrarian societies are by definition more ecologically determined than modern industrial societies. In Rwandan society, with more than 90 percent of the workforce in agriculture, competing land laws and land-tenure systems proved to be crucial factors in terms of various conflict formations in microcosms of intergroup power relations. Since the fault lines between ethnopolitically motivated actors

---

23 Edit du 25 April 1962 relatif aux associations sans but lucratif, J.O. du 15 May 1962.

and competing land systems coincided, the ruling elite could play the ethnic card to establish a strong sociopolitical hierarchy, subordinating most of the population - before 1960 chiefly Bahutu, afterwards Bahutu and Batutsi.

Ethnization through Colonialism and Regionalism

As I have suggested in the previous sections of this chapter, Bahutu and Batutsi are social and political categories based more on privilege, unequal distribution of property, and positions of power than on sheer blood.

The colonial powers accentuated the hierarchical order they found by *ethnopoliticizing* it systematically.[24] Under the German colonial administration, but even more markedly under Belgian rule, the Batutsi were regarded as the more highly developed race. The Governor-General of the Belgian Congo and Rwanda-Urundi, Pierre Ryckmas, considered the Batutsi as quasi-Europeans. They were perceived as *aristocratic black people* and thus as a born *Herrenrasse*, while the Bahutu and Batwa were despised as *negroes* or *primitive pygmies* respectively. The *imaginary* racism of colonial times, which was based on European race theories of the 19th century, promoted the sociocultural difference and froze the existing power relationship on an ethnic basis. Physical characteristics were rated as signs of racial superiority or inferiority, and thus the privileged social and economic position of the social caste of the Batutsi was consolidated. Specification of ethnic descent in the passport created administratively two ethnic groups. This division was artificial as only the genealogy of the father was crucial in the classification, the mother's lineage was neglected. By such colonial measures the continuous amalgamation of the two groups, for instance through intermarriages, was undone by the colonial powers as *exogenous* forces. Instead, ethnic fragmentation was artificially accentuated.

In addition, the colonial administration had no interest in democratization of the mandated territory. On the contrary, in the first half of this century it tried hard to support the expanded power of the king *(umwami)* over the whole country by military means. It provided consolidation of Batutsi privileges in the lower administration and court departments and introduced new schools reserved mainly for children of the Batutsi elite. On the other hand, the size of landed property of each Bahutu farmer was doubled (at least in principle), the *socage* shortened, and religious freedom introduced. However, these measures could stop neither polarization of society nor overuse of land. In 1958 the Batutsi accounted for more than 95 percent of chiefs and 88 percent of the whole bureaucracy. The latter ruled like an alien element over the Bahutu,

---

24 Belgium took over Rwanda, which had belonged to German East Africa since the Berlin Conference of 1893-94, twice as a mandate power: in 1919 for the League of Nations and after World War II for the United Nations.

which at that time made up 90 percent of the population. The growth of Batutsi political power in the course of colonization changed the patron-client relationships fundamentally: as stated above, old alliances with some elements of personal solidarity transformed into instruments of pure exploitation (Newbury 1988:209).

Another high potential for ethnization of politics can be found in the firmly established *regionalism*. Relationships between the north, the center, and the south of the country in particular are marked by different access of the military and political elite to power (Gasana 1995:212-237):

- The *Forces Armées Rwandaises* (FAR), supported by the ruling party, the *National Republican Movement for Development and Democracy* (MRND) and extremist Bahutu of the *Coalition for the Defense of the Republic* (CDR), recruited their officer corps chiefly from the northern provinces of Ruhengeri, Gisenyi, and Byumba. After Major General Juvénal Habyarimana from Gisenyi in 1973 revolted successfully against the president of the first republic, Grégoire Kayibanda, a southerner, the Bahutu from the north established their rule. The northern regions, where the regal clan stems from and where only few Batutsi were living, based their privileges basically on long resistance against the people from the south (*banyanduga*). It did not matter whether they were Bahutu or Batutsi. The northern Bahutu even regarded southern Bahutu generally as Batutsi (Nsengiyaremye 1995:241-248).
- The *Front Patriotique Rwandais* (FPR) was founded by descendants of Batutsi who had escaped to Uganda to get away from pogroms during the Bahutu revolution in 1959-60. Later exiled Batutsi from Burundi, Zaïre, and Belgium and even Bahutu from the political opposition to Habyarimana joined the FPR. Most of these Bahutu came from the province of Ruhengeri, that is, from the same area as the president of the FPR, Colonel Alexis Kanyarengwé. He had been a popular minister under Habyarimana until 1980, when he had to flee because he had been accused of conspiracy (Africa Confidential 1994:5).
- The political opposition was chiefly assembled in the *Republican Democratic Movement* (MDR), the *Liberal Party* (PL), the *Social Democratic Party* (PSD), and the *Christian Democratic Party* (PDC). As was shown by the elections of 1993, the opposition parties recruit their supporters mostly in the south, while the MRND and CDR have their base in the north (Africa Confidential 1994:5; Reyntiens 1995: 265-273).

The presidential guard (GP) and the two militias loyal to the government *(interahamwe* and *impuzamugambi)*, which began the massacres after the plane crash in April 1994, defended the supremacy of the north against the south. The first victims were opponents of the Habyarimana government and the *akazu*, which was dominated by the the president's wife.[25] The planned genocide of the Batutsi served to legitimize the use of force against all critics of the regime. Against the historical background of the Batutsi rule, it was not difficult for the FAR and the militias to stir up the population against the Batutsi. Mass media, above all the private station *Radio-Télévision libre des Mille Collines*, had already disseminated malicious and racist campaigns against the Batutsi since the end of 1993 (Braeckman 1994:161-168).

Channeling to Overcome Discrimination

In the second half of the 1980s it became apparent that the economic growth of the 1960s and 1970s - the heyday of Rwanda - and the ecological capacity of the land fell far behind demands of the rapidly growing population, of which more than 90 percent lived in rural areas.

For the inheriting sons the land property was barely enough, while the non-inheriting descendants had hardly any alternatives. A rift developed in the traditional social establishment in the rural arenas on the hills, and subsequently quasi-modern social mobility began, although without corresponding destinations. In Rwanda a rural livelihood was not - as, for example, in Europe since the 16th century - counterbalanced or at least moderated by alternatives. Since labor was hardly available, migration of the unemployed and those without inheritance to the capital was unwelcome. The state restricted the influx into town centers which, in turn, heightened social tensions in rural arenas. Crime and violence became rural social phenomena, whereas these concerns are usually known in countries with a high degree of urbanization.

Observers explain the disposition of the marginalized population for this »unconditional friend-foe attitude« in the days of the 1994 massacres[26] with reference to this previously exceptional rural phenomenon.[27] In 1989, for the

25 See African Rights (1994:131). Thus the first victims a few hours after the plane crash were moderate Bahutu, who were ready to compromise with the Batutsi; among them Prime Minister Agathe Uwiligiyiamana; the president of the constitutional court, Joseph Kavaruganda; and the minister of information, Faustin Rucogoza. The militias deliberately spread the rumor that the FPR would systematically massacre the Bahutu inhabitants in the areas controlled by them. Those prepared to compromise in their own ranks and the political opposition could be eliminated simultaneously under the guise of »ethnic revenge« against the hated former »lords«, *i.e.*, the Batutsi ruling elite before 1959/60.

26 Thus for instance Peter Molt, coordinator for development cooperation between Rwanda and the State of North Rhine-Westphalia (Molt 1994).

27 Michael Franzke writes: »If one asks 'Batutsi' why they did not organize themselves, why they did not defend themselves, the replies sound like those from Jewish survivors: we did not take this agitation seriously, we were good neighbors, we went to the same school, our children played with one another. And if one asks captive Bahutu why they slaughtered their victims with machetes, drowned them in cesspools, threw them bound into rivers, burned

first time in the history of the republic, severe regional famines occurred, which led to unrest similar to civil war. This social and political unrest had little to do with the historical Bahutu-Batutsi conflict formation. Nonetheless, in view of the dramatic ecological situation »the ethnic group« became once more *the* imaginary quantity along which the acts of violence could be *channeled* and at the same time deliberately *intensified*. The ethnic group served as a political channel to maintain the conflict latently, to stir it up when required, and to initiate a rapid escalation if considered necessary. Obviously »the environment« or »the economy« were poorly suited for this medial function, even though they played a major role in the background of the conflict causation process.

Finally, the Roman Catholic Church also contributed to shaping identity groups. Between 1930 and 1959 the church transformed itself into a force against the monarchy, not at least because of the growing pressure of European missionaries. Bahutu consciousness is not thinkable without the education and the moral support of the Roman Catholic Church. The missionaries encouraged Bahutu in mission schools to focus on social justice, which brought them automatically in opposition to the feudal rule of the Batutsi. Many national Bahutu leaders of the revolution had studied at the Catholic *Grand Séminaire* at Nyakibanda. They acquired the skills and contacts necessary for articulating protest and for communication with the world abroad. In rural arenas mission-educated schoolteachers played a key role in channeling the anger of the rural population (Newbury 1988:181).

Finding IV

The colonial administration accentuated the social lines of division by introducing stiff ethnic criteria by means of the passport law and granting of political, economic, and educational privileges to the *Herrenrasse*. Such exclusive criteria had not existed previously, despite the fact of the historical process of ethnogenesis in the Rwanda arena. The intensified polarization of society and channeling of political unrest along ethnic fault lines proved itself in the appointment of the Batutsi elite to the power positions of the central state. The Bahutu countered this elitist ethnopolitical concept of the Batutsi with ethnonationalism, which drew its strength and broad support from the homogeneously populated Bahutu strongholds in the northern provinces.

---

them alive or killed them like rats, the answer of farmer Francis Sibamana is valid for many others: he had to kill his neighbor Isaac Kimonyo on the orders of Army Colonel Celestin Rwagafirita as a proof of loyalty towards the dead Habyarimana. 'Kimonyo was 74 years old. I knew him well. He could not defend himself, I hit him so long with a stick, until he fell down and died. If I had not done it, they would have killed me'« (Franzke 1994).

5.2.4. TARGETING: FREEDOM FROM OPPRESSION VERSUS ELIMINATION

The liberation of the Bahutu from centuries of Batutsi rule brought about a radical change in terms of both interethnic relationship as well as the access to power.

Social Revolution of 1959/60

Transformation of society-nature relationship in rural arenas - inducing the transformation of land-tenure systems, land-use patterns, and clientship structures - fostered political consciousness among rural youths. However, in Rwanda - as in other places (*see:4.2.4.*) - the coalition between peasants and other classes was crucial to the emergence of a revolutionary situation (Wolf 1973:290). In the 1950s an inter-class alliance developed between peasants, landless laborers, local elites, such as traders, shopkeepers, etc., Bahutu *évolués*, *i.e.*, educated white-collar workers, as well as poor Batutsi. One of the more prominent leaders, Grégoire Kayibanda, fully recognized the necessity and importance of an alliance between *évolués* and rural population. Through articles in and editorship of several publications (such as *L'Ami, Kinyamateka, Temps Nouveaux d'Afrique*) he and his closest allies (a.o. Aloys Munyangaju) stressed this issue systematically - and rather successfully (Newbury 1988:180, 188-189).

In the mid-1950s the Belgians, well aware of the growing revolutionary potential, introduced some modest reforms. Two of them were the introduction of councils at the levels of subchiefdoms and territories in 1952 (decree of 14 July) and of subchiefdom elections in September 1956. Although the elections still brought a powerful result for the Batutsi, it was clear for the first time that the growing influence of ethnic voting blocks in direct popular elections could not be ignored anymore. In 1955, anticipating the possible end of colonial rule, the Belgians appointed a new Governor-General of Ruanda-Urundi. Jean-Paul Harroy turned out to be a sharp observer of the ethnopolitical fragmentation of the Rwandan society. In his monograph rich of details, which he dedicated to the people of Rwanda, he points out: »*On pourrait alors admettre qu'en 1955 je trouvai un Rwanda où régnait une forte tension sociale entre Tutsi et Hutu, plus forte qu'en 1925. D'une part, les abus avaient augmenté et de l'autre on osait de plus en plus les condammer publiquement (...) Mon diagnostic personnel (...) est que sur les collines la dite tension était à ce moment beaucoup plus grave que ne le faisaient apparaître les rapports officiels*« (Harroy 1984:91-92).

In 1957 a group of Bahutu had published the »*Manifesto of the Bahutu*«, which demanded an end of Batutsi rule, improved access to key positions in economy and politics as well as a fairer distribution of arable and pasture land. The Manifesto was an answer to a »Statement of View« by the High Council of Rwanda, which, on the one hand, called for accelerated progress towards broader political participation and, on the other hand, completely failed to address the Bahutu-Batutsi conflict or the discrimination against the Bahutu majority. Against the background of the ignorance of King Rudahigwa, who was mainly backed by conservative and elitist Batutsi in Kigali, the Manifesto denounced the unfair social and political conditions and blamed the discrimination against the powerless and rural majority. The Bahutu experienced the Batutsi colonial rule as a feudal-racist system. Therefore, the Manifesto treated social aspects under the »indigenous racial question« and attacked the political, socioeconomic, and educational monopoly of the Batutsi as well as the arbitrary use of power by the chiefs. To quote one crucial passage:

> »The problem is above all a problem of political monopoly which is held by one race, the mututsi; political monopoly which, given the totality of current structures, becomes an economic and social monopoly; political, economic, and social monopoly which, given the *de facto* discrimination in education, ends up being a cultural monopoly, to the great despair of the Bahutu who see themselves condemned to remain forever subaltern manual laborers and still worse, in the context of an independence which they will have helped to win without knowing what they are doing. The buhake is no doubt abolished, but it is replaced even more by this total monopoly which is largely responsible for the abuses about which the population is complaining« (Nkundabagenzi 1961:21-22; Murego 1975:760-764).

Whereas Kayibanda founded the *Mouvement Social Muhutu* - later converted into the *Parti du Mouvement de l'Emancipation Hutu* (Parmehutu) - in order to politically articulate the sociopolitical issues stressed in the Manifesto, the ruling Batutsi elite reacted to the unrest by ethnopoliticizing the conflict, condemning the Manifesto with reference to a centuries-old »natural master-servant relationship« between superior Batutsi and inferior Bahutu and Batwa. The *Herrenrasse* mentality, cultivated since colonial times, of the »fairer-skinned and taller Batutsi« towards the »shorter black Bahutu«, made ethnization of the social struggle relatively simple. At the same time, it revealed the deliberate ethnic charging of the conflict by the naturally superior ruling elite as ideology for the preservation of power. The social and political reasons for conflict were disguised and the blurred boundaries between the two ethnic groups were accentuated by means of superficial trivialities. The crisis intensified when the ruling Batutsi minority was accused of having murdered Bahutu Subchief Dominique Mbonumutwa on 1 November 1959 near Kabgayi Mission. The first massacres against Batutsi occurred only a few months later in

June 1960.[28] Violence spread from arenas with Parmehutu strongholds, such as the provinces Gitarama, Ruhengeri, Gisenyi, and Byumba. Although the violence - mainly burning and looting Batutsi dwellings - seemed to be spontaneous, the rapidity with which violence spread from certain chiefdoms indicates at least a strong »desire« if not tough organization behind the »spontaneous« upheavals.

The Bahutu revolution caused the Batutsi to brutally strike back. The aim of the Batutsi was to seize independent control of the state apparatus because they anticipated correctly that, once the Bahutu had time to reorganize their forces, the Belgians may cross the front line and begin to support - openly or not - the Bahutu. While the Batutsi regime intended to eliminate Bahutu leaders and to get rid of the colonial power, Governor Harroy asked Colonel Guy Logiest, at that time officer of the *Force Publique in Congo*, to send troops in order to prevent violence and to reestablish public order. After the November conflicts the population refused to obey Batutsi chiefs any longer. As a consequence, Logiest did his best supporting Bahutu to take over subchiefdoms and chiefdoms. The result was quite impressing: by the beginning of March 1960 Hutu interim authorities already headed 22 of Rwanda's 45 chiefdoms and 297 of the 531 subchiefdoms. Colonel Logiest prevented the Batutsi from returning to power, while he put the Bahutu in a favorable position to campaign for the communal elections scheduled for June/July 1960 (Lemarchand 1970:167-168; Harroy 1984:319-320).

The revolution ended *de facto* with the Bahutu winning in nearly all districts in the local elections of June 1960. In September 1961 the oppressive system was also legally abolished. The Parmehutu won the elections held on 26 September 1961 under supervision of the United Nations with 79.5 percent of the votes. In a referendum the monarchy was abolished with a majority of 80 percent. The schoolteacher and founder of Parmehutu, Grégoire Kayibanda, became the first president of the so-called first republic, which obtained its independence from Belgium on 1 July 1962 (Harroy 1984:299-310; Reyntiens 1985:267-271; Newbury 1988:196-198).

Betrayed Revolution

The revolution of 1959/60 was often compared with the French Revolution of 1789 (Chrétien 1989). Does this comparison hold? Of course the objectives or targets of the two revolutions can be compared, for all the differences in conditions and protagonists: breaking of royal power and abolition of the feudal system in favor of republican conditions. However, here the comparison ends. The

28 Mbonumutwa was only assumed to be dead. How far his »death« was staged to create a reason for the Bahutu revolt remains a matter of speculation. I owe this information to Richard Friedli, professor at the University of Fribourg.

analysis of the post-revolutionary period seems to be far more meaningful. Here a comparison with the rule of the Jacobins, with Napoleon and the *»Thermidor«* of 1848, perhaps even with the period after the October Revolution of 1917 in Russia would definitely have contributed to sensitizing the »betrayed revolution« theme.

First of all, the Parmehutu regime of President Kayibanda lacked the experience of how to rule a republic and how to manage an elected government. Thus the tradition of monarchic centralism was maintained in the form of a patriarchal centralist republic, now only *»Hutuized«*. As a matter of fact, an ethnic bias was an attribute of the political campaigns of Kayibanda from the very beginning. Already in the mid-1950s the *Mouvement Social Muhutu* (MSM) and the *Association pour la Promotion Sociale de la Masse* (APROSOMA), founded by Joseph Habyarimana Gitera, represented two divergent approaches concerning the question of ethnic relations. Whereas MSM stressed the liberation of the Bahutu by campaigning against all Batutsi, APROSOMA called for the liberation of all oppressed people in Rwanda - Bahutu or poor Batutsi. Although the social question was important during the mobilization phase of the 1950s, ethnic identity became dominant during the first elections as well as the first Bahutu rule in the history of Rwanda (Murego 1975:760-764; Newbury 1988:193).

The refugee crisis developed into a warning sign for the Bahutu revolution. According to figures provided by the International Committee of the Red Cross in 1964 there were about 336,000 Batutsi refugees in Burundi, Uganda, Tanzania, and Zaïre (Guichaoua 1995: 339). The great number of refugees indicated that for Parmehutu the goal was not only »to abolish tutsi political and economic privileges and control of promotion but also to chase away the minority, made up of 'hamite invaders', and to return the country to its sole legitimate owners, the Bahutu. The new republican ideology slipped easily into inherited political patterns, based on politico-ethnic exclusiveness henceforth used against the former élites and, in fact, all the members of the banished ethnic group« (Guichaoua 1992:9).

The refugees and expatriated Batutsi (nicknamed *inyenzi,* cockroaches) organized militant resistance, first in Burundi and after 1962 in Uganda as well. The target of the exiled resistance movement was restoration of the monarchy and the former Batutsi rule. The time between 1962 and 1966 was marked by repeated attempts of armed Batutsi invaders, mainly from Burundi, to reestablish power. After major intervention attempts in July 1962 and in December 1963 the FAR reacted in January 1964 with massacres and crossed the frontier to Burundi to act against the Batutsi population in the border area. Batutsi were virtually hunted throughout the country.

The main effects of the invasions were rather counter-productive because they kept the most radical Bahutu elements in power. As a consequence, the Batutsi stopped with their militant actions within Rwanda. However, a second wave of interethnic struggles reached Burundi in 1972. Tens of thousands of deaths in the neighboring country induced the escalation of anti-Batutsi hatred anew. In July 1973, after several months of ethnic crisis, the Parmehutu regime of Kayibanda fell victim to a bloodless coup by Major General Habyarimana. He banned Parmehutu and in 1975 founded the MRND, a united party of which all inhabitants were members from their date of birth. While Rwanda attained a level of stability from the point of view of domestic policy and experienced significant economic growth, Habyarimana established a personalized rule, centered on the person of the head of state and based on the »automatic majority« of the Bahutu (Molt 1994). *Divide ethnically and rule politically* became the central instrument of the government. What counted in the »presidential mwamiship« (Lemarchand) was loyalty and proximity to the »court«: at the core of the power were the so-called OTPs, *i.e.*, »les originaires du même territoire que le président« (Gasana 1995:218).

Finding V
The tremendous historical lack of political experience of the Bahutu government and even more the paternalistic one-party rule in an ethnically and socially fragmented society contributed considerably to the early racist drift of the social-revolutionary Bahutu movement. The Batutsi were forced to remain far from power in the countries to which they had fled. In fact, the social and political goals of the revolution were betrayed by the ethnopolitical program of the Parmehutu from the very beginning of the liberation process.

The year 1960 was the turning point in the history of Rwanda. For the first time the Bahutu majority took over power. At the same time, it marked the historic birth of two completely different ethnopolitical projects: the strife of the Batutsi refugees for recapturing power in Rwanda and subordinating the revolutionary Bahutu again. The formerly oppressed Bahutu majority, on the other hand, had the single goal of avoiding the renaissance of the old regime. Always having been oppressed, the only security guarantee to preserve state power seemed to be the expulsion of the former aristocracy. Intervention, war, and genocide in 1994 are the most visible expression of two irreconcilable concepts: *Batutsi oppression of the majority versus Bahutu elimination of the minority.*

5.2.5. DEVELOPMENT WITHOUT DEMOCRATIZATION

President Juvénal Habyarimana, the founder of the second republic (1973-1994), presented himself to the world as a paternalistic but well-minded leader who was primarily interested in development of the country to the benefit of all Rwandans. At the same time, he created the image to be a moderate facilitator concerning interethnic disputes. In fact, the government's position was relatively moderate compared to the racist goals of the Bahutu extremists. However, the maneuvering between democratic rhetoric and populist paternalism is said by observers to be rather a result of external pressure from donor countries and the intervening Batutsi than of Habyarimana's political insights. The president behaved more or less apolitically, worried about his image abroad, and placed the revolution skillfully under a banner with the inscription *endogenous development*. This ideology hit its target, that is, non-governmental organizations and development agencies: during the 1970s and 1980s there was almost no hill without at least one foreign aid project. *»Tout se passe donc comme si l'Etat rwandais voulait prolonger l'illusion coloniale fondamentale, celle d'une 'ingénierie sociale' consistant à assurer la 'mise en valeur' économique tout en figeant la société dans un 'équilibre' 'adapté' à sa culture et à ses traditions. La méconnaissance systématique du changement et des contradictions de la société réelle justifie la négation du pluralisme politique«* (Chrétien 1992:129).

The vision of an integrated society adapted to its culture and traditions led to a rigorous system of politically enforced conformity, social stratification, and ethnic quota-fixing. For example, only 9 percent of the places in schools, industry, and commerce were reserved for Batutsi, even though they made up 11-19 percent on account of their earlier key positions in the economy. According to numerous observers the political situation was characterized above all between 1970 and 1980 by political banality. For example, the candidates for parliament were chosen rather for their diploma, their profession, or their ethnic affiliation than for their political standpoint. However, political exclusion of the Batutsi was more total than the economic one. For instance, none of the 143 mayors in office belonged to the Batutsi. Thus it was obvious that the clashes between Bahutu and Batutsi in the early 1990s had been provoked for the most part by the administrative, military, and local political actors in close cooperation with the media (Braeckman 1994).[29]

Under pressure from the FPR, which could establish itself in the northeast of the country, a constitutional reform was initiated and various parties were allowed in 1990/91. The president's pro-development party MRND added the

---

29 Among others the massacres of Kibirira, Gisenyi in October 1990, Bagogore, Ruhengeri in January 1991, Kanzenze near Kigali in October 1991.

programmatic term *democracy* to its name. Democratization *(demokarasi)* became a passionate as well as demagogic theme. It served to consolidate the ethnocracy with the help of other means without making any fundamental cutbacks in the emerging *racist and anti-Hamitic program.* Of the remaining Bahutu parties, that is, the *Republican Democratic Movement* (MDR) as successor organization to the Parmehutu, the social democratic PSD, and the liberal PL, only the latter admitted representatives of the Batutsi into its leadership team. It was therefore regarded as a concealed Batutsi party (Becker 1993: 132).

James Gasana, a former minister with the Habyarimana regime, described the political situation during the second republic as *»un mélange de moderne et de traditionel, dans lequel on retrouve des éléments hérités de la monarchie tutsi, des traditions des royaumes autochtones hutu du nord, et du modèle nord-coréen auquel la deuxième République a emprunté certaines méthodes de contrôle de la population«* (Gasana 1995:215).

Finding VI

The category Batutsi was banished from the ensemble of the national community, and Habyarimana with his »automatic majority« of about 90 percent of Bahutu voters advanced to the top of a government, which could no longer be voted out of office. The racist anti-Batutsi movement was supported by the privileged and *parvenus*, the lower middle class and qualified people, the new political class of educated white-collar Bahutu and army officers. The »centralisme excessif«, the OTP-strongholds in the central and northwestern provinces, and the control of the rural arenas through Bahutu mayors led to a tough social control system in favor of the ethnocracy.

5.2.6. CATALYSTS: THE INVASION OF THE FPR AND THE ARUSHA PEACE AGREEMENT

In the course of the 1990s the situation intensified noticeably. There were two catalysts, which significantly enhanced the tensions and contributed to the escalation of April 1994, which was triggered by the plain crash. These two catalysts were the invasion of the FPR and the Arusha peace process (not the agreement as such).

In October 1990 the FPR intervened with indirect support from the Ugandan army and took up its position in the northeast of Rwanda in the province of Byumba.[30] According to James Gasana this war declared by the FPR

30 The influence of Uganda on the FPR has not yet been clarified adequately. The media often considered the invasion of the rebels as controlled by Museveni. In fact, there seem to have been weapon deliveries from Uganda for the FPR. However, they were obviously not

encompassed elements »d'une aggression externe entreprise par le couple FPR/ NRA, d'une contre-révolution des exilés rwandais, et d'une insurrection, qui ont convergés dans une seule crise« (Gasana 1995:212). The Front pursued the aim of hurrying along the repatriation of the exiled Batutsi, which had been blocked politically again and again, and, if necessary, overthrowing Habyarimana's regime. In contrast to 1960, when Batutsi perceived their power as threatened, the FPR tactically denied that there was any question of an ethnic conflict. They accused the government »only« of incompetence, corruption, and dictatorial behavior. The consequences of the FPR intervention were new waves of persecution by the government.[31] The invasion once more aggravated the ethnic polarization. At the same time, it became clear that the second republic was going to fail because of internal contradictions, mainly the ethnic and regional asymmetry of power. »La guerre d'octobre 1990 éclata donc sur un fond de profonde crise socio-économique et politique produite par la faillite du parti-Etat. Elle n'est ni la cause ni le point de départ de cette crise, mais une initiative opportuniste du FPR qui, convainçu que l'occasion de chute du régime était venue, voulait éviter que d'autres groupes n'accèdent au pouvoir par les voies démocratiques, en dehors de sa tutelle militaire« (Gasana 1995: 220).

After massacres, war, flight, return, and renewed in 1990, 1991, and 1992, the cease-fire agreement of 12 July 1992 was reached after prolonged mediation by Tanzania between the government of Rwanda and the FPR. On 4 August 1993 the *Arusha peace agreement* was signed.[32] In its six protocols, the agreement laid down a quota plan to divide power in a broad transition government until elections planned for 1995.[33] According to Article 7 of the agreement the transition government was to be installed on 10 September in Kigali, on the condition that a neutral UN blue-helmet contingent would be stationed there until then. Furthermore, it was decided to relocate a FPR battalion in Kigali in order to protect the Batutsi population. While the latter decision was carried out quickly and was a main reason for the later military victory of the FPR, the *United Nations Assistance Mission for Rwanda* (UNAMIR) arrived later than scheduled by the UN because of recruiting problems in UN

dependent on Ugandan financial help. In the »Strategic Comments« IISS London writes: ».... Museveni is in no position to dictate to the highly independent-minded RPF. In fact, Ugandan officials have complained that the RPF is an unguided missile, refusing to take advice from Museveni and his officials« (IISS 1994).

31 The United Nations assume that there were about 900,000 refugees (displaced) in March 1993, that is about 13 percent of the total population (Security Council 1994:para. 49).

32 See Security Council (1993a, 1993b, 1993c).

33 A procedure which was applied successfully in South Africa.

member states. As a consequence, Habyarimana also delayed appointment of the transition government.

The high expectations attached to the agreement reached under the auspices of the UN, the OAU, and Tanzania were not fulfilled.[34] In fact, it turned out to be the unintentional prelude to the elimination struggle and genocide, which raged between 6 April and 7 July 1994. The indecisive attitude of UN - together with the rather obscure policy of France - had a negative influence on the dynamics of the conflict.[35] Hence, *»(l)es accords de paix ne constitueraient qu'une étape de lutte, car ils ouvrent la possibilité de paralyser l'organisation de la défense du pays et de mener la lutte sur l'ensemble de celui-ci et non seulement au nord«* (Gasana 1995:224).

Further Accelerating Events

Besides the FPR invasion and the Arusha peace accord there was one external event, which also had an accelerating impact on the crisis in Rwanda: the military coup of 21 October 1993 in Bujumbura, Burundi, against the democratically elected President Ndadaye who was killed by Batutsi forces. On 26 October, after a new Batutsi regime had taken over power, there were about 270,000 refugees in neighboring countries; on 29 November the total number increased dramatically to 659,000, with 375,000 in Rwanda alone. These events definitely enhanced the great fears among the Bahutu population in Rwanda that the FPR could organize a similar coup and establish an oppressive regime in Kigali again.

After the Arusha agreement was signed the Bahutu opposition in Rwanda split along ethnic and political lines, which was only the expression of a further channeling: on one side were those supporting a strategic alliance with the FPR. This political faction was led by Faustin Twagiramungu, a Bahutu who later became prime minister under the victorious FPR. On the other side were Habyarimana allies, the so-called presidential movement, which denounced the FPR as »snakes« and »beasts« and castigated the moderate Bahutu as deceitful people and traitors. The latter were kept on the death lists of the militias and were among the first victims on 6 April. Even during the negotiation phase in Arusha so-called local security meetings were organized by extremist Bahutu,

34 Boutros-Ghali spoke in his interim report of a »stable peace« which could be attained thereby: »Both sides seem determined to achieve a durable peace through disarmament, demobilization and national reconciliation« (Security Council 1994:para. 62).

35 The harsh criticism of human rights organizations was directed in particular at the attitude of the UN Special Envoy of the General Secretary, Jacques Roger Booh-Booh. He was characterized as a rather timorous and obliging helper of Habyarimana. A main goal of French policy was to keep the country in the francophone sphere of influence. Leaders of the FPR were educated in English in Uganda and »French diplomats felt extremely uncomfortable when English was spoken around them« (African Rights 1994:664; Verschave 1994).

often in cooperation with communal mayors who served to identify Batutsi. At the same time, the two militias were strengthened, death squads were formed, and death lists were made. The genocide was well-prepared and strictly organized.

Traumata

The population of Rwanda was badly traumatized by war and repeated flight and was therefore extremely mistrustful. While the trauma of the Batutsi goes back to the expulsion of 1960 and the definite and hence existence-threatening exclusion from the national association, the force of the Rwandan population's collective memory of Batutsi-rule proved to be unbroken. The rural population could be deceived by radio and television about the intentions of the Batutsi. The fear of a revenge campaign by the FPR was intensified; the slightest rumor could lead to mass flights. The deep-seated traumas on both sides, the war crimes, the incessant refugee tragedy, and high tensions led to a permanently risky situation.

Accelerating Military Expenditures as Indicator for Changing Threat Perceptions

The military expenditure of the country - although starting at a relatively low level - increased alarmingly in terms of annual growth rate, namely from $43 million (US) in 1985 (1993 prices) to $113 million in 1994. Military expenditure as percentage of GDP increased from 1.9 percent in 1985 to 7.7 percent in 1993, that is from $7 to 15 per capita. The fast growing military expenditures more than doubled to the average of 3.6 percent of all developing countries in terms of per capita GDP. MILEX was also high compared to 1994 total expenditures of neighboring states: Zaïre's expenditures amounted to $114 million and Uganda's to $87 million.

Finding VII

Both the invasion of the FPR and the Arusha negotiation process are catalysts to the conflict. They are not the reasons as such. However, they contributed massively to enhance tensions and to spread fear on both sides. While the FPR feared to be definitely excluded from power through democratic elections, the extremist Bahutu perceived the power-sharing model of Arusha as a first step towards Batutsi rule. The Bujumbura events added fuel to an already critical situation. They only confirmed the diffuse fears in large parts of the Rwandan population.

Against the double backdrop of pessimistic medium- and long-term economic outlook as well as the accelerated increase in military spending between 1985 and 1994, one can only assume the Rwandan government's threat perception to be alerted by recent politico-military developments both within the country and in the region.

5.2.7. TRIGGER: PLANE CRASH AND FIRST MASSACRES

The shooting of the presidential plane arriving in Kigali from Arusha is usually considered the trigger of the war that followed. Indeed, the crash was a powerful symbol, an event that shocked the people. On the other hand, a trigger is a more complex phenomenon than usually assumed. It is - as a rule - composed of several elements, the plane crash being the most visible of them. However, it was not the plane crash that triggered violence throughout the country. There are at least two other elements. Firstly, it was crucial that the plane arrived from Arusha where the decision on the power-sharing model was taken. The agreement - signed under some pressure put upon the negotiating parties by the international community - was a signal for the actors prepared to fight militarily. The perceived consequences of the agreement and the related fears triggered them to cross the threshold of violence. The shooting of the plane was the first step in the use of violent means and as such closely connected to the other element of the trigger: the first wave of massacres only a few hours after the crash. The first massacres were against Bahutu ministers, the Bahutu opposition, and against those who sympathized with the FPR. These events indicated that everything had been carefully prepared. They triggered the massacres throughout the country, even on remote hills far from the capital.

Finding VIII
The trigger is a complex phenomenon. The most visible events - such as the plane crash - are not always the core of a triggering process. Instead, they are often the first step after the decision to use violent means has been taken. The Arusha agreement concerning power sharing, the plane crash, and first massacres are closely intertwined.

5.3. Preservation of Power at All Costs: Conclusions

The 1994 war in Rwanda had its specific and complex anatomy. One would fail to explain it with monocausal, biologistic, or holistic concepts. Instead, the causes and their roles have to be put in the framework of a differentiated conflict analysis. In this area study five interrelated roles have been conceptualized (*see: fig. 5.1.*). It proved to be meaningful to separate reasons from targets, channels, catalysts, and triggers. The result is more clarity regarding the difference between causation processes and what causes - or triggers - actors to use violent means. This procedure contributed to highlight adequately the role played by environmental as well as ethnosocial discrimination against the rural majority throughout Rwandan history. The following diagram reflects the individual findings illustrating the causal role of influencing factors (*see: fig. 5.3.*).

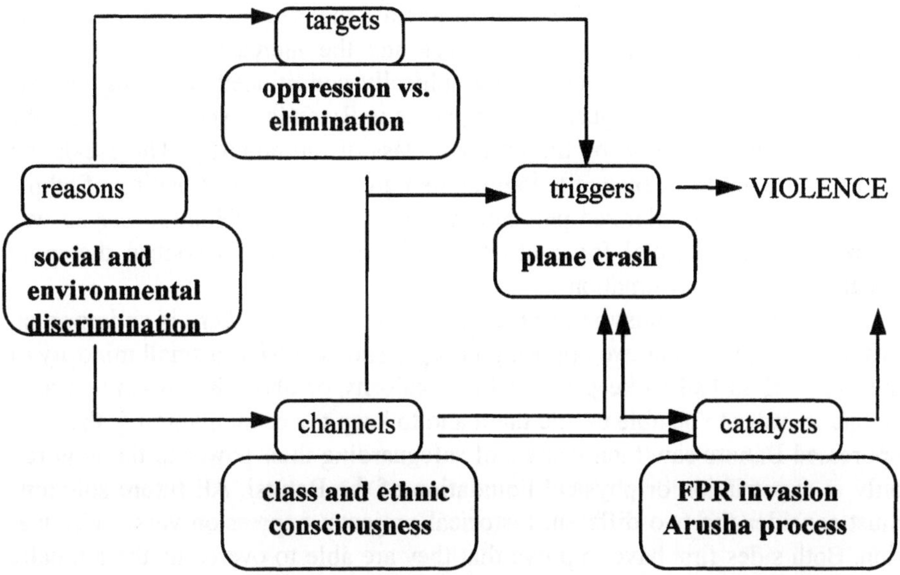

Figure 5.3. Multiple causal roles concept

Even though genocides of the Rwandan type and extent are quite unique in history, Rwanda is not a single case; it can be compared with other contemporary internal wars. The current violent conflicts in Africa, of which there are at least 20, display at least partly similar causal pathways. If the various causes and their roles are related to one another, it is noticeable that so-called ethnic conflicts are not spontaneous or indiscriminate pogroms. All actions are concerted, planned, systematic, and carried out by a political elite with method and strategy.

If the variables are combined, a picture of the Rwandan crisis emerges, which corresponds to the hypothesis formulated at the outset. A battle for suppression between two elites is discernible, which aims at political exercise of power in a hierarchical interethnic order. The social classes form the channels, which contribute to intensifying the fault lines in the country; reducing differentiation to two identity groups was achieved by ethnization. Not only the conflict lines but also the duration and intensity of the conflict were thereby easier to manipulate. Through the influence of exogenous factors since colonial times the power struggle turned into an actual elimination struggle between Bahutu and Batutsi elites, on the one hand, and a regionalist and chauvinist battle between northern and southern Bahutu elites, on the other hand.

The revolution of 1959/60 and the counter-revolution of 1994 were the peaks of this development. The growing social and ethnic tensions between these two dates can only be explained against the background of scarce and unequally distributed economic resources and the increasing overuse of the various ecoregions due to the demographically conditioned shrinking of subsistence economy. The majority of the people in Rwanda - except between 1962 and 1994 - felt *colonized* by the political class of the minority. The escalation of the 1990s led to a sense of belonging to a group, to a strengthening of ethnic consciousness and polarized perceptions, so that ethnic affiliation *per se* was stylized into a potential for conflict and thereby began to control the inner dynamics of the confrontation.

Nevertheless, it must be emphasized at this point that there is an important difference between the two opposing groups. The Batutsi as a small minority of the upper class had to keep the Bahutu majority of about 85-90 percent as *a people* in order to be able to rule them and to keep the economy going; the once oppressed Bahutu saw their chance of safeguarding their power in the long run only in expatriation or physical liquidation of the Batutsi. All future solutions must consider the two different historical projects: oppression versus elimination. Both sides first have to prove that they are able to overcome the mutually exclusive projects. *»Rwanda pour tous«* so far is a political slogan of liberal and moderate exiled Bahutu in Europe and not a political program of the parties to the conflict.

Despite the relative cultural unity of the country in the long period from the 15th century to colonial times, the ethnic amalgamation had not progressed far enough to avert accentuation of ethnic identity once and for all. The latter had happened three times in the history of Rwanda in a lasting manner. The first time the existing ethnization served the Belgian colonial masters to carry through their indirect rule efficiently. The second time it served the seizure and consolidation of power by the Bahutu government and exclusion of the Batutsi from the national unity between 1960 and 1994. The third time it served an extremist Bahutu minority inside the government to seize power through massacres of opposition Bahutu and genocide of the Batutsi.

The removal of traditional forms of conflict management within the framework of the family and lineage institutions did not lead to reliable new mechanisms. Eroding social and economic conditions developed into a potential for conflict, which led to famine revolts, clashes between two ethnopolitical groups, on the one hand, and degenerated patron-client relationships, nepotism, and dictatorial conditions, on the other hand. While the respective rulers pursued an aggressive ethnopolitization in order to preserve and consolidate power (Batutsi during colonial times and in 1959/60; Bahutu in 1962-66, 1973, and in 1990-94), the respective opposition simultaneously pursued a politicization of social contrasts (Bahutu during colonial times and in 1959/60; Batutsi in 1960-1994).

After the liberation in 1959/60 Bahutu copied former Batutsi-style ethnocracy, which became the guiding sociopolitical system. The ethnic antagonism proved a most effective lever against democratic pluralism. »Bahutuized« Batutsi practiced nepotism and corruption in the center of power, »Batutsized« Bahutu and opposition intellectuals sympathized with the FPR. Against this background there are in principle two possibilities to end a boundless and permanent power struggle: either the power imbalance forces interested parties into mutual acceptance of the *compromise* as a distribution norm, or a stable hierarchy forms, which allows an individual or an elite to dictate the conditions of distribution. The latter means that the monopoly on the use of force can be permanently *usurped* by factions within a group (Münch 1992:266). Habyarimana with his automatic majority seemed to strive for the latter, but intervention of the FPR and the political opposition in the south and within his own ranks forced him to sign the peace agreement of Arusha and with it a compromise on the division of power. An extremist group sacrificed Habyarimana and tried to continue the usurpation by falling back on force. The genocide against the Batutsi, who were decimated to about 500,000 or a third of the population, served to maintain the undistributed power. This plan was shattered by the military victory of the FPR, which seized power on 18 July 1994 to reestablish Batutsi rule.

It is still an open question how the deep-rooted socioeconomic and ecologically determined power conflict will develop in the future. As long as the despotic and the exterministic concept confront each other, neither the MRND government in exile nor the FPR intervention government can claim to be the legitimate government of Rwanda.

In 1996 and 1997 political tensions in Rwanda remained very high. Observers assumed an *inter-war period* rather than the beginning of a period of peace. The situation is now more polarized than ever. If the present government allows the refugees to come home, it is confronted with the situation as it existed before the revolution of 1960: it must establish an oppressive system without power sharing and broad participation of the Bahutu population. Otherwise it risks becoming a minority that could be expelled once again.

Only a process of *equalization* towards more horizontal intergroup as well as intra-group relationships, as ideally asserted for the mythical origins of early Rwanda, could lead to a step forward towards a government in a democratic and pluralistic state. The broad participation of the rural population and innovative capacity building - in terms of new political and legal institutions as well as of economic alternatives - are pure necessities in order to overcome in a sustainable way what finally caused the war: environmental and social discrimination against the majority of the rural producers on the *mille collines*.

CHAPTER 6

MODEL

Causal Relationship between Environmental Transformation and Violent Conflict

New theories normally start with attempts at explaining puzzles because they deal with social or political phenomena which are not subjects of existing theories. Now, after having collected and described so many puzzles and after having created several general patterns with the help of the single pieces I proceed to a systematic explanation of environmental conflicts in terms of why and how environmental conflicts escalate into open conflict between different actors. In order to provide a systematic comparison of different cases reaching general conclusions, I will introduce criteria enabling me to falsify the hypothesis of this study. Tentative rather than definitive conclusions have to be reached. The propositions are therefore essentially probability statements describing likely relationships.

The environmental conflict program does not lead to monocausal explanations of violent conflicts or war. Rather, environmental disruption is embedded in a syndrome of factors complicating any conflict analysis. The ENCOP cases have shown that transformation of society-nature relationships do not automatically and necessarily lead to the use of violent means. Since there are many countries with severe degradation of renewables that are confronted with violence and many countries with similar degradation that are not, the correlation between environment and conflict is spurious. The same causes do not always have the same effects because there are distinct intervening variables, which urge different outcomes. However, both the area studies review as well as the in-depth case study indicate that there is ample evidence for the repeated appearance of environmental change in the genesis of the conflicts the studies focused on. The natural environment is not just a resource in the sense of a contributing or (c) condition as other socioeconomic factors as well. In some cases this may be true. In others - which the study concentrates on - the transformation of society-nature relationships is part of Polanyi's Great Transformation of rural societies (Polanyi 1995). Such transformation is in any case a painful and often - not always - a violent process. In these cases transformation is considered to be a necessary or (n) condition for the genesis of the conflict.

This status of the independent variable will be reflected in the model proposed in this chapter.

6.1. Model Building and Constraints

There are several challenges concerning development of a model aimed at explaining a rather complex reality. One major challenge is the data problem, which will not be dealt with in greater detail. It should only be mentioned that an empirical test of the model is not intended. It may be tested by other scholars as soon as adequate measurable, comparable, and metric-form data will be available. Nevertheless, some indicators will be proposed in order to better grasp issues concerning transformation processes. Since more and more actors (and not only insurance companies) are interested in data on drought, flood, water availability, etc. there is hope that the UN agencies will provide more and accurate data on environmental issues on subnational and local levels.

Another challenge is that modeling urges conceptual clarification of the topic under consideration. It necessarily leads to reconsidering arguments more rigorously and, above all, it helps to put individual case studies into a general framework. Confronted with the choice to develop a model that is either comprehensive and close to the configurative case studies or that is empirically testable and therefore as narrow as the basis of the quantitative data available, the first option is chosen here. Unsatisfied with a pure case-studies approach - however important the single case may be - and concerned about the »data market«, a model is designed, which has to be seen as one further step in a cumulative research effort of a rather small community of researchers.

6.1.1. EXPLAIN A LOT WITH VERY LITTLE

The correlation between violence and environmental degradation is relatively weak to the extent that other factors such as, for instance, poverty and poor state performance are also causally relevant. There is evidence for the following basic assumption:

Intractable poverty of discriminated rural populations and environmental transformation become more and more closely intertwined in particular geographic zones with fragile environmental conditions; these are above all dry zones and sensitive mountain arenas with high rural population, which depends heavily on renewable resources. If poverty and maldevelopment lead to violent conflicts and wars, on the one hand, and if poverty and maldevelopment lead to environmental transformation in those arenas where violent conflicts occur, on the other hand, there is supposed to be a correlation between environmental transformation and violence. Poverty-driven violence and poverty-driven degradation may have a common denominator. Indeed, as we have seen above, countries most affected by violence (44 out of 48 of the least developed countries and 31 out of 69 medium-developed countries) belong to those countries with most serious environmental problems. Moreover and above this, societies with a large part of the population in the agricultural sector or living in the countryside are considered to be either politically unstable or undemocratic, or both. The specific sociopolitical circumstances - and not the degradation of the environment as such - determine what kind of environmental conflict the observer is confronted with. According to the typology provided above (*see: 4*) the following conflict types have to be grasped and covered by the model:

(1) in *center-periphery conflicts* the fault lines are between the modern national and international segments in the center of the state and traditional sectors of societies in which the environment is transformed by industrial projects (large-scale farming, mining activities, dams; *e.g.*, Bougainville);
(2) in *ethnopolitical conflicts* the fault lines are between ethnic groups, which either share one ecozone with degraded and thus scarce resources or which settle in neighboring ecozones of highly distinct support capacities and thus different degrees of productivity (*e.g.*, Jebel Marra in Western Sudan);
(3) in *interregional migration conflicts* migrants of the same ethnic group but stemming from a different and often remote part of the country settle in fertile rural or (peri-)urban arenas where they challenge local populations; (*e.g.*, mountaineers settling in the plains of Central Asia);
(4) in *cross-border migration conflicts* migrants from one country want to escape an unproductive ecozone and settle in a productive zone across the border (*e.g.*, the Senegal-Mauritania conflict in the Senegal Valley);
(5) in *population pressure conflicts* demographic development - above all high population density in marginal rural arenas - either triggers local conflicts or forces migration, which leads to conflicts in the arenas of destination (*e.g.*, Bangladeshi in Assam Province, India);
(6) in *international water conflicts* the asymmetric dependence of upper riparians and lower riparians sharing an international river basin triggers tensions, international bargaining, military threats, etc. (*e.g.*, Turkey vs. Syria and Iraq in the Euphrates-Tigris basin);
(7) in *globalized* or *«long-distance« conflicts* the producer or »sender« of environmental damage and the »recipients« of it may be located in geographically distinct parts in the world. Environmental damage is perceived by the recipients as foreign environmental aggression (*e.g.*, small island states vs. industrial air polluters) or as environmental neocolonialism (e.g., French Polynesia/Greenpeace vs. France at the Mururoa Atoll).

To further reduce the high complexity created by the diversity of fault lines and actors, the seven conflict types were assigned to specific causation patterns, which share a small number of common indicators. Thus I got three general patterns that accelerate socioeconomic, environmental, and political heterogeneity either within states or internationally:

(I) *marginalization of poverty* as a pattern drawn from cases, where poverty segments of societies in transition are forced to produce in marginal ecozones (arid zones, steep slopes, remote mountain arenas, etc.);

(II) *national and international sacrifice arenas* drawn from cases, where central governments - often together with multinational firms - develop islands of modern industry degrading remote ecozones used and traditionally owned by indigenous groups, subsistence farmers, smallholders, or other coherent groups;

(III) *high international dependence* drawn from cases, which reflect international power asymmetries resulting from (a) unequal access to freshwater resources by upper and lower riparian states or (b) unequal distribution of environmental pollution effects between producers and recipients.

These three patterns are driven by the factor of environmental discrimination against actors forming part of the conflict structure. Environmental discrimination is the common denominator not only of all conflict types but also in view of the linkage between the structural problem of environmental degradation with the actors' perceptions regarding this very problem. Environmental discrimination is the entry point into the model.

6.1.2. PREREQUISITES OF A MODEL

The FAO early warning system concerning food crises, for instance, is well established and accurately functioning to date. But how to predict human behavior? And even more problematic: how to predict human behavior if ecological disasters are not disasters in terms of sudden or stochastic events but rather resemble invisible system forces? The only way is to link an environmental change model with a conflict model by elaborating on the complementary or overlapping issues.

Robust Model Building

The model proposed here is designed to explain a lot with relatively little - without being too simplistic. On the other hand, even the most encompassing model cannot deal with all factors involved in a conflict syndrome within a given arena. If a multi-factor analysis tries to bring too many variables into the picture, explanatory power declines.[1] If the model, by contrast, is too narrow in terms of variables, it fails to address different types of environmental conflicts. A compromise follows the path of a given conflict without getting lost in the jungle of countless intervening factors.

If there is a lack of theoretical input, even the most complex model will fail in predicting events. The FUGI model (Onishi 1996), for instance, is a model of global interdependence, which can also be used as an early warning model for refugee and displaced persons crises stemming from destruction of the environment. The model is well-developed, complex, and empirically tested. However, it has tremendous shortcomings concerning theoretical assumptions and transformation of qualitative data into metric terms. To highlight two aspects in order to justify the criticism: Onishi states, for instance, that there is a risk of »ecologically displaced persons« emerging in accordance with the »destruction of environment«. This statement is correct. However, in order to get some measurable data, he takes some problematic steps forward. He proposes, among other things, that in Africa »there are continuing large increases in population.« Thus, »(a)s human beings increased in number (...) (m)ore firewood becomes necessary, and so forests are cut to get it« (Onishi 1996:4). He assumes that population growth is responsible for deforestation, and, hence, deforestation for migration and displacement. Although this may be one plausible pathway, it is just one out of countless other possible pathways. In today's Africa it is not even the most probable one. Neither population growth nor deforestation (in many areas there is no forest left) are or will be major causes for displacement and flight.

The next step Onishi takes is even more critical. Since there is a lack of accurate data on the »destruction of the environment«, he introduces a point system to rank six different environmental problems (water pollution, air pollution, etc.) divided into five-year periods: 1995, 2000, 2005, 2010, 2015 (1996:12). According to this system, China, for instance, is rated a »natural disasters» factor of 60 points in 1995. Onishi predicts a decline in China to 55 points by the year 2000 and 47 points by 2015, etc. The Chinese government will be pleased to hear from Onishi that water pollution will decline from 50 to 48 points by 2015, and ecological imbalances - deforestation, desertification - from 60 to 46 points by 2015. At the same time, economic growth will decline from 11.1 percent in 1995 to 5.4 percent in 2015; per capita income stagnates, whereas foreign debt service ratio will increase dramatically from 9.5 percent to 29.1 percent; MILEX will grow from 2.3 percent (GDP) to 7.0 percent, and basic human-needs satisfaction will decline from 48 to 35 points (Onishi 1996:

1 Overcomplexity may, for instance, constitute a major problem for the »complete model« of Wieder-Goodrich/Brecke (1996:20).

13). The question arises regarding the data basis of predicting China's environment to be far better off while the economy stagnates and human needs are more difficult to fulfill. The only explanation for this result is his prediction about less abnormal weather conditions in China, such as floods, droughts, etc. Despite the fact that nobody really succeeds in predicting long-term climate change and weather conditions, the Onishi approach provides some conceptual shortcomings over the society-nature relationship in China - and elsewhere - due to more than unpredictable weather conditions.

Good Theoretical Propositions Are Indispensable

Against this background it becomes clear that each model must be based on clear assumptions formulated in *propositions* that deal with major issues in terms of causal factors, which are based on empirical evidence and plausibility assumptions. The following propositions are drawn as much as possible from empirical results provided by the ENCOP cases, but also from the conceptual background highlighted above (*see: 1-3*).

Barbara Harff stated that even if abundant information is reliable and thus potentially useful, »we still need means to isolate and interpret essential messages. We argue that theoretical models are the essential filters by which we can detect the patterns that can guide us through the maze of information available« (Harff 1996:3-4). Propositions must lead to a set of linked hypotheses or sentences, which put the proposition into an empirically testable form. »All models provide frameworks through which taxonomies can be derived, similarities and dissimilarities be discovered, and particular configurations detected. Rarely are we fortunate enough to figure the precise constellation of factors that lead to a certain type of conflict« (*ibid.*:4). On the other hand, there is some hope because »(t)he fact that the same variables in different configurations appear in different models shows that various types of conflicts have common dimensions« (*ibid.*:1). If individual models are good enough, they should be integrated in a broader context in order to compare both the propositions and the results in terms of predictability - which is only an indicator for the extent of generality.

The Quality of Indicators

Different types of conflicts need different models and - as a consequence - different sets of indicators. With the help of indicators we are able to describe a certain stage of events. An indicator points to something, which it is not on its own. It indicates complex interrelationships, which cannot be measured or described directly. Therefore, an indicator has to represent adequately a topic or issue. In other words, a good indicator expresses relationships one wants to explore as accurately as possible. It is not just a statistical unit of measurement.

A quantitative unit is only an indicator if it is put in a theoretical context (Nohlen/Nuscheler 1993:77). In development theory an indicator marks the attempt to quantify certain concepts of development in the light of a specific approach. Indicators should point to those elements supposed to be relevant for this very conception of development. If, for instance, improvement in the health of a population is a goal, then the rate of child mortality measures its extent and indirectly the state of health. This means: child mortality is a health indicator for the population in general. Probably it is even more. As mentioned above, child mortality highly correlated with the phenomenon »failed states«. Obviously this well-documented indicator stands for a society's stability or instability, which can hardly be measured in quantitative terms.

Since there are numerous individual indicators that can point to a certain complex state of affairs - such as poverty, maldevelopment, or discrimination - one has to find and select the most representative indicators (Taylor/Jodice 1994). The sheer quantity of indicators is not crucial, but their quality and the data they are based on are important in terms of analytical results.

Nohlen/Nuscheler (1993:80) distinguish between three different kinds of indicators. The three types should be used altogether in a considerable mix to avoid major biases, misperceptions, and miscalculations. They are:

(1) per capita indicators designed to measure a statistical average;
(2) percentage indicators, which point to distribution of a certain asset or value; and
(3) structure indicators, which are also measurable in percentages but do not point to a distributive aspect.

Indicators serve different purposes, such as diagnosis of human development, basis of planning, evaluation of certain goals, etc. In the present context they are used to systematize a certain amount of knowledge accumulated in several case studies. They indicate socioeconomic and environmental problems that lead to crises and violent conflicts or wars. Beyond that, they help to develop a model based on the different area studies. The model itself can serve to identify and explain similar cases that are not part of the sample. Therefore, modeling should lead to a degree of generalization over the findings, which allows a certain predictability, *certain* because the results are probable in any case.

Problems related to indicators are not only selection but also availability of data at the national level, which allows cross-country comparison, as well as disaggregated data, which allows looking into subnational arenas and comparing heterogeneity in a single country. Besides data consistency and completeness, validity and significance of data are also important. *Validity* asks if the indicators really represent the theoretical problem a researcher deals with. *Significance* asks if the indicator really points to the important aspects of the problem. Moreover, the question arises if the indicator really discriminates between those factors an analyst aims to focus on. Indicators should show trends over time and simultaneously be responsive to changes in underlying conditions. Therefore, the selection of valid and significant indicators is crucial: they should focus on structural data, such as land/energy use, access to freshwater, rural-urban poverty gaps, environmental discrimination, crop failures, growing dependence on scarce/degraded renewable resources, etc. At the same time, they should be designed to be linked easily to data on events or sudden changes - such as drought, livestock decline, flood, or acute water shortages, etc.

A major shortcoming in the indicator development is the lack of a defined set of indicators for sustainable or unsustainable resource use. There are still disputes over the most representative bunch of indicators (Nohlen/Nuscheler 1996:88-89; OECD 1991). In the best case work in progress may lead to a sustainability index. The catalogue drafted so far encompasses environmental monitoring in order to understand and predict habitat changes and the socio-economic and demographic forces behind it (FAO, WRI, WCMC, TREES project of the EU; World Bank 1995:24-26). The World Resources Institute (WRI) and its partners developed a map-based indicator approach for assessing human pressure on ecosystems and the degree to which arenas are at risk from continued human activity. The *natural capital index* (NCI) is based on a country's *remaining natural areas* (RNA) in hectares, adjusted by the *biodiversity indicator* (BDI) so that NCI = RNA x BDI (*ibid.*:23). The NCI indicates different degrees of dependence on natural capital. However, it does not focus on environmental degradation as a *social* stress factor as does FAO with prediction of crop failure and starvation - based on rather accurate models.

UNDP (HDR 1995, 1996) focuses more and more on trend analysis than on single indicators. The annual reports provide helpful data on growing rural-urban gaps concerning human development or deprivation (*see: app. 13a, b et al.*). All these attempts should be accompanied by further experiments dealing with subnational environmental disparities, which could indicate fault lines evolving between different groups. Such a pilot study done by UNDP led to a significant but somehow not surprising result: as mentioned above, in the Philippines the National Capital Region has quite a high HDI, whereas Mindanao is ranking next to Zambia (*see: 1.1.4.*). New ideas UNDP is experimenting with, such as an encompassing »capability poverty« index or a »natural resource dependence« index, combined with a »gender disparity index« could provide many insights of importance for an early recognition of a crises system aiming at integrating environmental key factors.

A future task is to devise a sustainable development index. There are indicators provided by OECD (1991) to highlight links between development and ecology. Preliminary results are similar to those referred to in previous sections: poverty groups are most affected by soil erosion, pollution of surface water, etc. Experiments of the WRI to define criteria to measure environmental productivity and the sustainability matrix of the World Bank (1995:76) lead in the same direction. UNEP with its established monitoring systems (GRID in Geneva and INFOTERRA in London) should be at the forefront of the movement to provide an index based on their own global monitoring capacities.

Although environmental indicators for global cooperation have been published quite recently, they focus on international comparative studies. They are not designed for subnational levels and use of disaggregated data (Rodenburg/Tunstall/Bolhuis 1995). For studies that are dealing almost exclusively with domestic affairs induced by subnational environmental impacts, those indicators may be useful theoretically but do not provide an adequate set for practical purposes.

This brief overview demonstrates that development of indicators is moving in the direction of grasping transformation processes on subnational levels.

6.2. Propositions and Hypotheses

Propositions are preliminary arguments about a certain state of affairs or a kind of relationship. The set of propositions begins with the common denominator or what was called the *entry point*. This is *environmental discrimination* in the sense that distinct actors - based upon social, ethnic, linguistic, religious, or regional identity - experience significant inequality in terms of material well-being, livelihood security, and societal participation concerning environmental issues compared to other actors. This means data on degradation of renewables - such as, for instance, land degradation - are only interesting in terms of conflict analysis if and when these data indicate discrimination against actors.

Members of the group constituting an actor or actors as an entity are systematically restricted in their access to natural capital relative to other groups in their societies. This is an adapted definition of discrimination provided by Taylor and Jodice (1994:71), who measure scope and intensity of economic discrimination. Since it is not possible to draw a direct line from the state of the environment to political decisions, discrimination is designed to link the natural environment and the sociopolitical environment by introducing valid and significant indicators, which differentiate in two directions: firstly, concerning environmental and socioeconomic trends; secondly, concerning political/institutional mechanisms as well as opportunities and preferences of actors. The first hypothesis is as follows:

The greater the environmental discrimination actors experience in their arena, the more likely they are to organize for action against the sources of discrimination (hypothesis 1).

Environmental conflicts mainly become violent in developing countries. Why is this so? First of all, there is a general trend towards violent conflicts and wars being a phenomenon of societies with low human development in general and low state stability and poor performance in particular. Secondly, violence occurs in the context of rapid social and environmental transformation of agrarian societies. This process does not lead immediately or smoothly to modern societies. On the contrary, it has been pointed out that the environmental crisis feeds structural heterogeneity in developing societies. These societies depend heavily on their natural capital to form a substantial part of the nation's wealth. The Natural Capital Indicator (NCI) points to dependence of a national economy as such, but also addresses unsustainable use of renewables. The degree of dependence differs enormously - and not only internationally. Also at the subnational level various actors' dependence on natural capital may differ significantly in terms of economic activities, modes of production, and productivity.

The more dependent victims of environmental discrimination are on degrading renewables, the more likely they are to organize for action (hypothesis 2).

Environmental discrimination most likely affects more or less coherent types of actors in a certain arena. These actors may consist of community-based social groups, such as small rural producers (peasants, landless, herders, etc.), ethnic minorities, indigenous or religious groups. It may affect individuals, families, clans, tribes, or any other kindred groups in many ways. Intra-group competition for scarce resources is overwhelmed by intergroup struggles. Both the perception and fact of environmental stress enhance identification of individuals with their own group, which is viewed as an *individual* actor against other individual actors. Thus group cohesion triggers intergroup struggles about degraded resources along different fault lines: interethnic strife, immigrants vs. residents, farmers vs. nomads, indigenous peoples vs. representatives of the modern sector, rural vs. urban dwellers, etc. If individuals, families, and lineages seek support by joining or building larger social or kindred groups, they tend to create »strategic groups«. Strategic groups are actors with political leadership, organizational ties, common goals, and more or less developed capacities to fight. Thus strategic groups create a political arena where use of violent means is most likely to happen. The arena of environmental conflict lies in marginal ecoregions, in national sacrifice areas, and in internationally shared river basins (according to the three patterns).

Group cohesion increases to the extent that discriminated actors are concentrated in sensitive arenas, such as marginalized ecoregions, national sacrifice areas, or shared (transboundary) river basins (hypothesis 3).

The concept of group cohesion is drawn from Gurr and Harff (1994:83). It has been modified according to the topic of this study. Concerning international water-basin conflicts group cohesion is not a necessary condition; however, behavior of an upper riparian can induce group cohesion within a lower riparian state, which is confronted with many internal fault lines (as does the behavior of Turkey in the Syrian society).

The cohesion of »strategic groups«, which tend to organize action in a given arena, is related to two other key factors. The first is the decline of traditional mechanisms and institutions and the lack of modern ones to regulate resource use and tenure conflicts where they occur. Most projects focus on weak states, state failures, poor state performance, etc. Although this is an important factor in most countries with low human development, the nonexistence of a state is only one aspect of modernization that got stuck. The reverse of failed modernization is the weakening or even dissolution of traditional or local and provincial regulatory mechanisms. Rural communities are either confronted with a weak central state, which makes decisions without participation, or they suffer great pressure because of environmental and socioeconomic discrimination; this often results in collapse of traditional communities, atomization, migration, and flight.

The weaker the regulatory mechanisms, the more likely is the formation of »strategic groups«, which prefer to resort to violence (hypothesis 4).

The second key factor concerns population dynamics in a given arena. Local or regional rural as well as peri-urban clusters of high population density influence scope and intensity of discrimination against groups, which depend highly on renewables or natural capital. One consequence of this pressure is either the unsustainable use of productive arenas or the movement to marginal arenas. Another consequence is intergenerational subdivision of already scarce resources. There is a coincidence between high population density and clusters of rural poverty, which must also be seen in the light of structural heterogeneity and a lack of socioeconomic opportunities:

The higher the population pressure in terms of growth rate and density, the more likely »strategic groups« are to organize for action (hypothesis 5a).

However, population pressure may have inverse effects. It may inhibit identity group shaping and organizational strength. Instead, it may contribute to intragroup tensions, to atomization (anomie), to high migration rates, and to the collapse of traditional institutions:

The higher the population pressure, the less likely that organized action occurs (hypothesis 5b).

Besides structures the model also covers institutions and actors. Environmental conflict does not occur without actors »acting« in a certain political environment and institutional framework. Present violent conflicts and wars occur in non-democratic political environments, that is, generally speaking, outside the OECD segment of the world. Violence occurs in countries where state stability and performance is low and local mechanisms have not sustained the pressure of social change induced by imperfect modernization. This does not necessarily mean that violence occurs mainly in countries with strong military and/or repressive regimes; there are different options or nonlinear relationships at work:

The weaker the state stability and the poorer the state performance, the more likely coherent »strategic groups« are to form, which aim at organizing violent action (hypothesis 6a).

However, state authorities that have used extreme force (such as massacres, torture, and genocide) are less likely to be challenged either because groups cannot organize open resistance or fear the consequences of doing so.

The more extreme force is used, the less likely the chance of open violence (hypothesis 6b).

This hypothesis corresponds to the model provided by Gurr and Harff (1994: 85). The authors also suggest that a curvilinear relationship exists between state violence and the extent and level of violence of political action taken by challengers. I conclude that extreme state repression has the same effect as extreme marginalization: atomization, livelihood insecurity, social fears, apathy, and - in the best case - migration or flight. Thus there are weak non-repressive states that »invite« groups subject to discrimination to organize themselves. There are also weak repressive states that inhibit marginalized groups from doing so.

The influence of the external factor is as complex as the weak state factor. First of all, if environmental transformation occurs in the context of historically violent conflicts - be it subnationally or regionally - a »culture of violence« exists. Therefore, the threshold to use force is perceived as being quite low in general. This perception of actors, in turn, bears violent reactions at several front lines. A historical conflict may spread directly to arenas where societal relationships are already at stake because of environmental discrimination. The latter may also be harnessed to add fuel to a historical conflict (channeling):

The lower the (perceived) threshold of violence in general, the more likely »strategic groups« tend to violent reactions - particularly in their own arena (hypothesis 7a).

The same external effect may be triggered by joint ventures between multinational firms and central governments, which are designed to exploit natural capital in a certain arena. Such projects often symbolize the »evil« *per se* because they sometimes destroy livelihood systems irreversibly without any compensation, let alone participation at the beginning of the planning period. Mainly dams and mines create national and international sacrifice areas where degradation of renewables is significant and where coherent groups are built:

The more likely environmental disruption by international industrial projects, the more likely is a violent response by »strategic groups« (hypothesis 7b).

International environmental organizations, such as Greenpeace, WWF, IUCN, and Friends of the Earth may add fuel to an open fire because they contribute to environmental awareness and knowledge. Awareness encourages affected groups to create strategic groups aimed at protecting their environment against foreign »aggressors«. Since such organizations aim at protecting the environment with peaceful means, they may contribute to nonviolent settlement of an escalating conflict - for instance, through enhancing global awareness and/or improving the atmosphere for cooperation and resource management, negotiation, mediation, or arbitration:

The more prominent the role of INGOs, the less likely the resort to violence (hypothesis 7c).

6.3. Environmental Conflict Model

The preliminary theory of environmental conflicts encompasses seven linked hypotheses. The following model is designed to illustrate the causal relationships:

Key to the model:

independent variable	political and institutional key factors
dependent variable	hypotheses
socioecological key factors	no hypotheses developed

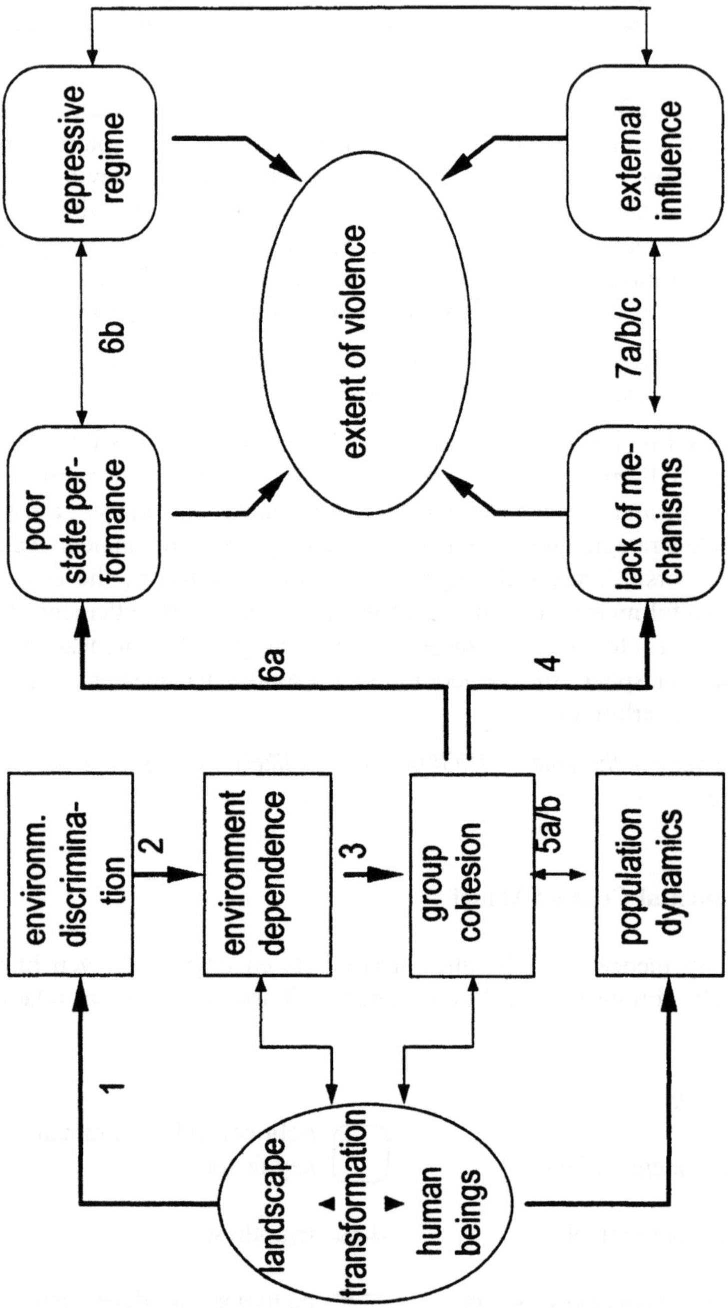

Figure 6.1. Environmental conflict model

6.4. Indicators

The concept of indicators serves to grasp the key factors and their properties.

Concept 1

The variable property is the degree or extent of environmental discrimination. It consists of two subconcepts, the first one addressing the socioecological dimension and the second one the politico-ecological dimension of discrimination.

Major forces of socioecological and political transformation in dry zones and mountain arenas - and partly in tropical regions - are: marginalization of poverty, increased demand for marketed products, mechanized farming, increasing lifestocks, the challenge of local political authorities by national governments, a short growing season, strong risk for recurrent droughts, severe soil degradation due to vulnerable soils subject to agricultural overexploitation, and limited water surplus after evapotranspiration to feed aquifers and rivers. Structural heterogeneity can be shown by the size of the »informal sector« and rate of employment or unemployment in this sector.

Indicators of socioecological discrimination:

- limited group access to freshwater (also time needed to haul water);
- high livelihood insecurity in terms of high »capability poverty« (HDR 1996:27);
- high under-five mortality;
- food production per capita (or rural-urban gap concerning calorie supply per capita);
- low environmental productivity per hectare compared with the national average and/or with privileged groups (degree of marginalization in terms of ecosystem vulnerability).

Indicators of politico-ecological discrimination:

- degree of vulnerability induced by limited access to the resources a group depends on (*e.g.*, 95 percent of Egypt depends on the Nile water);
- low degree of participation in politics and/or administration compared with other groups in the society;
- disaggregated HDI addressing marginal and national sacrifice areas in heterogeneous societies.

In addition, the Gini-coefficient could be used to show social justice and unequal distribution of income and wealth in terms of rural groups, which become victims of discrimination.

Concept 2
The variable property is dependence on degraded renewable resources. A new indicator is proposed here, which addresses different degrees and intensities of environmental stress. It is a collective indicator, which points to all kinds of high dependence on natural capital and thus use and overuse of renewables. The so-called *Resource Availability Index* (RAI) consists of *total hectares of arable land under production per capita* times *the annual commercial energy use per capita.* RAI is able to highlight the gap between North and South and the heterogeneity within societies concerning dependence on renewables as well as lack of alternatives.

Another argument arises from the strong correlation between the *level* of affluence or wealth of a state (not growth rate or relative deprivation!) and its stability and performance (Lane/Ersson 1994:197). According to a new approach to the problem of the source of national wealth, the World Bank suggests that natural capital accounts for nearly 44 percent of total wealth for exporters of primary goods; compared to 28 percent for all other developing countries and 17 percent for high-income economies (human capital: 36 percent to 56 percent to 67 percent for the latter). Sub-Saharan Africa, for instance, depends up to 52 percent on its natural capital, whereas human capital accounts for only 31 percent (World Bank 1995:62-63). It is evident that the source of wealth is not only fragile but actually a shrinking pie. As H.J. Leonard (1989:5) puts it: »In many rural areas characterized by traditional technology and few off-farm opportunities, growing populations continuously subdivide a resource whose potential to yield food, fodder, and fuel is relatively fixed«.
Indicators of environmental dependence are:

- soil erosion/desertification risk (index in: World Bank 1995:76; only partly measurable because of lack of data);
- scarce arable land resources (hectares per capita) combined with low commercial energy consumption (tons per capita) compared with other groups (RAI);
- remaining natural areas as part of the natural capital index (NCI);
- HDI as an indicator of high dependence on natural capital. Disaggregated HDI may serve as an indicator of the gap between groups discriminated against and the HDI of the average population. In the near future the »natural capital index« (Rodenburg *et al.* 1995) may provide a further indicator;
- importation of food vs. exportation of cash crops;
- high rural poverty (index: GNP/population growth rate).

Concept 3
The variable property is the degree of cohesion among leaders and followers. »Cohesive groups are those with dense communication and interaction networks linking leaders with followers. The more factions exist within a group, the less cohesive it is. Strong leaders generate a climate in which people willingly subordinate personal preferences to group preferences« (Gurr/Harff 1994: 89).
Indicators of group cohesion are:
- degree of acceptance of traditional leadership roles;
- degree of acceptance of established social order within the group;
- number of factions within the group;
- perceptions of environmental discrimination (using a scale).

Concept 4:
The variable property is the degree of decline in traditional means or mechanisms and/or the lack of democratic ones to deal with intergroup - and to some extent intra-group - conflicts. There is no significant indicator. One could therefore use the proportion of employment in the formal and informal sectors as an indicator. Another indicator may be the unemployment rate as well as the extent of displacement and rural-urban migration. A high proportion of inhabitants in the informal sector combined with high unemployment and high urbanization rate may serve as an indicator for the decline of local and traditional cultures and regulatory mechanisms.
Indicators of the decline in traditional methods in societal conflict regulation are:
- unemployment rate compared with employment in the formal sector and urbanization as a statistical (percent) index;
- availability of legal instruments in order to sue (judges and/or lawyers per capita);
- degree of political participation in environmental affairs (for instance, realization of the goal: strengthening »other groups« according to UNCED Agenda 21 in terms of financial support or resource allocation);
- degree of communal self-determination and centralization of decisions concerning the environment (participation as an indicator);
- degree of political marginalization in traditional producer communities (access to jobs in the administration).

Concept 5:
The variable property is the population pressure on the landscape or renewable resources in an ecogeographical context.
Indicators of population pressure are:

- trends in the demographic development indicated in annual growth rates (probably integrated in the natural capital index in the future);
- disaggregated population density (regional distribution in terms of capita per hectare of arable land);
- rural-urban population distribution and rural-urban migration;
- high rate of agricultural workforce combined with high rural unemployment rates.

Concept 6a:
The variable property encompasses dimensions of state performance, such as polity durability, civil order, legitimacy, and decisional efficacy, mainly concerning environmental and socioeconomic goals.
Indicators of state performance are:
- type of government and degree of civil and political rights;
- consultation of local leaders, *e.g.*, concerning traditional resource management and biodiversity preservation;
- regime stability (duration in years);
- general government consumption (sectoral distribution);
- expenditure for the environment, rural development, and poverty alleviation, *e.g.*, as percentage of GDP, MILEX;
- quality of habitats;
- percentage of protected areas (parks);
- implementation of national environmental action plans.

As a test of the impact of climatic conditions on state stability and performance suggests, »there is confirmation for the climate and the family structure models. The correlations here are so pronounced that these factors have to be looked into« (Lane/Ersson 1994:147).

Concept 6b:
The variable property is the severity of force used by governments against societal groups of all kinds. The most severe kinds of violence range from genocide, massacres, and systematic rape to torture, execution, detention without due process, forced displacement, and relocation of groups or people.
Indicators of state repression and violence are:
- number of arrests;
- forcible displacement and/or relocation of groups;
- torture and executions;
- massacres;

- political mass murder and genocide (concept 6b is drawn from Gurr/Harff 1994:90-91).

Concept 7:
The variable property is the degree and kind of external influence.
Indicators of external influence that contribute to an escalation are:

- the presence of multinational firms and degree of their environmentally destructive activities in rural areas (*e.g.*, Ogoni vs. Dutch Shell in the Delta State, Nigeria);
- the support these companies get from the national elite;
- the control of large reserves of scarce resources (including biodiversity);
- the spill-over from historical conflicts (culture of violence);
- the degree and kind of compensation affected groups get;
- the status of relationship between strategic groups and INGOs.

The model consists of seven key factors I considered as important predictors of environmentally induced violence. The set of indicators is composed of per capita, percentage, and structure indicators. They are designed to provide a consistent picture of reality. If the indicators are valid and significant will be examined in the next chapter. Most factors are interrelated or even interdependent in influencing one another. They differ in importance, depending on the area study we have in mind. However, when a rapid increase is observed in one or more of the variables in the model, we can assume that conflict *is likely to intensify*.

Thus the general hypothesis is: conflicts caused by the environment of society-nature relationships - in terms of degrading the landscape as a life-support system - generally manifest themselves in socioeconomic crisis regions of developing and transitional societies - if and as far as social, ethnopolitical, regional, and international fault lines can be harnessed by actors who prefer violence to other methods of dispute settlement. Groups who depend heavily on degraded renewables and suffer from environmental discrimination tend to establish coherent strategic groups along those lines. Population pressure and external intervention (industrial projects) may also aggravate conflicts.

The model enables researcher to further explore if the relatively few cases of major environmental conflicts and wars occurring today are only harbingers of a global trend, which may lead to more environmental conflicts and probably more severe ones. The capability to make suggestions concerning future developments by systematically analyzing conflict formations at an early stage may also lead to preventive action countering the causes that are supposed to urge actors to use violent means.

CHAPTER 7

EMPIRICAL EVIDENCE

Six Area Studies and Six Control Cases to Check the Model

The model is designed to serve as a general explanation of environmental conflicts. Seven different types of transformation conflicts were distinguished, according to the actors involved, the societal fault lines, and the attributes of the conflict arena. Given the fact that type II, *center-periphery*, and type VII, *global/long-distance*, are variations of the same conflict formation, six conflict types remain driven by different actors in order to check the model. The term „checking" is used to distinguish it from quantitative „testing".

To make sure the model is as robust as necessary and as little biased as possible not only ENCOP studies were checked but also case studies provided by other scholars or teams such as ECACP. Since ENCOP also analyzed control cases where transformation did not result in some sort of organized violence, I will also examine six control cases of nonviolent environmental conflicts. The list of the 12 cases which will be checked is found in figure 7.1.

The second pair, for instance, considers the violent conflict in Chiapas, Mexico, *vs.* the nonviolent conflict in the Okavango Delta in Botswana. It will be neither possible nor necessary to check *each* indicator discussed in the previous chapter. Such a detailed procedure would be beyond the scope of this work. At the same time, focusing on quantitative data would only disturb the main lines of arguments drawn here. Clear statements of evidence that support or falsify the hypotheses are crucial at this stage of work and not the mere statistical comparison of the cases.

A problem of representation must be mentioned. In order to avoid this chapter growing too long, I will go through all 12 cases in some detail for the first hypothesis only, dealing with the crucial concept of environmental discrimination (*section 7.1.*). As far as the other hypotheses are concerned (*sections 7.2.-7.7.*) only Chiapas in Mexico - including the Botswana control case - and South Africa will be dealt with in the same way as in section 7.1. Results concerning the other five pairs will be presented in cursory form.[1]

1 The check of the other hypotheses in tabular form is part of Baechler (1997): Empirical Evidence to Support the Environmental Conflict Model (manus.).

violent conflicts	*nonviolent conflicts*
intrastate	
A.I. ethnopolitical conflicts	
South Africa (Percival/Homer-Dixon, ACP)	Nigeria, Northern (ENCOP A.II.15)*
A.II./C.VII center-periphery / global conflicts	
Mexico; Chiapas (Howard/Homer-Dixon, ACP)	Botswana; Okavango (ENCOP A.II.7)
A.III. internal migration conflicts	
Sudan (ENCOP A.I.4)	China (Smil, ACP)
transboundary dimension	
B.IV. cross-border migration conflicts	
Senegal/Mauritania (Dokken; ENCOP C.VI.39)	USA/Mexico (Rogers; ENCOP C.VI.35)**
B.V. demographically caused conflicts	
Bangladesh (ENCOP B.V.25)	China (Smil, ACP)***
interstate	
C.VI international water conflicts	
Central Asia, Fergana (ENCOP C.VI.30)	Colorado/Rio Grande (Rogers; ENCOP C.VI.35)****

Since all six ethnopolitical conflicts of the ENCOP sample resulted in some sort of violence I use the Northern Nigeria water conflict as a control case. It has elements of both center-periphery and ethnopolitical fault lines. **There is no separate control case available; I therefore use here the Colorado/Rio Grande study. This case not only has international river-basin dynamics but also transboundary migration as a major problem. ***There is no control case available. I use the China Study of Smil (ECACP) here because it also has a demographic component with transboundary implications. ** The numbers behind the ENCOP cases refer to the appendix (see: app. 11).*

Figure 7.1. 12 cases to check the environmental conflict model

7.1. Environmental Discrimination

The first hypothesis (*see: 6*) is: *The greater the environmental discrimination actors experience in their arena, the more likely they are to organize for action against the sources of discrimination.*

7.1.1. ETHNOPOLITICAL CONFLICTS

This pair consists of South Africa as an example of a violent conflict and Northern Nigeria as an example of a nonviolent conflict.

South Africa: The Homeland and KwaZulu Natal Arenas
The South African ecosystem is characterized by little rainfall, water scarcity, and soils susceptible to erosion. Approximately 65 percent of the country receive less than 500 mm of annual precipitation, a threshold widely regarded by experts as the minimum required for rain-fed cropping (Percival/Homer-Dixon 1995:7). »The severe environmental problems confronting the country have been eclipsed by the negative social impacts of apartheid, opposition to minority rule, and the effort to build a post-apartheid political order (...) Apartheid has left behind a grim ecological legacy that will influence political, social, and economic conditions for decades« (*ibid.*:1).

It turned out that the system of apartheid, established after the 1948 victory of the National Party, was designed to be the perfect system of environmental discrimination - among other forms of discrimination that were among the strongest throughout the world: »Apartheid provided whites with 87 percent of the land, while blacks - almost 75 percent of the country's population - lived within the 10 Bantustans or homelands, which accounted for only 13 percent of the land. The black population in these areas sustained themselves through agriculture, local service industries, and as migrant labor in white-owned mines and industry« (*ibid.*:3). The homelands became »a dumping ground for the black South African population. From 1960 to 1980, the government forced 1.75 million people into the homelands to clear what it called 'black spots' - squatter settlements in urban areas and rural villages consisting of blacks whose labor-tenant contracts on white farms had been canceled. The population density of the homelands increased dramatically as a result. Quasi-urban communities emerged on homeland borders as the labor force commuting to neighboring cities, mines, and industries grew rapidly in the 1970s« (*ibid.*:3).

Black women belong to the most marginalized groups in general and environmentally discriminated against in particular. Since they are responsible for providing food, water, and fuel and face a lack of paid employment, they cannot but degrade the environment they depend on. High infant mortality rates, combined with the usefulness of children for labor under subsistence or poverty conditions, lead to high birth rates, as families have no guarantee that their children will survive to adolescence (*ibid.*:9). At the same time, population dynamics indirectly contribute to the marginalization of black women.

The apartheid system both institutionalized and accelerated the uneven distribution of natural capital in South Africa. This caused acute structural scarcity of resources for black South Africans. Land ownership by blacks was limited in both rural and urban areas, curtailing socioeconomic development. Racially denied access to productive land affected at least 15 million black inhabitants living either in the homelands or as tenants and laborers on land of whites (*ibid.*:8). Not only did blacks suffer from bad distribution of the overall quantity of land. They also often received the most marginal and least productive land (*ibid.*). Widespread landlessness among blacks existed even in Transkei, having been the homeland with the most fertile land. Fewer than 50 percent of the rural dwellers were allocated a field, and 60 percent of them owned no cattle whatsoever. As other homelands, Transkei was overpopulated and short of housing. Infrastructure was inferior with a lack of services, such as electricity and running water (*ibid.*:9).

Thus a major aspect of racial discrimination was that the apartheid regime situated the homelands in fragile environments with thin topsoil unsuitable for supporting the level of agricultural production required by their populations (*ibid.*:10). On the other hand, South Africa is generally a water-scarce country. Between 12 and 16 million people lack potable water supplies, and 21 million people, *i.e.*, half of the country's population, lack adequate sanitation. Concerning access to water discrimination works as well: 70 percent of urban blacks do not have access to running water and are forced to rely on heavily contaminated river systems for their daily water needs (*ibid.*:11).

Hence, in South Africa environmental discrimination was extremely accentuated in the context of a state policy in which racism became part of the system. Environmental discrimination is a strong argument to explain upheavals against apartheid.

The Northern Nigeria Dammed River Basins Arena

Chiefly after the Sahelian drought of 1972-73 the flat savanna and arid Northern Nigeria emerged as an arena most vulnerable to environmental conflicts caused by a vast network of dams and their impact on rural producers, mainly because of their huge sizes. Although dam construction policy predated the drought, the latter had a catalytic impact on subsequent implementation of far-reaching water development programs. These also included exploitation of underground water resources through the sinking of wells and bore-holes (Okoh 1996:233-234).

Conflicts erupted particularly because of the Bakolori dam in the Sokoto-Rima Valley, about 100 km southeast of Sokoto, but also because of many others in the 15-dam network. The dams, instead of supporting indigenous peasant communities affected by the drought periods, were designed to produce cash crops in large-scale irrigation areas at the southern edge of the Sahel. The Bakolori dam alone was envisaged to feed a terraced irrigation area of 27,000 ha. It was planned to produce 75,000 MT of wheat, 50,000 MT of sugar cane, 30,000 heads of cattle, as well as large quantities of rice, maize, onions, tomatoes, etc. (*ibid.*:235-236). Besides the fact that these figures were never realized, the adverse effects on the peasant communities were tremendous. Not only was the goal of damming seasonal rivers to irrigate land and provide farmers with two certain crops a year instead of an uncertain one not reached at all. The dams also altered the natural flow cycles throughout the year. The closure of dams in anticipation of annual rainfall and flood periods often induced crop failures of downstream rural communities.

Dams and irrigation forced displacement of people if not communities as a whole. Displacement and resettlement created major problems for authorities of the Northern Nigerian states. The Kainji dam, for instance, led to 44,000 displaced peasants, the Tiga dam to 12,000, the Bakolori dam to 14,000, the Goronyo dam to 20,000, and other dams in Kano to 40,000 (*ibid.*:238).

Farmers and local producers perceived the dams as discrimination by the central government together with foreign companies - such as FIAT-owned Impresit - against their mode of production. The dams - instead of broadly mitigating effects of the drought - led to the accelerated disruption of the people's habitat and thus of their livelihood security. Peasant farmers displaced by dams were not compensated for the loss of productive land. »Those resettled are often poorly treated and marginalized in their new settlement. In many project arenas, there is a high level of resentment amongst the local population« (*ibid.*:238). Between 1978 and 1980 mainly in the Bakolori dam arena, the discontent of the peasants resulted in militant actions against companies handling the project (*ibid.*).

Large-scale dam building in an arid and drought-prone rural arena did not favor the indigenous population. On the contrary, it contributed to further marginalization of self-reliant food crop production, heightened rather than mitigated livelihood insecurity, and induced widespread migration of people from their ancestral communities. Protesting groups experienced the discriminatory policy through massive use of coercive and violent measures undertaken by central government forces. No doubt, the indicators illustrate a precarious situation.

7.1.2. CENTER-PERIPHERY CONFLICTS

Mexico: Violence in the Chiapas Arena

On New Year 1994 when the *Ejército Zapatista de Liberación Nacional* (EZLN) took over San Cristóbal and other urban centers in the Highlands of Chiapas, Mexico, it brought »the plight of Chiapan peasants to the attention of Mexicans, foreign investors, and the international community« (Howard/ Homer-Dixon 1996:1).

The principal target of the EZLN was supposed to be »relief from escalating environmental scarcities that have impoverished their communities« (*ibid.*: 1). The two authors stress neither orthodox political-economic nor pure environmental explanations. They do not connect the rebellion with deforestation, soil erosion, and loss of biodiversity. In search of the root causes of the conflict in the Chiapas arena they state: worsening »environmental scarcity« and »maldistribution of natural resources« were to be seen as both reasons and channels of rural dwellers and their sympathizers victimized by discrimination.

Concerning *hypothesis 1* almost all the proposed indicators apply to the Chiapas case. The authors mainly emphasize relatively low environmental productivity, high degree of vulnerability due to livelihood insecurity, low degree of political participation of the affected groups, and extreme marginalization through the central government. Discrimination against the local or indigenous population is the major topic addressed by the authors. Discriminatory policy of the center against marginalized arenas and their mainly rural population is expressed by the fact that the southern states are rich in oil, natural gas, forests, and farmland and, at the same time, the poorest region in Mexico (*see also: 6*).

The only indicator which was not particularly mentioned in the study is *limited access to freshwater*. This may be a minor issue in the Chiapas arena. It may also be underestimated by the authors who focused on overuse of land. This single factor notwithstanding, the Chiapas arena confirms the discrimination concept extremely well.

Botswana: Nonviolent Conflict Management in the Okavango Arena

The Botswana arena consists of the Okavango River Delta in the northeastern part of the country. This largest inland delta on earth is a unique ecosystem endangered by the Southern Okavango Integrated Water Development Project (SOIWDP). The key players were the central government and the beef and mining industries (diamond), on the one hand, and the local population, Greenpeace International, and other international environmental organizations, on the other hand.

Water is among the scarcest resources in Botswana. Since most of the surface water evaporates in the dry climate and Botswana is affected by protracted drought, 80 percent of the population as well as stockbreeding and mining depend on ground water, which is a nonrenewable resource. Botswana has one of the fastest growing economies in Africa, and population growth and urbanization rates are high (*see: app. 8, 9, 13a*), which means that the pressure on limited water resources is increasing. Therefore, the major concern of SOIWDP opponents - besides the unique environment of the delta - was denial of access of small rural producers to freshwater. Large-scale beef production and diamond mining would have profited most from the disputed water-development plan.

As in the Chiapas case, environmental discrimination against marginalized groups holds as well. Why it did not lead to violence will be addressed later in this chapter.

7.1.3. INTERNAL MIGRATION CONFLICTS

This pair consists of case studies dealing with particular arenas within Sudan, on the one hand, and China, on the other hand.

The Jebel Marra Arena in Western Sudan

The case study in Sudan covers the Jebel Marra mountain arena in Darfur, the remote western part of the Sudan bordering on Egypt, Libya, and Chad. Darfur, which is as big as France, has historically been marginalized. From 1650 to 1917 it was an independent state. Since 1917, when it was annexed to the Sudan by the British, and after independence of the Sudan in 1956 only minor efforts have been made to develop the region economically. »This negligence by the central government has certainly contributed to the exclusion of Darfur from the rest of the country, not only economically, but also in political and cultural terms« (Suliman 1996:155). According to Suliman, the Sudanese regional government system *(wilaya)*, in place since 1982, has been a major factor for economic underdevelopment as well as relative political and cultural isolation of a number of provinces in the vast country. The Darfur arena is more a victim of indirect discrimination in terms of its isolation and *imperfect penetration* than the object of an intended discriminatory policy of the central government against its population of about 4 million people.

Against the background of remoteness there is an ethnic as well as an ecogeographical division of the Darfur arena, which induced major conflicts between groups discriminated against environmentally and groups relatively privileged in terms of access to natural capital, the latter no longer willing to share their resources with the former after the Sahel drought of 1972-73 and increasingly so after the 1984-85 drought. The ethnic division of the Darfur arena into groups of Arab origin and the non-Arab population is consistent with the occupational and ecogeographical structure of the region. Despite the official *wilaya* division the region is actually divided into three tribal arenas known locally as *dar*, which means homeland. The center of Darfur consists of the huge and fertile volcanic mountain massif Jebel Marra; an arena twice as big as Switzerland and with the highest peak reaching 3,071 meters. It is inhabited by the Fur and other non-Arab groups of settled small-scale farmers. The nomadic Arabs live at the foothills and in the semi-arid savanna and arid semi-desert surrounding Jebel Marra. Camel nomads belonging to the Arab Zaghawa, Majaryia, Bani Hussien, and Irayqat tribes used to live in the *dar* north of Jebel Marra. In the southern *dar* there are the cattle nomads of the Arab Reziegat, Taisha, and Bani-Helba tribes (Suliman 1996:155-157).

The central Jebel Marra *dar* consists of the richest zone in the arena in terms of abundance of surface and underground water, regular rainfall, soil productivity, and other natural resources. Unlike most other parts of the two other *dars*, the mountainous region has ecologically favorable conditions. As Suliman describes, »the pattern of life in the agricultural communities of the Fur central heartland is characterized by peaceful coexistence and harmony within its ethnic groups. Not a single major incident of open ethnic warfare has occurred within this group. The recorded conflicts that took place in the most susceptible upper and lower zones were mainly due to intrusions by the cattle nomads coming from the south, or camel nomads moving from the north into the farmlands of the Fur peasants« (*ibid.*:160).

The northern *dar* is the most disaster-prone and environmentally fragile arena within Darfur, which, as Suliman suggests, is the explanation why the camel nomads have always been active participants in violent conflicts of the region. The Arab groups perceive themselves as superior to the sedentary farmers. Loaded with nomadic self-esteem, they perceive the other rural dwellers as inferior, not only ethnically but also culturally by virtue of occupation. Simultaneously, they experienced serious pressure with their livestock economy, not only through environmental transformation but increasingly through the central government - and the Arab business class (*jellaba*) - which support absentee landlords with large-scale mechanized agriculture against the nomads and their land-use patterns. As a result of the nomads' perception of being more and more discriminated against, the violent intergroup conflict on the territory of the central *dar* expanded in intensity and frequency after the mid-1980s (*ibid.*:167).

The Northern Interior in China

In *China* internal migration conflicts chiefly affect the northern interior part: the arena consisting of the provinces of Shanxi, Shaanxi, Ningxia, and Gansu. These are not the poorest provinces in the country. These are Yunnan and

Guizhou in the remote south. The fault line that matters is constituted by the greatest contrast between the northern interior and the coastal provinces of Jiangsu, Zhejiang, Fujian, and Guangdong. Concerning the drought-prone northern interior arena, Smil points out that »no other large region in the country combines the unusually high environmental vulnerability with relatively high population density to such an extent as the basin of the Huang He between the western Hebei and the deserts and high mountains of interior Asia« (Smil 1995: 77). Whereas the country's 100 economically most advanced rural counties in 1991 were in the southeastern coastal provinces, about one quarter of China's poorest counties are in the northern interior arena.

Environmental disadvantages obviously coincide with a syndrome of sociopolitical deprivation factors, a phenomenon having produced significant environmental discrimination against the local population since the early years of Mao's rule. With the highest population density, the arena's infant mortality and chronic malnutrition rates are considerably above the national mean. At the same time, the average life expectancy and literacy of its people are far below average (*ibid.*:79). There is further contrast between China's coastal and interior provinces worth mentioning (1990 figures): the grain yield amounts to 4,800 MT per hectare at the coast compared to 2,700 MT per hectare in the interior. The net agricultural income is 920 Rmb per capita at the coast and 500 in the interior, whereas the gross industrial output with 3,300 Rmb per capita is more than double than that of the interior, which is only 1,500 Rmb per capita. In the basin of Huang He less than 20 m^3 of water runoff are available for each hectare of cultivated land and no more than 600 m^3 per capita. Compared to the figures of the Yangzi basin: available water amounts to 170 m^3 per hectare and 2,800 m^3 per capita (Smil 1995:79).

Following the argument of Smil, the combination of resource mismanagement chiefly under Mao's rule, widespread human hardship in a marginalized arena of low agricultural productivity, and droughts aggravated by extensive deforestation led to conflicts over ancient property rights and the desperate search for ever more distant resources. Migration to nearby towns or to more distant coastal cities went hand in hand with massive, state-organized resettlements: »Between 1983 and the end of 1990 320,000 people were resettled from arid, eroded, desertifying, poverty-stricken and clearly over-populated mountainous areas in the middle of Gansu province and in southern Ningxia. (...) This is clearly one of the world's largest organized moves of environmental migrants, and according to the Chinese plans the total number of relocated peasants should reach 900,000 by the end of the century« (*ibid.*:83).

To conclude: transformation of a vulnerable environment in the interior arena led to double discrimination. Firstly, mismanagement of resources in terms of indiscriminate conversion of forests, shrubs, and orchards to grain fields induced a discriminatory policy against the relatively poor rural dwellers in sociopolitical terms. Mainly under Mao's rule these were systematically neglected by rural development programs. Secondly, the marginalized and impoverished population is the passive object of forced resettlement to other arenas, which did not only cut the cultural roots. It also contributes to their further discrimination in the new arenas where they are confronted with hostile attitudes of the indigenous population challenged by its own problems.

7.1.4. CROSS-BORDER MIGRATION CONFLICTS

This pair consists of case studies dealing with particular cross-border arenas shared by Mauritania and Senegal as well as the US and Mexico.

Cross-border Violence between Mauritania and Senegal
The violent conflict between *Mauritania* and *Senegal* emerged in the Senegal River Valley, which forms the border between the two neighboring countries in West Africa. Besides the fact that the Senegal River Valley has historically been an arena where white Moors and black Africans clashed, escalation of the conflict in 1989 was due to the following reasons: transformation of the landscape in general and the Sahel drought in particular locally enhanced the interethnic tension between Moors and black Africans, the latter chiefly consisting of the Wolof, the Soninké, and the HalPulaar (Wegemund 1996:288-289). Following explanations of several scholars (*see:2.3.4.*) the effects of severe and persistent drought, water scarcity, and soil erosion caused the governmental decision to build a dam and a barrage in the Senegal River Valley. As a consequence of river development carried out by the *Organisation pour la mise en valeur du Fleuve Sénégal* (OMVS), land prices in the region increased significantly. The Mauritanian Arab elite did not hesitate to change the land-tenure system in the so-called land reform of 1983 after becoming aware of future economic benefits. Both river development and increasing land prices led to denying black African small-scale farmers, herders, and fishermen to use land and water resources as they had done before.

Enhanced pressure on black Africans by the Moor elite and growing parts of the Mauritanian population being concentrated in the Senegal River Valley favored channeling ethnopolitical consciousness and strengthening extremists on both sides. In the 1980s Moorish extremists defended an Arab nationalism inspired by Iraqi Baathism: »For them Mauritania has to be a white country, black Africans have to be expelled« (Wegemund 1996:291) - similar to what happened before colonial times when the Moors drove the black African population from the Mauritanian to the Senegalese side of the Senegal River (*ibid.*:289). Promotion of Arab nationalism was part of a policy to become independent from France and French-dominated Senegal, which was the political and administrative center for Mauritania and its economy during colonial times. Extremist black Africans organized themselves into the *Forces de libération des Africains en Mauritanie* (FLAM) founded in 1983. In a manifesto published in 1986, they described how they felt discriminated against by Moorish nationalists. They blamed the oppression against blacks by Moors in terms of denied access to politics, administration, and economic resources; they even compared the racial politics in Mauritania with South African apartheid (*ibid.*: 292). They were supported by Senegalese nationalists who blamed their own government for not defending its population »against aggressive Moors who try to enslave the Senegalese population as centuries before« (*ibid.*:293).

According to Dokken (1995) and Wegemund (1996) the high intensity conflict in 1989 was triggered by a transboundary clash between Senegalese peasants and Mauritanian nomads in April (*ibid.*:293). The immediate background was that Senegalese fishermen were more and more persecuted in Mauritanian territorial waters while Senegal tried to control the wandering of Mauritanian nomads because of their impact on the environment.

Thus in both countries drought, resource scarcity, and resource development (OMVS) caused ethnopolitically or racially motivated oppression of either black Africans in Mauritania or Moorish nomads in Senegal. Chiefly the systematic economic, environmental, and political discrimination against the blacks in Mauritania and support of the latter by neighboring Senegal triggered the large extent of transboundary violence in 1989 and beyond.

The Migration Conflict between the United States and Mexico

The cross-border migration conflict between the *US* and *Mexico* is a conflict between a northern industrialized country and a southern developing economy. The migration conflict must be distinguished from the transboundary water conflict, even though both overlap to some extent. Bilateral relations are influenced by geographical proximity, power asymmetry, and economic and technological disparities (Rogers 1996: 35). Social issues, such as decline in agricultural productivity, low incomes for small-scale farmers, population growth, and changing consumption patterns, enhance the flow of migrants from Mexico to the US.

Environmental discrimination is more *subtle* than in other case studies. However, it is a real issue in the cross-border arena that should be seen in the broader context of the overall relationship between the two countries. Mexico, as an economically dependent partner, does experience »a sense of inferiority« (*ibid.*:34). Historical experiences, such as the loss of about 50 percent of its territory in the 19th century or repeated interventions and punitive actions undertaken by the US until Franklin Roosevelt established his nonintervention policy in 1936, may still play a certain role. However, recent developments are much more influential. Mexicans are well aware of the fact that »the US is more important to Mexico than Mexico is to the US« (*ibid.*:34). Mexico lives next door to one of the largest countries in the world - a country with a much higher living standard and better socioeconomic conditions than Mexico.

»Environmental problems are just one more aspect of the tensions underlying Mexican-US relations. This tension is subtle and becomes evident periodically over problems such as drug trafficking and migration, both of which are partially facilitated by an extremely porous border« (*ibid.*:34). The American side of the border is generally less prosperous compared to the rest of the US, while the Mexican side is more prosperous compared to the rest of Mexico (*ibid.*:46). Therefore, discrimination is part of a complex pattern of superiority and inferiority, of welfare and relative deprivation. The pattern of the divided border arena is, however, completely different than the pattern of discrimination against the rural population in other developing countries. The US/ Mexican cross-border arena dramatically shows that rural dwellers affected by water shortage, pollution, soil erosion, and livelihood insecurity *do use* the opportunity of off-farm capacities, *if available*. Some figures may illustrate this: in the border arena there is the rare phenomenon of eight pairs of fast growing twin cities (among them Tijuana, Ciudad Juarez, Chihuahua, and El Paso). Mexican cities have been growing at double the rate of their US twins; on the Mexican side 19 percent of the 1980 population of the six border states used to live in the border cities, whereas on the American side only 3.8 percent of the population of the four border states live in this border arena. Beyond this, in Mexico the population in these border states grew from 10,691 million in 1980 to 13,221 million in 1990, most of the growth having occurred near the border. The overall growth rate for border states is 3.6 percent, thus far above the Mexican average of 2.7 percent. The high rate is not only due to a high fertility rate but also due to immigration to urban arenas. In a state with more than 50 percent of its population in the rural sector, eight urban communities account for 80 percent of Mexico's frontier population (*ibid.*:31).

Rogers does not provide figures on illegal immigration of Mexicans to the US. However, we know from other sources that the migration flow provokes major political friction in US domestic affairs.

To summarize: because of a lower standard of living and relatively high livelihood insecurity of the rural population immigration to the northern border regions, high rural-urban migration, and cross-border migration to the US (which is better off than Mexico in most terms) have always been options for Mexican migrants and are increasingly so today.

7.1.5. DEMOGRAPHICALLY CAUSED CONFLICTS

This pair consists of case studies dealing with Bangladesh and China.

Population Pressure on Scarce Resources in Bangladesh

Bangladesh - along with China and to some extent also Rwanda - belongs to the few countries where population size in relation to the natural capital available causes major social friction and conflict. Within a territory of 144,000 km^2 Bangladesh hosted a population of 110.8 million in 1991 with an annual growth rate of 2.16 percent. By the year 2000 the expected population will amount to 141 million. »While poverty and decreasing land-man as well as resource ratio are causing the environment to degrade at a faster rate, ecogeographic setting and various physiological and morphological characteristics have made the country extremely vulnerable to natural disaster« (Hafiz and Islam 1996:3). Access to safe water is one of the crucial problems in Bangladesh. Mainly the rural population is deprived of freshwater because either there is seasonal drought between October and March, when only 9-11 mm of precipitation falls in most parts of the territory, or there are heavy floods in the summer months, affecting large parts and causing displacement and destroying harvests, agricultural stocks, and livestock (*ibid.*:54-55). Beyond this, most available water is far below drinking water quality.

With a density of 793 persons per km^2 the arable land per capita was 0.117 ha in 1990 (Hafiz and Islam 1996:37). Rural population pressure induced large migration flows towards large cities. Various surveys in Dhaka indicate that 57.3 percent of the migrants left their homestead due to economic reasons - landlessness, poverty, unemployment, etc. - while 25 percent migrated for environmental reasons (*ibid.*:54).

Urban migration notwithstanding, overpopulation, mainly in the Chittagong Hill Tracts, led to major fault lines. This arena is located in the southeastern part of the country, close to the border with India and Myanmar. It comprises the only mountainous region of the country. While the conflict in this arena is often seen as an ethnopolitical conflict, according to Hafiz and Islam it is also caused by environmental, demographic, and migrational factors.

The first serious tensions between the indigenous people and the central government have been triggered by construction of the Kaptai Hydro power project, which forcibly displaced a huge tribal population from its land. In the aftermath of these events, further confrontations were caused by the declining capacity of nature in different regions of the country. Mainly disastrous cyclones, which affected vast parts of Bangladesh in 1970 and 1991, together with protracted environmental problems, such as seasonal floods, river bank erosion, declining soil fertility, produced millions of economic migrants and environmental disaster refugees. Migrants and refugees were attracted by the comparatively favorable environmental conditions in the hill tract. The government allowed and sometimes actively encouraged such migration and settlement. Thus these people created extra pressure on the limited resources and job opportunities (*ibid.*:58, 72-75).

Land-use and distribution issues ended in most cases in favor of the immigrants supported by the government. This contributed to further discrimination against the indigenous people in terms of access to resources. In 1951 the ratio of indigenous vs. immigrants was 90.9 to 9.1 percent, while in 1980 it already amounted to 67.1 to 33.9 percent. Against this backdrop, Hafiz and Islam stressed the fact that immigration and mutual mistrust fueled the perception among the indigenous »that the aim of the Bengali leadership was to reduce them to an abject minority« (*ibid.*:58).

Population Pressure in China's Interior North

In some *Chinese provinces* internal migration conflicts are closely interrelated with population dynamics, low agricultural productivity, and overuse of natural capital. These are reasons why »large numbers of desperate Chinese peasants may be forced to leave severely degraded, food-short interior provinces, becoming environmental refugees and seeking succor and economic opportunity in the nation's richest destinations, in its large eastern and southern coastal cities« (Smil 1996:127). The syndrome consisting of China's large and extremely diverse territory, huge population, limited transportation links, and spatially heterogeneous economic structures has produced »complex patterns of relative affluence and continuing poverty, as well as notable differences in prevailing environmental degradation and pollution. Significant numbers of people living below the statistically defined poverty line can be found in every province of the country, and larger areas of persistent misery and ecosystemic decay which might in future become significant sources of environmental refugees are located in 20 of China's 26 provinces« (*ibid.*:128). In 11 of the poorest provinces of the interior the population amounted to about 400 million people, over 80 percent of them having been rural in 1992. High density and intensive farming on small plots led to overuse of water, heavy soil erosion, desertification, and rapid deforestation (*ibid.*). In central Chinese Gansu province, for instance, the natural population increase (without immigration) was over 300,000 people annually during the 1980s, while losses of arable land amounted to 25,000 ha per year; given grain yields of less than 2,5 MT per hectare, the loss is the equivalent of losing food for about 300,000 people (*ibid.*: 135).

Smil estimates that displacement of peasants amounts to 20-30 million due to population pressure and land degradation during the 1990s. In 1994 the vice minister of public security estimated the floating population of job seekers in large cities at over 80 million people (*ibid.*:145).

In China population growth and migration are interrelated. Not all migration movements are caused by population dynamics. However, high population pressure almost always leads to migration. In this context, rural and urban areas form one major arena: while conflicts caused by high population growth and density most often happen in the marginalized arenas of origin, conflicts caused by migration occur in the arena of destination. These are mainly the large cities.

7.1.6. INTERNATIONAL WATER CONFLICTS

This last pair consists of case studies dealing with transboundary river basins in Central Asia (Syr Darya and Amu Darya) and the United States (Rio Grande and Colorado).

Upstream-Downstream Conflicts in Central Asia

After the collapse of the Soviet empire, the new interstate borders in *Central Asia* »created classical clear-cut upstream-downstream situations between different independent countries« (Klötzli 1996:287). Since the Central Asian arena consists mainly of arid and semi-arid vegetation zones, precipitation and usable ground water resources were insufficient to meet the dramatically growing demands of agriculture and private households in the recent past. Most of the river water originates in the high mountain ranges of Pamir and Tien Shan in the eastern part of Central Asia, feeding the two main rivers Amu Darya and Syr Darya, flowing westwards and northwards to the Aral Sea. Most agricultural activities in the downstream arenas were only made possible by enormous irrigation and sophisticated water distribution system withdrawals. Thus the Aral Sea has become one of the world's worst anthropogenic environmental catastrophes. Its drainage basin, an area of roughly 2 million km^2, is shared by the newly independent states of Turkmenistan, Uzbekistan, Tajikistan, Kyrgyzstan, and the southernmost provinces of Kazakhstan.

Since the 1960s »depletion of river flows and groundwater reserves, as well as degradation of water and soil qualities have become widespread. The consumption of water has tripled mainly due to significant extension of irrigated agricultural land and to the rapid population growth which has more than doubled and is expected to reach 50 million by the year 2000« (*ibid.*:249). In 1988, water withdrawals in the Aral Sea basin for all purposes were 125 percent of the average annual water resources; withdrawals were more than 100 percent because the return flows are used repeatedly downstream. Water resources are unevenly distributed among the countries: about 80 percent of the water flows in the relatively small upstream countries Kyrgyzstan and Tajikistan. The lower riparians Turkmenistan, Uzbekistan, and parts of Kazakhstan - being much larger, more developed, and containing most of the arable land - control only 20 percent of the total surface water flow (*ibid.*:290-291).

In the Syr Darya basin the upstream-downstream conflict forms part of a classical highland-lowland formation (*see:3.3.*). In this case, the mountain state Kyrgyzstan is the actor most discriminated against: poorly developed, economically marginalized, and politically relatively powerless, despite being the water tower in the arena. After independence the country faced serious economic problems, mainly because of a shortage of fossil energy supply from Russia and its neighbors. Downstream Uzbekistan significantly enhanced prices for oil and gas exports for its upstream neighbor. On the other hand, since hydroenergy is actually the major source of income, Kyrgyzstan wants to get the maximum profit out of its abundant water resources. It tried to prevail upon Uzbekistan to purchase hydroenergy as compensation for the amount of water usually flowing free into Uzbekistan. But the latter is only interested in getting the necessary amount of seasonal water for irrigation, while it can produce its energy with fossil resources more cheaply (*ibid.*:291).

The primacy of energy production over irrigation needs in the middle and lower courses of the basin has produced bilateral tensions between Kyrgyzstan and Uzbekistan: in the summer, when most of the water is needed for irrigation, water is stored in the large reservoirs upstream, such as Toktogul and Naryn, to be released in winter, when energy demand is at its peak. In summer 1993, for instance, Kyrgyzstan retained about 50 percent of the water allocation granted to Uzbekistan in the Toktogul dam, whereas it was blamed for releasing too much water from the Toktogul dam in the winter of the same year (*ibid.*:291-292).

Especially complicated is the upstream-downstream situation in the densely populated and ethnically extremely mixed Fergana Valley where Kyrgyzstan, Tajikistan, and Uzbekistan have common borders. In this arena of the Syr Darya, where 45 percent of all irrigation systems of the basin are located, »(a)ll countries making up parts of the valley have territorial claims against each other mainly because of large minorities living in districts bordering their own republic« (*ibid.*:295). Ethnopolitical confrontations with water-related disputes occur between Kyrgyz and Uzbek populations. Uzbeks living in Kyrgyzstan reside mainly in the Osh district bordering the Uzbek part of the Fergana Valley. Although strained relationship between Kyrgyz and Uzbek communities are historical, tensions increased after the Soviet withdrawal. Kyrgyzstan, with the self-perception of being discriminated against by its more powerful neighbor, started by discriminating in turn against the Uzbek minority. Klötzli reports on clashes between Uzbek and Kyrgyz people over land tenure in the city of Osh in 1990, resulting in at least 300 deaths. The trigger for the violent fights »was the official permission for a Kyrgyz cooperation to use irrigated land of an Uzbek *kolkhoz* to build residential buildings on it. After the clashes the Uzbek minority in Kyrgyzstan claimed an autonomy status in the Osh region«, adding fuel to the bilateral territorial conflict in the basin (*ibid.*).

In June 1989 Uzbeks clashed with Meskhetian Turks - deported by Stalin from Georgia to the Fergana Valley. According to the study's author, »(t)he economic frustration of one group, the Uzbeks, is reflected by a coincidence of socio-economic stratification with national or ethnic divisions, leading to conflicts between the indigenous and economically privileged immigrated Meskhetian Turks, often having better income opportunities as little traders. This cleavage is not only obvious on the countryside but also in the cities, where a mostly Slavic immigrant population have the qualified jobs« (*ibid.*:296). Since the share of young people out of work is 40 percent in some regions, it does not seem surprising that a lot of young men participated in clashes in the Fergana Valley: »The indigenous youth has turned its anger against those who have jobs, high wages and decent material and living conditions, *i.e.*, against those who are doing well« (*ibid.*).

The bilateral water conflict between Kyrgyzstan and Uzbekistan shows the complexity of environmental discrimination against different groups of the neighboring countries. Whereas the downstream country is in a position to discriminate against the upper riparian, the latter uses its central authority to discriminate against the Uzbek minority in the economically important, densely populated, and ethnically divided cross-boundary subarena of the Fergana Valley.

River Basin Conflict between Mexico and the United States

The *US* and *Mexico* share the transboundary Colorado and Rio Grande river basins. Despite improved cooperation over water resources during the last decade or so, there is a considerable potential for conflicts over water resources. As Rogers points out: »The environmental problems of increasing scarcity of surface and groundwater and growing degradation of water supplies are likely to produce social consequences that in turn could lead to higher levels of conflict« (Rogers 1996:27). Both increasing scarcity and pollution of water as well as increasing demand of water by a rapidly growing population in the fragile desert ecoregion that characterizes the Mexico-US border arena »is the essence of and the basis for environmental conflict and cooperation« (*ibid.*:28). Transformation of water resources is perceived by both sides as a serious issue, particularly concerning salinization, which has an adverse impact on agricultural productivity in a sensitive ecozone.

The Colorado is a successive river, flowing from the Rocky Mountains through several US states and - after 80 miles through Mexican territory - to the

Mexican Sea of Cortes. The Rio Grande is a so-called contiguous river forming the boundary between the two countries, 60 percent of its tributaries coming from Mexico. The US as an upper riparian of the Colorado and using 60 percent of the Rio Grande not only has more favorable geographical conditions than its neighbor. It also has more appropriate enforcement mechanisms in place, a more highly educated citizenry, superior technological and scientific capacities, better mobilization of public opinion which facilitates awareness of environmental problems along the border, better health care facilities along the border, etc. (*ibid.*:36).

The reverse of the coin is that water degradation occurs mainly on the US side, whereas Mexican users' access to freshwater resources is restricted by expansion of irrigation, overuse of fertilizers, insecticides and pesticides, discharge of industrial waste, chemical pollution, and untreated sewage (*ibid.*:55). In Mexico, for instance, water is considered fresh if it contains salt under 1,000 ppm, while the main US polluter sets drinking water standards at 500 ppm or even less (*ibid.*). With an average of 1,870 m^3 water per capita and year the US is the third largest water consumer worldwide; in Mexico, with 899 m^3 per capita, the average is much less. Serious health problems - such as increased cancer rate, dysentery, diarrhea, etc. - result from degraded water quality and air pollution on both sides of the border. Chiefly on the Mexican side the *maquiladora* industry is a main source of pollution, substandard housing of attracted workers, no running water, and sewage disposals (*ibid.*:57).

In the US-Mexican relationship denied access to freshwater resources on the Mexican side has been and still is a major topic on the bilateral agenda.

7.2. Dependence on Natural Capital: The Cases of Mexico, Botswana, and South Africa

The second hypothesis was: *The more dependent victims of environmental discrimination are on degrading renewables, the more likely they are to organize for action.*

Mexico: Violence in the Chiapas Arena

One reason for the Zapatista insurgency was the »rising grievances among peasants« in the Highlands »caused largely by worsening environmental scarcity« (Howard/Homer-Dixon 1996:2). High dependence is indicated by scarce land availability combined with low commercial energy consumption, a factor stressed repeatedly by the two authors of the study: »Particularly in and around San Cristóbal, the growing population has consumed much of the forest and occupied most of the potentially arable land, greatly changing the local landscape« (*ibid.*:8). The critical resources are forests and cropland. »Land degradation often begins with forest removal, continues with unsustainable agricultural practices, and ends with overgrazing by cattle, sheep, and goats« (*ibid.*:9). In the Central Highlands, deforestation has led to firewood shortages, so that many communities have exhausted nearby supplies and must travel high into the hills for wood. Other indicators are soil erosion and a comparatively low HDI. Soil erosion and desertification are supposed to be strong factors in some highland areas contributing to the decline of sensitive tropical soils productivity. Most of Mexico's arable land is »rain-fed, undercapitalized and of poor quality« (*ibid.*:10). This applies especially for deforested arenas in the Highlands and eastern foothills where degraded land is converted into pasture land for cattle. The disaggregated HDI for indigenous groups in Mexico is 0.700 and for non-indigenous 0.887 (HDR 1996:31). Other factors are food shortage and export of cash crops (see below).

There are no figures on importation of food to the Chiapas arena vs. exportation of cash crops from there. However, the problem is tackled by the co-authors. While food crop areas were shrinking, the *latifundistas*, a small class of landowners, controlled vast territories in the Eastern Lowlands where most of the land is devoted to capital-intensive cash crops for export. As a relatively new group of entrepreneurs the *rancheros* have taken control of huge land areas supported by state subsidies (*ibid.*).

Botswana Control Case

The SOIWDP water development plan was originally designed to cope with the hardships of local people induced by high dependence on scarce water resources and relatively poor soils - besides having to cope with persistent drought and erratic rainfall. It should help to ensure the food production of the country and to increase the living standard for people living along the Thamalakane fault and the Botetis River (Claus 1996:271). It is assumed that the project would have discriminated against more people downstream than would have benefited from it: »Evidence is that during recent drought periods, grazing availability has been the major constraint rather than water supply in the vicinity of the Boteti. Furthermore the two reservoirs (...) would flood some of the most valuable land in Botswana on which thousands of people were also dependent for molapo cultivation and grazing« (T. Scudder, article for the Okavango Observer, manuscript dated 26 Sept., 1994. Scudder led the IUCN team. In Claus 1996:272). Claus considers the dependence of rural producers on scarce and maldistributed arable land to be high. The author does not mention other aspects of high dependence on natural capital. However, some further figures are to be found in the appendix of this study (*see: app. 7-9, 13a*).

South Africa: The Homeland and KwaZulu Natal Arenas

Concerning land erosion and scarce arable land, the authors pointed out that only 16 percent of all used arable land (about 16 million ha) are considered suitable for growing crops, while the rest is used for pasture only. About 4 percent of the land used is high-potential agricultural land reserved for commercial farming. Out of the total area of cropland, about 13 million ha are situated in large-scale commercial farming areas, while 2.5 million ha of less productive land are in small-scale farming arenas in the former homelands (Percival/ Homer-Dixon 1995:8).

The co-authors also emphasize the lack of commercial energy supply in rural arenas. Because of insufficient energy services, estimated 40 percent of the South African population depend on fuelwood for cooking and heating. They consider the annual volume of fuelwood consumption to be about 11 million MT. This is the reason why »(i)n the past 50 years, 200 of KwaZulu's 250 forests have disappeared« (*ibid.*: 10). In addition, in KwaZulu, in 1993, 80 percent of all rural households and 40 percent of urban households were supposed to be below the poverty line, whereas formal jobs were available for only 50 percent of the overall South African population. Poverty is »endemic and deep within the rural black community«. In 1980, when apartheid was perceived to become counterproductive for the system as a whole, »81 percent of the rural population earned less than the minimum subsistence wage« (*ibid.*:13).

7.3. Marginalization and Group Cohesion: The Cases of Mexico, Botswana, and South Africa

The third hypothesis was: *Group cohesion increases to the extent that discriminated actors are concentrated in sensitive arenas, such as marginalized ecoregions, national sacrifice areas, or shared (transboundary) river basins.*

Mexico: Violence in the Chiapas Arena

Howard and Homer-Dixon leave no doubt about the economic and social problems of Mexico and the marginalization of the rural population widely neglected by the ruling PRI (*Partido Revolucionario Institucional*). The EZLN »inspired campesinos, indigenous people, labor, and the urban poor of central and northern Mexico to express discontent with the PRI regime by engaging in violent protest and grassroots democratic campaigns« (1996:2). In the 1990s, group cohesion among EZLN supporters - despite ethnic differentiation - was quite strong and factions insignificant. On the other side of the warring parties were the *latifundistas*, the *rancheros*, and the political »bosses«, the *caciques*, who mobilized support for the PRI and were perceived as particularly oppressive in the *Cañadas*, in the Eastern Lowlands, the basic mobilization zone for the Zapatistas (*ibid.*).

A group of intellectuals, clerics, and opposition leaders helped organize *indígenas* and *campesinos* and encouraged them to express their dissatisfaction. »The Catholic and Protestant Churches have provided key venues for the discussion of political repression in Chiapas« (*ibid.*: 25). The bishop of the diocese of San Cristóbal, for instance, has drawn heavily on liberation theology to inspire the struggle of the peasants. Both churches encouraged *indígenas* and *campesinos* to set up non-PRI social organs working for land reform, labor rights, and fair credit programs (*ibid.*:26).

The probability of civil violence is higher if groups are already organized around clear social cleavages, such as ethnicity, religion, or class. The Zapatistas not only exploited long-standing social and class cleavages and mobilized *indígenas* and *campesinos* against elitist groups. With Subcomandante Marcos they also provided charismatic leadership. Marcos argued: »It strikes me that what most radicalized our companions were the changes to Article 27. That was the door that was shut on the Indian people's ability to survive in a legal and peaceful manner. That was the reason they decided to take up arms, so that they could be heard, because they were tired of paying such a high blood tax« (Marcos in *La Jornada*, 7 February 1994. In Howard/Homer-Dixon 1996:26).

Art. 27 was an amendment to the Mexican Constitution which in 1992 introduced an agrarian law that drastically changed the traditional land-tenure system, a state of affairs strongly criticized by *campesino* and *indígena* organizations (*ibid.*:19). With this step the Salinas government obviously triggered the »cognitive liberation« of the peasants when it abolished the institution of *ejido land* titles introduced in 1917. The small *ejido* plots have always been seen as the last guarantee to reach the goal of livelihood security, however precarious it may have been. The peasants' perception of environmental discrimination significantly increased after the new land law of 1992: »The Zapatista leadership articulated and channeled peasant grievances so as to focus blame for the peasants' hardship on the PRI regime« (*ibid.*:24).

Botswana Control Case

In Botswana, group cohesion was relatively high in terms of organizing a political opposition against a single project of the government. In the city of Maun, just after the first preparations for construction had started, local actors raised protests against the water project as early as 1990. In January 1991, more than 500 people participated in a traditional *kgotla meeting* (*see: 7.4.*) to discuss the project (Claus 1996:277). The participants, supported by the influential local tourism firms in Maun, strongly rejected the governmental project. Regional parliamentarians of the ruling Democratic Party were also strongly opposed to the plan, as were opposition members of parliament from the Botswana National Front (BNF). The mobilization notwithstanding, the opposition movement did not consist of one coherent group with a strong intra-group identity. The project brought together political groups with different programs but with one goal in common, comparable to »single-issue movements« in industrialized countries.

Leadership was reduced to subgroups of the movement. Only the chief of the largest ethnic group in the region, the BaTawana, was identified personally with the anti-dredging campaign. Chief Mathiba Moremi said in an interview in Gaborone on 10 May 1991: »I am happy that all BaTawana are opposed to dredging. BaTawana say to dredge a river is to destroy it. They therefore refuse. I am happy that my people have taken such a stance« (In Claus 1996: 278). Thus there was a coalition of groups, more or less well-organized, with pluralistic leadership which challenged the democratically legitimized government.

South Africa: The Homeland and KwaZulu Natal Arenas

In the KwaZulu arena, the homelands, and some of the larger settlement areas warlords often tried to build coherent groups. Warlords are »local leaders who control their own paramilitary forces and owe only nominal allegiance to any higher authority. Warlords establish patron-client relations with settlement residents: the residents support the warlord in return for essential resources and services, including policing and dispute settlement. Paramilitary forces allow warlords to exercise strict control over the right to conduct business and over environmental resources, such as land and water. Resource control multiplies warlord power and wealth, permitting extraction of surpluses in the form of taxes, rents, and levies« (Percival/Homer-Dixon 1995:15).

Warlords, at the same time, »have limited ability to provide their communities with infrastructure or services. They have no real control over electricity, refuse removal, roads, and other services, since these services - to the extent that they are available - are provided by the local municipality« (*ibid.*:15). Nonetheless, political leaders on all sides as well as warlords aimed at manipulating conflicts among different actors over access to resources, such as land, housing, water, and services, for egoistic reasons (*ibid.*:20). As a result, warlordism not only led to a coherent opposition movement against the apartheid system. It also induced segmentation and societal fragmentation. On the one hand, these phenomena contributed to a collapse of social networks, to the weakening of community norms, and to erosion of trust, undermining the ability of state institutions to function. On the other hand, segmentation often took the »form of divisions among ethnic groups, among family-based clans, and among the residents of townships, informal settlements, and work hostels« (*ibid.*:15). The ethnic groups were not only divided among themselves; they were also politically fragmented in their interior, such as the Zulu population. The internal cleavages had »political overtones«; this means that they often manifested themselves as conflicts between the ANC and Inkatha. »Group divisions are constructed and manipulated by political leaders and warlords on each side« (*ibid.*:18).

Against the background of the plurality of more or less coherent groups with strong and competing leadership, the co-authors maintain that »(i)dentification with one's ethnic group was necessary for survival and advancement within apartheid's ethnically divided political system. (...) A survey taken just prior to the 1994 election shows the salience of ethnic identity: When asked to name their nationality, 16 percent of blacks replied South African or black South African, while 63 percent gave the name of their tribe. Much of the recent violence in South Africa has been between supporters of the ANC and those of the Inkatha movement«, that is, mainly between Zulu and Xhosa (*ibid.*: 18).

Indeed, the ANC under Nelson Mandela and the *Inkatha Freedom Party* under Mangosuthu Buthelezi provided two strong and charismatic leaders who could mobilize their supporters. At the beginning of the 1990s, the political reform process »radically changed the opportunities for political violence« (*ibid.*:19). As the ANC and Inkatha became fully-fledged political organizations battling for control of the institutions of power, they mobilized huge numbers of people and substantial financial resources for their political ends. »At the same time, the capacity of the state to do anything about the violence between the ANC and Inkatha was debilitated by the reform process» (*ibid.*:19).

7.4. Lack of Regulatory Mechanisms

The fourth hypothesis was: *The weaker the regulatory mechanisms, the more likely is the formation of »strategic groups«, which prefer to resort to violence.*

Mexico: Violence in the Chiapas Arena

Howard and Homer-Dixon more or less directly address most indicators dealing with the decline of traditional methods in conflict management.

The *indígenas* in the Highlands and Eastern Lowlands still have their own language and culture. The main groups are the Tzeltal, Tzotzil, and Chol. Through resource scarcity, voluntary migration, and forced displacement to the *Cañadas* their internal structure is at stake. A lack of communal self-determination and decentralization is expressed in the facts that *indígenas* and *campesinos* have neither a lobby to fight against the problematic change of the land law at the central government level nor do they have access to agricultural credit at the local level. Both »the state and *municipio* governments try, but often fail, to contain and manage these groups« (Howard/Homer-Dixon 1996: 4). As a result, the degree of political marginalization is supposed to be higher than elsewhere in Mexico. The political elite has rarely consulted *indígenas* communities when planning welfare programs. In contrast to this, both *latifundistas* and *rancheros* »have withstood federal attempts at political reform and land redistribution and have retained control of state politics« (*ibid.*).

Botswana Control Case

Lack of local regulatory mechanisms in Botswana was not a big issue at all. Marginalization and dissolution of traditional societies is not as pressing as in Chiapas. An important channeling element for the opposition was the *kgotla meeting* in Maun mentioned above. A *kgotla meeting* is a traditional regional parliament led by the local chief. »It plays a significant political role in Botswana's democracy«, mainly concerning issues that affect local people (*ibid.*:277-8). This statement gives us the impression that traditional mechanisms still work, although within the framework of the democratic state with its legal institutions.

South Africa: The Homeland and KwaZulu Natal Arenas

The lack of discussion of traditional mechanisms is a weak point in the Percival and Homer-Dixon study (1995). Passages on local collapse of the state's institutional order and on identification with one's own identity group as a necessity for survival may be interpreted in the way that traditional mechanisms had to be there as a kind of self-help capacity. The co-authors only stressed the fact that the Zulu population was based more on traditional structures than other groups. At the same time, they also addressed internal division and the relative weakness of traditional society, partly induced by apartheid, partly by interethnic strife: »The ANC promoted a revolt against tradition among Zulu youth, who expressed their opposition to traditional tribal structures through participation in ANC recruitment drives« (*ibid.*:20).

It seems that traditional mechanisms were not worth mentioning because they stopped playing a major political role under the apartheid regime decades ago. They were supposed to be a victim of the strong segmentation of society, the state's intervention at all levels, as well as of manipulation if not occupation of local power by warlords and political party leaders.

7.5. Population Dynamics

The fifth hypothesis was: *The higher the population pressure in terms of growth rate and density, the more likely »strategic groups« are to organize for action (hypothesis 5a).* However, population pressure may have adverse effects. It may lead to intra-group tensions, to atomization (anomie), high migration rates, etc. *The higher the population pressure, the less likely that organized action occurs (hypothesis 5b).*

Mexico: Violence in the Chiapas Arena

Population dynamics in the Chiapas arena accelerated rather than inhibited formation of strategic groups. Regional high density in the populous arena of the *Cañadas* at the eastern foothills of the Central Highlands is supposed to having played a crucial role: »It is in this frontier region between the Highlands and the Eastern Lowlands that people have been most severely affected by environmental scarcities, and it is from here that the EZLN draws its support« (Howard/Homer-Dixon 1996:4). It was not the high population growth rate alone - among the *indígenas* up to 4.6 percent - that caused tremendous population pressure but also immigration of different groups. Some came from other Mexican states in search of fertile land, some from Guatemala seeking refuge from the civil war in the mid-1980s. Since several huge hydroelectric projects flooded high-quality farmland in the Grijalva basin, »the government forcibly relocated tens of thousands of smallholders into the Eastern Lowlands. These people now live in one of the most marginal parts of the state, often without potable water, electricity, or infrastructure« *(ibid.*:8).

As a result of the population dynamics, rural unemployment rate is extremely high among the *indígenas*: in Chiapas it is 32.5 percent, while the average is 21 percent among *indígenas* all over Mexico and 19.0 percent in the national average (*ibid.*:29).

Botswana Control Case

Since the dependence on nonrenewable fossil water is high in Botswana, population dynamics play an economic role. Politically speaking, the population factor was not a crucial issue in the protest against the Okawango water project. With 1.2 million inhabitants the average population pressure is rather low. At the same time, the annual growth rate of 3.4 percent is quite high. This leads to a rapidly growing urban population. An annual urban growth rate of 9 percent will push the population of the capital, Gaborone, from 130,000 inhabitants in 1994 to 550,000 in 2020 (Claus 1996:270).

It is not the population growth rate as such that poses a major problem in Botswana but the growth rate with increasing demands of people in a fast growing, tourist-oriented urban economy. This factor may have a major impact on the handling of future water development. The government decision concerning SOIWDP may be reconsidered.

South Africa: The Homeland and KwaZulu Natal Arenas
According to the study of reference, the rapid growth of South Africa's black population results in more severe scarcity of arable land. Under the apartheid regime, as the co-authors pointed out, the average population density in the former homelands grew by up to ten times the density of rural 'white' South Africa. In the context of the decline in labor requirements in commercial agriculture, »apartheid ensured that black South Africans could not move to cities when they were expelled from rural white areas, police forcibly moved blacks to the homelands; partly as a result of this forced migration, the population of the homelands grew from 4.5 to 11 million between 1960 and 1980« (Percival/ Homer-Dixon 1995:9).

Although rural-urban migration rates are vague, they give the number of blacks moving into cities as 750,000 a year (*ibid.*:14). This may be a conservative estimate. In KwaZulu Natal, population growth rate in urban areas is estimated to be three times higher than that in rural arenas, the latter growing fast as well. An expression of rapid urbanization, on the one hand, and marginalization, on the other hand, is the increase in informal settlements. »In the early 1990s, these represented more than one third of the total urban population and more than half the total black peri-urban population, exceeding growth rates within formal urban arenas« (*ibid.*:14).

The analysts suggest population growth, overpopulation of the homelands, and black rural migration to white urban areas to be crucial causes for the collapse of the apartheid system. High regional density in homelands, KwaZulu Natal, and peri-urban areas had also adverse effects on group cohesion and mobilization power of parties representing black Africans. As mentioned above (*see: 7.3.*), decline of the social order and interethnic tensions contributed to the build-up of factions among opponents of the apartheid system which, to some extent, survives the old regime today.

7.6. State Instability and Poor Performance

The sixth hypothesis was: *The weaker the state stability and the poorer the state performance, the more likely coherent »strategic groups« are to form, which aim at organizing violent action.*

Mexico: Violence in the Chiapas Arena
»To cause civil strife, economic crisis must be severe, persistent, and pervasive enough to erode the legitimacy or moral authority of the dominant social order and system of governance. System legitimacy is therefore a critical intervening variable between rising poverty and civil conflict« (Howard/Homer-Dixon 1996:24). The fragile legitimacy of the Mexican sociopolitical order was reflected in the fact that the ELZN not only succeeded in mobilizing marginalized groups in the Chiapas arena but also attracted many sympathizers across the country: »Mexican authorities have been forced to devote substantial resources to keep similar insurgencies from flaring up elsewhere« (*ibid.*:2).

Another element of poor performance of the Mexican state is that Chiapas is both a marginalized arena as well as a typical national sacrifice area exploited by the center and in favor of the center. This factor is well taken by the co-authors. As they specify: »Chiapas produces 5 percent of the nation's oil, 12 percent of its natural gas, 46 percent of its coffee, and 48 percent of its hydroelectric power, yet only a tiny portion of the wealth generated from these resources is returned to the state for development programs« (*ibid.*:4). And poverty statistics reveal both discrimination and poor state performance connected with some racist elements: »In *municipios* where the *indígena* population is less than 10 percent, 18 percent of the people are at or below the poverty line; for *municipios* where the *indígena* population is between 10 and 40 percent, 46 percent of the people are poor; and for those where more than 70 percent of the population are *indígena*, over 80 percent are poor« (*ibid.*:6).

The »structural scarcity« of fertile land for impoverished farmers has arisen from the steadily growing domination and manipulation of the land-tenure system by a wealthy elite of large-scale cash-crop producers. The scarcity is reinforced by »the political hegemony of this elite, by the economies of scale and commercial success of large agricultural producers, and by corrupt and inequitable credit and social-spending programs managed by the state« (*ibid.*:13). In 1990, 87 percent of the rural smallholders had no access to agricultural credit whatsoever. And this group had only very limited access to infrastructure provided by the state. Wealthy beef producers and exporters, on the other hand, increasingly benefited from subsidies (*ibid.*).

Poor state performance is also indicated by badly realized revolutionary goals. In the aftermath of the 1910 revolution the new Constitution of 1917 was chiefly designed to enhance livelihood security of the broad rural population by land redistribution and introduction of the *ejido* tenure system. However, as the co-authors state, good lands were seldom subject to such efforts. Instead, »the government often met its land reform obligations by promising, usually without actually granting, titles to unused lands in the rain forest with difficult market access« (*ibid.*:17). The years of »economic liberalization« from 1982 to 1994 brought a further retirement of the state, largely because of huge privatization projects. By minimizing controls of the markets, by substantially reducing subsidies and public credits, the deregulation policy further shifted power from labor and peasantry to private capital and entrepreneurship (*ibid.*).

Besides events such as forced displacement, incidences of police violence, and heavy corruption state repression is not seen as a major issue. On the contrary, rising grievances and the policy of »economic liberalization« further weakened political control of the ruling PRI throughout Mexico. Deregulation of the »corporate state« also significantly enhanced the political opportunities of opponents. Therefore, in the early 1990s, *de facto* declining state repression politically encouraged potential challenge of the regime (*ibid.*:23). In other words, EZLN activities must be seen as part of a larger »wedge« splitting open the authoritarian character of Mexico's central and local governments. Thus the Chiapas insurgency had a profound impact on the corporatist regime, catching it at a particularly »sensitive moment of transition«, disrupting »traditional patterns of presidentialism, party dominance, patrimonialism, and clientelism« (*ibid.*:27).

The region of Chiapas is still heavily militarized in the mid-1990s. However, President Zedillo of the ruling PRI has chosen to negotiate with EZLN leaders and to keep the dialog open. The violent uprising of the Zapatistas has indeed brought new hope in large parts of the society for democracy, participation, and improvement of livelihood security in a less discriminatory environment.

As we can see, Howard and Homer-Dixon emphasized the issue of poor state performance and instability more than, for instance, population growth or group cohesion. Together with state repression or a lack of democracy, these two factors seem to indicate the relative importance of hypothesis number 6 in the framework of our model. As we will highlight in the context of the Botswana control case *it is indeed the hypothesis that makes a difference in terms of violent or nonviolent outcomes of environmental conflicts.*

Botswana Control Case

The political situation in Botswana is quite different from Mexico. The central government stopped the water development project before it provoked more manifest reactions. Although environmental problems should not be neglected, environmental discrimination is less accentuated in this case than in the Mexican or South African one. Botswana has a functioning multiparty system. It guarantees opposition movements sufficient freedom to channel their function in order to correct state decisions (Claus 1996:277). Political stability is surprisingly high, which is proven by the fact that the Botswana Democratic Party has been ruling since 1966. Although there are no well-organized political groups and the presidential system tends towards a paternalistic approach, the government is recognized as legitimate by the people. This was the reason why the central government had to withdraw its plans as the main promoter of the water development project after »meeting severe local resistance« (*ibid.*:277).

As IUCN confirmed: »The extent of the local opposition is striking. Within the SOIWDP impact region of Ngamiland District, there is virtually no support for the project among villagers, local government officials, other citizens, and non-citizens. Opposition was also surprisingly strong along the Boteti« (IUCN Review, Executive Summary, Geneva 1993:11. In Claus 1996:278). The government faced the problem that the initial reasons for the project - increasing the cultivated area for traditional and commercial irrigated farming - were considered unrealistic. On the other hand, to establish the project just for delivering a huge amount of water to the diamond industries was perceived as an illegitimate goal. Such as step could have undermined the legitimacy of the present government. It would have found itself in a position of defending the diamond industries against large parts of the population including the tourism industry.

There were neither physical confrontations nor ethnopolitical conflicts in Botswana. Each side tried hard not to let the situation escalate. Claus mentioned that »(d)uring the historical kgotla-meeting in January 1991, some residents reportedly pledged to sabotage the machines of the company doing the dredging work if the government did not back-track on the project» (*ibid.*:279). Even sabotage - also carried out against disputed projects in industrialized countries - was not a measure that had to be taken. State stability, a legitimized government, and better state performance were factors that contributed to mitigating the water development conflict and to finding a political solution - at least for the time being. This enhances the chance that similar solutions will be found, even if the government revives the plan in the future..

South Africa: The Homeland and KwaZulu Natal Arenas

»Both townships and settlements receive few services and lack infrastructure. The inadequacy of the infrastructure, such as sewage systems, water supplies, and energy sources, means that the urban black population relies extensively for its day-to-day needs on the local environment - including small vegetable plots and local streams, trees, and brush. Because many of these communities are located in fragile environments close to hillsides and river valleys, the environment quickly deteriorates (Percival/Homer-Dixon 1996:4).

On 1 February 1990 the white South African President de Klerk lifted the 30-year ban on the ANC and a number of other political organizations. He released Nelson Mandela on 11 February 1990 and allowed the return of political exiles from abroad (*ibid.*:4). The analysts interpret these events as follows: »The government began to shift away from apartheid when it realized that it was impossible to stem the tide of Africans moving to the urban areas in search of employment, signaling that the homeland system did not work« (*ibid.*).

Weak state and poor performance were also visible at the community level: »Because community institutions are debilitated by the processes just described, they are unable to provide infrastructure - including sewers, electricity, and running water - to keep residents from wrecking local environmental resources« (*ibid.*:15).

Above all the key element in development leading to the elections of April 1994 was Nelson Mandela, the charismatic leader of the ANC and the first freely elected black president of the new republic. After Mandela's release in February 1990 violence became pervasive because the failures of apartheid became obvious and competition among the distinct factions of the anti-apartheid movement increased. Percival and Homer-Dixon estimate that between September 1984 and the end of 1989 about 3,500 people died because of political violence throughout South Africa. From January 1990 until December 1993, political violence killed an estimated 12,000 people, whereas criminal and political violence combined produced more than 20,000 deaths in 1992 alone (*ibid.*:16). In July 1990, the so-called Reef Township War around Johannesburg began. Clashes broke out between hostel dwellers, often migrant workers who supported Inkatha, and residents of townships and informal settlements, most of them members of the ANC (*ibid.*).

They conclude: »The South African state was unable to control violence at the local level because communities were segmented and institutions were weak (...) The battle took place within a deeply aggrieved society, brutalized for generations, and well rehearsed in fighting the system of apartheid» (*ibid.*:19). And further: »Predatory warlords and opportunistic members of the ANC and Inkatha took advantage of a weakened state, debilitated local institutions, and an aggrieved population to mobilize group identities and instigate group rivalries« (*ibid.*).

In the late phase of apartheid, societal demands on the South African state increased tremendously - despite the apartheid system - as thousands of black people migrated to peri-urban areas. Simultaneously, the capacities of both national and local institutions to compete with the increasing demands of a fast growing population declined rapidly (*ibid.*: 21).

In South Africa - as in Mexico - political instability, poor state performance, and delegitimization of the central government indirectly accelerated the use of violence. In contrast to Mexico, state repression against the anti-apartheid movement and the marginalized poor was an inhibiting factor, at least until the early 1990s. After the 1990 events, use of force could not compete with the eruptive violence, since the political reform process has opened some room for maneuver.

7.7. Complex External Influences

The seventh hypothesis was complex, indicating curvilinear relationships: *(a) The lower the (perceived) threshold of violence in general, the more likely »strategic groups« tend to violent reactions - particularly in their own arena. (b) The more likely environmental disruption by international industrial projects, the more likely is a violent response by »strategic groups«. (c) The more prominent the role of INGOs, the less likely the resort to violence.*

Mexico: Violence in the Chiapas Arena
In the Chiapas case no external influence in the sense of international political intervention played a significant role in triggering the escalation. At the same time, political violence in the Central American neighborhood was declining in those years. The authors mention only the fact that the regime found itself »scrutinized by multiple external and internal groups, each with access to different pressure points and each pursuing diverse, but not always opposed agendas« (Howard/Homer-Dixon 1996:27).

Besides this, from the standpoint of the Zapatistas there is a historical component of quasi-external intervention worth noting. Indeed, they perceive the »resource capture« of the former colonial power as an important factor to the present conflict. Since the 15th century *indígenas* have fought in countless riots and rebellions against the Spanish and later the Mexican state. The demand for land has usually been at the core of each uprising (*ibid.*:15). Later, »(d)uring the tumultuous 19th century«, the Mexican center did not firmly establish sovereignty over Chiapas, »and the state remained quasi-autonomous because of its isolation« (*ibid.*:16).

The discrimination against the inhabitants of the Chiapas arena is deeply rooted in the colonial past, which led to imperfect penetration of the region by modern institutions. This is the reason why the central state policy towards the region - such as the land act - is perceived as a kind of foreign intervention with mostly unfavorable results for the region.

Botswana Control Case
The government canceled the water development project in the Okavango arena in May 1992 because of internal and external pressure (Claus 1996:269). In this particular case, hypothesis 7(c) applies. Apart from the strong reluctance of the local population, Greenpeace International launched a massive international campaign against the project in 1990 threatening the Botswana government with the call for an international boycott of Botswana beef. In October 1991, a team of experts engaged by the International Union for Conservation of Nature (IUCN) reviewed the water project SOIWDP; the final report was submitted in

October 1992 identifying serious errors in the hydrology of the project proposal (*ibid.*:275; 279). On the other hand, the Debswana mining company, jointly owned by the government and the South African De Beers Group also withdrew its support for the project - not least because the chairman of Debswana was the president of the Kalahari Conservation Society, supposedly the environmental watchdog of the delta arena.

However, the government did not comment on the IUCN report. Nor did it initiate a public debate on it. There is some evidence that the plan will be revived. In March 1994, for instance, President Masire stated that »the government is thinking of a revisit of the SOIWDP (*Midweek Sun*, 16 March 1994. In Claus 1996:281). This statement may also reflect that international disputes over water could top the future agenda of regional cooperation between Botswana, Namibia, Angola, and South Africa. »As all main water-systems of Botswana have an international character, any development or change of the water resources in the originate countries, specially in southern Angola or the Caprivi Strip of Namibia (...) would have significant impacts on Botswana (*ibid.*: 283).

South Africa: The Homeland and KwaZulu Natal Arenas

Percival and Homer-Dixon (1995) paid little attention to external influences on violence in South Africa. They seem to support the thesis that the internal events were much more determinant in terms of inducing change than external ones. They quote President de Klerk who stated in his Nobel Peace Prize speech in 1993 that it was neither international sanctions nor armed struggle that forced apartheid to change, but the movement of millions of people into the cities that created social upheaval and strained community and state institutions (*ibid.*).

Although this point of view may be disputed, if one tries to explain the change in South Africa one must put emphasis on the internal dynamics rather than the long-term effects of sanctions. The latter enhanced pressure on the white government to open the system, whereas the international community condemned the use of political and criminal violence. Actually, numerous governmental and non-governmental organizations - among them many church organizations - tried to mitigate the violence or mediate between the various actors. The role of the UN in preparing the monitoring of the 1994 elections was crucial in terms of a nonviolent transformation of apartheid (Bächler 1994: 54-64).

7.8. What Makes a Difference?

The model on environmentally caused conflicts is supported by the 12 cases used in order to check it. Not only the ENCOP case studies but also the analyses provided by the ECACP fitted into the framework of the seven linked hypotheses. The distinct area studies, which deal with completely different types of conflicts, contend that all factors build up the model.

To take the Chiapas case as an example: the co-authors of the study address the issue of environmental discrimination against marginalized groups; they deal extensively with high dependence on shrinking natural capital, and they mention the necessity of group cohesion following societal fault lines; they also stress the ability to organize marginalized groups by an accepted leadership as well as by mobilizing allies targeting on land reform and land tenure. Mobilization is seen as a consequence of poor state performance in one of the poorest regions of Mexico, whereas the rich natural resources are exploited by the center, which is perceived as incapable of organizing the return flow of resources to the region. Environmental degradation is seen as a reason underlying conflict formation. A major goal of the EZLN was to overcome the structural scarcity of resources. Both the issue of maldistribution and the lack of arable land helped mobilize a fast growing and ever more marginalized rural population. Whereas the land issue facilitated channeling and group identity building, Article 27 became a trigger accelerating the use of violence against a system perceived more and more as unjust. The relative weakness of the previously repressive government in the phase of its own transition towards economic liberation constituted a catalyst in terms of creating opportunities for protest and violent actions against the morbid state authorities.

If livelihood security is at stake and there is an everyday struggle for survival, relative if not total deprivation can be grasped by the discrimination approach. In the case of Mexico, groups already discriminated against felt even more discriminated against after the new land law of 1992 took effect. The growing gap between their own marginalization and the growing wealth of privileged landowners was perceived as an inevitable situation that forced them to react as soon as the political situation seemed opportune to do so.

In the conflicts analyzed above, the environment and its transformation are seen as a cause of violent conflicts. As far as the *intensity* of a certain conflict is concerned, *severity* of degradation in chemical or physical terms is only one side of the coin. It is neither exclusively nor chiefly responsible for the *extent* of violence involved. There is no equation: the more severe the environmental degradation, the more intense the related conflicts. Therefore, the reverse is that perception of the environmental situation determines both the interests actors have as well as the political opportunities they are aware of. Environmental transformation is a structural type of cause that works via channels, triggers, and catalysts, depending on the state of affairs as well as on perceptions of the actors involved. To explain this in other words: in some ethnopolitical conflicts, for instance, ethnicity may serve as a channel, whereas the structural reason is the scarcity of degraded resources leading to economic decline. In others, ethnicity may be a structural reason constituting a historical conflict formation, while an environmental catastrophe or a certain governmental decision (for instance a land act) can trigger ethnopolitical violence.

As noted, most of the case study authors emphasized the issue of poor state performance and instability more than, for instance, population growth or group cohesion. Combined with state repression or a lack of democracy, these two factors seem to be the *top indicators* for the relative importance of hypothesis number 6 in the framework of the model. As highlighted in the context of the Botswana control case, it is indeed the hypothesis that makes a difference in terms of either violent or nonviolent outcomes of environmental conflicts. This also holds for the other conflict pairs of the sample. In all control cases the role of state performance, participation, and institutional solutions was crucial in view of nonviolent resource conflict management. However, one has to differentiate this result because nonviolence can be a consequence

- of a relatively high PFI, such as in Botswana (74) (*see: app. 13a*);
- of a cooperative approach between neighboring states, at least one of them having a high PFI, such as between US (84) and Mexico (*see: app. 13b*);
- or of an autocratic regime being able to suppress violence successfully, such as in the cases of Nigeria (5) and China (8). In Mexico the Chiapas conflict was seen as the result of a formerly repressive state in transition (*see: app. 13a*).

The model takes account of this multicausation/multiple roles situation. It can deal with the environment as a reason or background condition (the independent variable) and with environmental scarcity, which may contribute to triggering events. However, the entry point to linking environmental transformation and the sociopolitical system is constituted by the concept of discrimination. It proves helpful to explain why various actors perceive themselves in an inevitable situation and why they prefer violence instead of other means.

Hence, not the degradation of the environment as such transforms the society-nature relationship but rather the discrimination against actors highly dependent on the landscape as a life-support system or the perception of being discriminated by those who profit from the exploitation of the nature's capital in one way or another. And these are deep-rooted sociopolitical conflicts.

CHAPTER 8

OUTLOOK

Conflict Potential, Sustainable Development, and Environmental Security

This study aims to elaborate on the violent transformation of society-nature relationships. The assumption of the coincidence of poverty-driven environmental transformation and ongoing violent conflicts and wars (*see: 1, 2*) could be confirmed empirically. Various methods adapted in this study to examine environment-related conflicts provided similar and complementary results. The cultural ecological background of dryland and mountain systems transformation has shown that a correlation does exist between environmental degradation and violent conflicts (*see: 3*). Moreover, as the synthesis of 40 case studies carried out by the ENCOP team indicates, there are different causal pathways of present violent conflicts and wars that could be traced to the environmental roots of the conflict genesis (*see: 4*). However, not all conflicts examined in the ENCOP became violent. Quite a number of them - although acute - remained below the threshold of organized violence, let alone war. This sample served as a control group in order to differentiate between it and those pathways, which ended up in violent confrontations among the actors involved. As the typology of environmental conflicts in general and the in-depth case study of the Rwanda arena in particular have indicated, environmental conflicts are related to socioeconomic, ethnopolitical, interregional, migrational, and demographic problem syndromes (*see: 5*). The significance of these factors notwithstanding, the reasons why these conflicts become violent are found in the sociopolitical network in terms of poor state performance, repressive regimes, lack of regulatory mechanisms, and unavoidable situations. This statement could be strengthened and systematized by designing a general model of environmentally caused violence (*see: 6*). The conflict model itself was checked by comparing it to a number of case studies provided by the scholarly literature. The linked hypotheses based on a set of different types of indicators turned out to be valid and significant for the cases examined (*see: 7*). This being said, one must add that the model applies more for internal conflicts than international ones, firstly, because it stresses sociopolitical aspects and, secondly, because almost all of the violent conflicts of the sample are domestic ones.

It was intended to go beyond using only one single method in order to overcome or at least compensate for necessary restrictions given by individual approaches. Research on internal violent conflicts in general and on environmental conflicts in particular depends heavily on in-depth area studies, hermeneutics, and a combination of explanation and understanding, that is, describing and analyzing historical and political events and their interdependence. A syndrome of factors, diverse trends, multiple-causation patterns, their change in the course of a conflict history, and erratic behavior of actors must be taken into consideration. The mere fact that the same causes do not always have the same effects makes any prediction extremely difficult. Only a mix of collective and cumulative research efforts may lead to a stage where in-depth studies and empirical testing enrich each other in enhancing prognosis. The empirical test - and falsification - of the conflict model provided here may therefore be a future goal.

However, it is impossible to build upon a single model covering all violent conflicts and ongoing wars. Since there are many key factors involved affecting causation, triggering, and escalating a conflict, several models must be designed. This model focuses on environmental degradation and conflict. It must be compared with other models, some dealing with different conflicts, others with the same one seen from another angle (*e.g.*, from the poverty point of view or focusing on ethnicity, refugees, the food crisis, etc.). Further research must concentrate on integrating different models and approaches to explain a conflict in an encompassing context.

In the following section different aspects will be summarized and discussed in the light of the overall results.

8.1. Multiple Causal Roles Concept

The study did not strengthen alarmist monocausal approaches dealing with environmental scarcity as the single cause of future all-out wars. Moreover, it has clarified that environment as a cause of conflict must be integrated in a concept, which takes account of the multiple roles and multifaceted character of the cause of sociopolitical conflict, such as discrimination against rural producers and accentuated structural heterogeneity under conditions of growing environmental stress.

Environmentally caused violence is hardly found in interstate relations of the present international system. Although climatic change, ozone layer depletion, and loss of biodiversity have created global conflict lines, which did not exist until recently, no major catastrophe threatens the world immediately, such as the prospect of a nuclear winter scenario discussed about ten years ago. »Climatic world wars«, mass flight due to rising ocean levels, military crusades by northern countries against destruction of tropical rain forests, and Atlantic

»fishery wars« - for instance, between Canada and Spain - may rather remain threatening blusters than lead to virulent conflict situations with a high risk of escalation.

However, such reluctant and hopefully realistic prognosis should in no way lead to underestimating the global environmental problems and their impact on development and security. The prognosis is rather aimed to draw attention to current and foreseeable crises. Because the biosphere transformation occurs rather unspectacularly, it emerges from many local processes, varies from region to region, differs in shape, and depends on numerous socioeconomic, cultural, and political factors. Meanwhile, in the context of many violent internal conflicts - partly less intense yet traceable to environmental destruction - society-nature relationships are locally at stake. An indication of this is that recent environmental conflicts arise ever more clearly from historical resource conflicts over land use and land tenure as well as access to fuelwood and freshwater.

Indeed, environmental conflicts are much more dramatic than conventional resource conflicts. While these result from inequitable socioeconomic distribution of scarcity and surplus, environmental conflicts are seen to be connected to historically unprecedented universalism, speed and extent of environmental changes, especially physical and chemical degradation of the landscape bearing renewables on which particularly rural societies in the South depend heavily. Figuratively speaking, the issue is no longer conquest of the *oases* but defending them against desertification and overuse. Therefore, it cannot be ruled out that currently observed armed environmental conflicts are only a harbinger of an international wave of violence, *e.g.*, due to conflicting international interests over regional impact of climatic changes.

So far, arenas of environmentally caused violent conflicts are most often subnational, that is, regional and local ones. They encompass distinct strategic groups using or overusing one or more ecoregions or subdividing ecoregions through political boundaries or migrating across such boundaries. Acute conflicts that tend to become violent constitute a kind of small-scale north-south conflict within the countries concerned. The issue is denial of access to the natural capital related to structural heterogeneity, political oppression, poor state performance, and a decline of (traditional) local resource management capabilities. In contrast to the national framework, on the international scene - especially in more integrated regions with cooperative elements - it is difficult to determine who is discriminating against whom and to what extent. Although in bilateral upstream-downstream conflicts the asymmetric relationship seems to be obvious, it is not as clear as it is supposed to be. There are several cases where the owner of the water tower is in fact much less powerful than the downstream riparians, which threaten the upstream country not to develop its water resources. What counts - as in domestic conflicts - is the quality of state performance as well as sociopolitical and military strength. That is why the model with its sociopolitical entry point covers international upstream-downstream and transboundary conflicts as well.

In principle environmental causes do not differ from economic, political, cultural, or any other causes of conflict. As any other cause the environment is part of a multilayered pattern or, as mentioned above, consists of a syndrome of factors, which - in an interplay of causes and effects, reasons and triggers - leads to violent conflict and war. In some cases environment may be just a contributing or (c) condition to a given conflict; in others it is a necessary or (n) condition either »co-causing« and/or triggering a conflict. Thus in the range of possible causation processes leading to a degree of conflict where violence is used, transformation of the society-nature relationship is no exception whatsoever. The ecosystem closely interacts with the socioeconomic and the political system. The society-nature relationship determines any cultural system and vice versa. This being said, one has to stress some specifics concerning the causal role of environmental transformation - any other cause has its own specifics as well; in this perspective the environment is no exception either.

Norman Myers made the point in »Ultimate Security« by concluding: »Regrettably, environmental problems do not often lend themselves to ready and precise analysis. The interlocking webs of cause and effect do not allow us to treat the issues in the way we are accustomed to when dealing with major issues, whether national or international. We cannot generally say that environmental problem *X* will necessarily lead to economic consequences *Y* that demands policy response *Z*. It cannot be predicted precisely how these factors will interact among themselves. (...) But it is realistic to suppose that environment-based tensions and conflicts, whether domestic or international, will escalate and multiply« (Myers 1993:227-228). That is exactly why a multiple causal role analysis based on case studies is a tool to define more precisely some crossroads in the interlocking webs of cause and effect.

Structures or background variables alone cannot explain why there is a violent conflict in one case and none in the other. Environmental degradation or socioeconomic decline may be a major reason for most of the ongoing violent conflicts. But these reasons alone do not lead to understanding the outbreak of violence between certain actors at a given period in time. Thus one also has to focus on the actors, their opportunities and preferences. The actor's perceptions of the severity of a problem and his preferences concerning strategic group building are crucial for the outbreak of violence. At the same time, with the exclusive focus on actors' behavior one may fail to explain conflict dynamics in the light of determining structures and underlying forces. Thus purely structure- and actor-oriented approaches do have their specific shortcomings. Analyzing the transformation of society-nature relationships, we had to look at the extremely interwoven character of environmental problems: their structure but also the actor's behavior competing with this structure as a prisoner of the society-nature interdependence - sometimes to the extent that an adverse outcome is perceived as unavoidable.

The study was aimed at integrating background conditions and dynamic elements in one single program. The environmental conflict model with seven linked hypotheses combines both elements. It provides an entry point in order to follow the integrated approach as vigorously as possible. As far as environmental discrimination and its indicators are concerned, the model deals with institutions and underlying structural heterogeneity. The hypothesis on group cohesion addresses the perception, preferences, and strategic goals of actors. Any group identity conflict - such as ethnopolitical conflicts - is about institutions and structures, and any war-proneness of such institutions and structures depends on how the actors in a certain arena perceive each other and how they behave on the basis of their perceptions.

The highway of integrating structures and actors leads beyond defining different roles of causes and correlates of violence. The cause of a conflict is not necessarily what causes the actors to resort to violent means. Not all causes are equally interesting: not all have the same impact and not all play the same role. Five major roles of causes have been distinguished: *reasons*, which form the background of the battlefield; *triggers*, which are responsible for resort to violence by actors that previously did not use violent means. Triggers are seen as the proverbial straw that breaks the camel's back; actors' *targets* focus on what the conflict is supposed to be all about; *channels* help to build strategic groups with an identity strong enough to be a tool in the struggle; and finally *catalysts* add fuel to the fire and work on the dynamics of conflict. The concept would be ill-understood if individual roles are seen in a linear time sequence.

Reasons, for instance, may reach far back in history, even to ancient or mythical times, but in order to be reasons for today's conflict they still have to be of some relevance. If they would have faded away in the collective memory, they would not play a powerful role in terms of causing actors to fight. Reasons may therefore be compared to stars in the universe: watching the sky, we recognize a great number of stars simultaneously, although some of them might be hundreds of light years away and others only ten or even less. The latter stars, though being much closer to the observer's eyes, do not necessarily have more impact in terms of brightness and size than the stars further away. The same holds for conflict causes. To give another example: land use was a reason for severe identity group conflicts in the Great Lakes arena already in the 17th and 18th century. In Rwanda as well as in the neighboring states land use, land distribution, and land tenure remain strong reasons and ever more so today.

A *trigger*, on the other hand, is always related to a relatively immediate event or various events running together. By definition a trigger for present violence, for instance, cannot be found with an inefficient hill chief distributing land in 18th century Rwanda. *Channeling* can be both part of a historical process of group identity building and a political process of relatively short duration - as, for instance, in terms of mobilizing and rallying in the forefront of an escalating conflict. The dynamics of channeling depend on several factors: above all availability of leadership, societal and cultural fault lines and cleavages, motivation, state performance, and means of mass communication (*e.g.*, Radio Télévision Mille Collines in Rwanda). *Catalysts*, finally, can be extremely different events, which do not necessarily have to be of acute nature. In Rwanda, as our study of reference, the myths of origin had the function of being a catalyst, first as element of the Batutsi-feudal system of oppression and after 1959/60 as an element of the anti-Hamitic Bahutu system of automatic majority, excluding the Batutsi from the nation.

8.2. Crisis Triangle: Maldevelopment, Transformation, Conflict

There is ample empirical evidence for the basic assumption that maldevelopment induces both environmental transformation and violent conflicts. The three issues must be seen in a triangular constellation, since intractable poverty and conflict-prone environmental transformation become more and more closely intertwined in particular ecogeographic zones. These zones encompass both dryland and sensitive mountain arenas with large rural populations being increasingly dependent on relatively shrinking natural capital in general and scarce as well as degraded renewable resources in particular.

8.2.1. RURAL TRANSFORMATION

As has been pointed out earlier, correlations between violence and environmental degradation are relatively weak to the extent that sociopolitical factors such as poverty and poor state performance are also causally relevant. The *violent transformation hypothesis* basically supports a prediction made by Barrington Moore (1966) and modernization theorists, even though the latter approach it from another angle than Moore: societies with large rural populations are highly unstable, undemocratic, or both.

Environmental conflicts probably indicate that transition from agricultural to industrial urban societies is at stake. However, this process definitely does not happen in the smooth way predicted by modernization theory but sometimes violently and always in the extremely painful way Moore suggested. The

»silent revolution« (Pradervand 1994) of the peasants discriminated against has become a noisy battlefield in many rural arenas, chiefly in African countries. The *Great Transformation* (Polanyi 1995) from rural to industrial societies did not only begin with remarkable delay, it also coincides - *horribile dictu* - with an era of accelerated human-ecological transformation (Turner *et al.* 1990). According to Polanyi, previously feudal and agrarian societies were transformed by development of the market and the nation-state. As a consequence, they created a subsystem within themselves, *i.e.*, the industrial system. On the basis of its expansionist tendencies the industrial system began to dominate or to colonize its social and natural environment. While Polanyi was in a position to emphasize the socioeconomic side of more or less successful modernization in the North, this study necessarily focuses on the global socioecological side of adverse impact of modernization and development in the South.

A major aspect of the transformation of society-nature relationships is the degradation of landscape as the arena of rural actors. The deterioration of life-support systems not only constitutes a loss of economic means of living. It is also a force, which contributes in the long term to livelihood insecurity and dissolution of traditional institutions with which societies regulated their access to and use of sensitive resources. The southern arenas under consideration are confronted with *imperfect* penetration by modernization. Landscape transformation - if not perceived by religious rural populations as divine punishment - is seen as a consequence of the modernization at stake. Large rural populations find themselves being discriminated against with limited off-farm opportunities while highly dependent on shrinking natural capital. Peasants, smallholders, landless people, rural workers, pastoralists, and nomads - the major victims of imperfect penetration - are both structurally and actively in permanent resistance against impoverishment, marginalization, and displacement by force and elimination. Decreasing productivity, lack of innovation, widespread fatality, starvation, migration, and flight belong to a syndrome of key factors, which - if there are triggers and strategically organized actors - cause violence in weak states with generally poor state performance. Environmentally caused conflicts should be understood as struggles of the land against the town, as competition among endangered rural livelihoods and invasive modernity. They are conflicts in marginal areas with migrants and refugees seeking their way in peri-urban arenas overflowing with former rural dwellers.

This study has basically confirmed a conclusion reached in our ENCOP final report:

> »Confronted with the fact that in many areas similar conflicts occur simultaneously with comparable causes, actor groups, and goals, the issue arises if these individual environmental conflicts (which we have indicated above to be the peak of an iceberg) foreshadow a larger disintegration process which especially covers the structures of developing and of

> transitional societies. If this is true, there are two possible interpretations for this phenomenon and its further development: Either environmental conflicts are at the heart of retreat skirmishes of an increasingly marginalized rural population in the South which sees itself maneuvered into a no-win situation because of capitulation to the modern and the environmentally caused dissolution of its lifestyles (a kind of a chanceless peasants' rebellion). Or they are vanguard skirmishes of a conflict in the making which will lead to sustainable change in society-nature relationships by basic revaluation and revival of rural structures« (Bächler *et al.* 1996:332).

The conflict-proneness of environmental transformation arises from the strong correlation between the level of a state's affluence and its degree of stability and performance; a low level of wealth relates to politically unstable and weak states. Since natural capital is a crucial part of the wealth of developing countries, its degradation enhances structural heterogeneity, leads to environmental discrimination, and thus correlates highly with unstable and conflict-prone states.

The crisis triangle can be empirically underlined in particular by focusing on the Sahel belt extending from West to East Africa encompassing dryland as well as highland-lowland arenas. This belt not only hosts most of the war-torn arenas in post-Cold-War Africa; it is also transformed by human action in large parts. Major forces of the Sahel syndrome are:

- marginalization of subsistence production, pastoralism, small-scale farming, and traditional trans-Saharan trade, whereas demands for marketed products have increased, with the effect of expanding mechanized farming and increasing stocking of pastures; high population growth rates leading to high-density pockets in low productivity areas;
- the challenge of local and/or traditional authorities and self-regulation by national governments, supranational and global developments, interethnic and intercultural conflict formations;
- short growing seasons and strong risk for recurrent droughts, strong landscape erosion due to vulnerable soils subject to agricultural overexploitation.

Violent conflicts and wars almost always occur at a domestic level. Typically, many actors are engaged: local communities (traditional families, clans, and tribes) *versus* provincial and central governments, as well as international companies worrying about loss of local privileges, land-order rights, and water scarcity; nomadic pastoralists *versus* central governments which try to settle nomads; or various ethnic groups using the same water points, pastures, or farmland; farmers *versus* herders, especially where permanent cultivation has been introduced in arenas previously used by herders for seasonal migration; conflicts between migrants, refugees, and displaced or indigenous groups, urban dwellers, and governmental forces; local population *versus* armed gangs, mercenaries, and dissatisfied ex-fighters.

The rural actors perceive their fights not necessarily as struggles against degradation of the environment, let alone to improve it. They defend themselves above all against further discrimination and finally sociopolitical marginalization within their societies. With their livelihood security at stake, they organize around political goals, which leads to channeling individuals along different societal and cultural fault lines in the region. One major fault line is between Christian- and Muslim-dominated arenas which, at the same time, are often areas of different production systems in distinct ecoregional environments. In the complex interaction between cultures resource scarcity and maldistribution as well as perceived discrimination or oppression can be major accelerators for identity group building and provide reasons for political struggles ranging from local self-determination to regional autonomy and secession.

8.2.2. ETHNOPOLITICIZED ENVIRONMENTAL CONFLICT IN THE RWANDA ARENA

In Rwanda - as in other agrarian societies - fertile land was traditionally scarce and led to distribution conflicts. Historically, land distribution conflicts led to clashes over fertile soils in relatively favorable ecoregions between actors in search of a better and above all richer life. The power struggle of the lineages and clans was entwined with a conflict between semi-nomadic livestock herders and settlers, which manifested itself periodically in land-use and land-distribution conflicts. The distribution conflicts over scarce resources - due to agricultural conditions and combined with demographic dynamics - led to agricultural overuse of land, forests, and available water resources and thus resulted in ethnopolitically motivated transformation conflicts.

In the Rwanda arena two factors cumulated and thus accentuated the conflict potential of environmental scarcity. Firstly, demographic dynamics led to high density clusters of small rural producers, which - due to lack of financial and human capital - were unable to escape the deteriorating society-nature relationships. Secondly, social stratification led to discrimination against the majority of small rural producers. This fact alone may not explain the escalation of violence well because the same phenomenon holds for many if not most societies throughout the world. The crucial point in Rwanda is that the rural-urban gap was overarched by clear-cut ethnopolitical segregation used by the ruling elite to exclude the other group almost completely from any resources, whether power, education, natural capital, or the entire country (the latter referring to Batutsi exiled in 1960 who were not allowed to return to their country).

With the emergence of absolutism, further exacerbated and strengthened by the colonial powers, a steep hierarchy developed with firm role assignments in which a participant was never supposed to be subordinate in one area and dominant in another. The integration of the two groups in the Rwanda arena already failed in the 19th century. Because of the dissolution of traditional mechanisms and kinship relations and the transfer of the *social hierarchy* to *politics* the expanding central state became an arena where ethnicity became a predominant channel to either hold power or compete with it. The central state itself - created only after 1860 - became an important tool in this struggle: the winner could easily turn the state into an agent of his own interest which was supposed to be ethnic interest - the latter following the logic of a winner-takes-all system in a fragmented society.

Demographic dynamics and transformation deepened the fault lines between Bahutu and Batutsi and contributed to formation of coherent strategic groups, which perceived that they were fighting each other for ethnic reasons and not because of economic, demographic, or ecological ones. Because of the ethnopolitical hierarchy and the environmental discrimination against most rural producers it did not prove possible to dissolve the socioeconomically and environmentally harmful agrarian structures, to overcome the spatial fragmentation and political weakness of the peasants and landless, and to induce sustainable development on a broad basis by means of a solid agrarian constitution. The environmental discrimination under the quasi-feudal Batutsi agrarian constitution and the related socio- and ethnopolitical conditions became a modernization trap, which could have been avoided only by means of innovative social strategies to overcome ethnization of the agrarian institutions.

The intensified polarization of society and channeling of political unrest along ethnic fault lines proved itself in the appointment of the Batutsi elite to the power positions of the central state. The Bahutu countered this elitist ethnopolitical concept of the Batutsi with ethno-nationalism, which drew its strength and broad support from the homogeneously populated Bahutu strongholds in the northern provinces.

After the Bahutu revolution in 1959/60 the paternalistic one-party rule in an ethnically and socially fragmented society contributed considerably to the early racist drift of the social-revolutionary Bahutu movement. The Batutsi were forced to remain far from power in the countries to which they had fled. 1960 marked the historic birth of two completely different ethnopolitical projects: the strife of the Batutsi refugees for recapturing power in Rwanda and subordinating the revolutionary Bahutu again. The formerly oppressed Bahutu majority, on the other hand, had the single goal of avoiding the renaissance of the old regime. Always having been oppressed, the only security guarantee to preserve state power seemed to be expulsion of the former aristocracy. Intervention, war, and genocide in 1994 are the most visible expressions of two irreconcilable concepts: *Batutsi oppression of the majority vs. Bahutu elimination of the minority.* Both the invasion of the FPR in 1990 and the Arusha negotiation process are only catalysts to the conflict. The Arusha agreement concerning power sharing, the plane crash on 6 April 1994, and first massacres are closely intertwined. They triggered a war which has its long roots in the history of land-use and tenure conflicts, environmental discrimination against the majority of the rural population, and a steep social and ethnic hierarchy.

8.2.3. POPULATION DYNAMICS AND MIGRATION

Population dynamics are an accelerating factor concerning both environmental degradation and conflict potential. Pressure on the biophysical basis of an underdeveloped economy is normally determined by the interplay of three factors: high population growth rate, changing consumer behavior through marketing of products and availability of cash, and low technical potential, which determines the extent to which a society or its subgroups depend on natural capital. Since the three factors function interdependently, a lack of technical resources indirectly enhances population pressure on a fragile environment, whereas large-scale mechanized farming, for instance, may have the same adverse effects. In rural arenas characterized by predominant or exclusive use of traditional technologies combined with little off-farm opportunities, growing populations subdivide land further and further while its potential to provide food, fodder, and fuel remains relatively fixed.

Environmental flight or migration is among the most visible and, at the same time, most serious consequences of transformation. Migration occurs when people are forced to abandon their traditional surroundings either temporarily or permanently because environmental damage (either natural or anthropogenic) has put their livelihood in danger and substantially impaired their quality of life.

Natural environmental damage, in the form of so-called natural disasters (*e.g.*, earthquakes) forcing human beings into sudden flight, have always occurred from time to time. However, nowadays it is not as easy as it once was to distinguish between natural and anthropogenic causes of disasters, since the transformation process has led to an increase in the number of natural events with disastrous consequences for an ever greater number of people. Environmentally induced flight must be seen in the framework of the crisis triangle - in connection with maldevelopment affecting violent conflicts, expulsions, state repression, and marginalization of minorities or ethnically oppressed majorities.

Violent conflicts of varying intensity are usually associated with the beginning and the end of flight and migratory streams.

Environmental problems accentuate conflict situations in which economic decline, political instability, or traditional conflicts are already present and/or receive additional fuel from population pressure. Pre-colonial forms of conflict often break out anew in such circumstances. Impoverishment processes in urban arenas promote social and political unrest potential. But conflicts can develop just as easily between nomads and settlers, between ethnic groups or neighboring states sharing the same ecosystem as between regions or states targeted by environmental flight.

At present, water scarcity in connection with soil erosion (desertification, overuse) is the most important cause of environmentally induced flight. Flight and migration may be caused by various phenomena: *i.e.*, periodic flooding, as in the central Chinese province of Henan; dramatic shrinking of great bodies of water, as in the Aral Sea and Lake Chad arena; persistent drought, as in the Sahel, the Middle East, or Central Asia; and immense dam constructions, as the Narmada project in India or the Three Gorges project in China.

The predicted rise of the sea level will possibly pose a major threat in the future. However, combined political, economic, and demographic factors might produce a gigantic migration potential in densely inhabited lowland coastal arenas - particularly since 60 percent of all human beings live on or near coast lines and delta areas.

8.2.4. FORCE OR COOPERATION?

The geographical taxonomy provided by Michael Renner (*see: fig. 2.2.*) may serve as a good tool to reach some final conclusions concerning the level of future conflict and its intensity in the international system. Transformation of the societal-nature relationships has profound impact:

- affecting *renewable resources locally or subnationally* (landscape, soil, vegetation cover, water) is most likely to lead to violence at a domestic level. These conflicts originate from both home-grown and internalized environmental disruption - mainly in arenas of developing countries. It is in this area that we find the highest proportion of conflict potential. If it will not be possible in the foreseeable future to reduce the adverse global environmental effects on these countries or to eliminate national sources of transformation, there remains little room for conflict resolution whatsoever;
- affecting *transboundary ecoregions in two or more neighboring states* (river basins, forests, arid lands) is perceived as »ecological attack«, which induces tense relations, diplomatic notes, verbal attacks, and military threats. However, use of military power is not usual, since the environ-

mental component of unilateral foreign policy today is more strongly anchored in international policy than ever before. Nonetheless, given the lack of regulatory mechanisms, paired with poor state performance and with repressive governments, the rebirth of classic interstate wars cannot be excluded (*e.g.*, in the Euphrates-Tigris Basin, the Jordan Basin, the Nile Basin, the Sir and Amu Daria Basins, or the Mekong Basin);

- affecting *not just the territory of one or several neighboring states directly but the environment in a wider regional context*, such as regional air pollution through emissions. These are not very violence-prone. So far, air pollution was supposed to be a typical problem of industrialized countries, such as acid rain in Europe. Nonetheless, rapid industrialization, mainly in East Asia and Southeast Asia, may also transform regional commons into a heavy international problem in these newly developed regions. Despite growing pollution, the conflict potential is relatively low because the fast industrializing economies must have an interest in a growing regional market. Export-oriented industrialization and »attacking« one's own neighbors environmentally are often mutually exclusive goals. China may turn out to be a notorious yet powerful free rider, creating some conflict potential that must be watched carefully;
- affecting *global commons with nondiscriminatory impacts on all human settlements as well as on the extraterritorial regions of the earth* (stratospheric ozone layer depletion, climate change) leads to conflicts over global commons, which could be waged successfully by political and/or legal means. It has been possible to pinpoint the most important environmental goods as subjects of international arrangements and conventions. The regulatory system in this respect is relatively well advanced, so there is no reason to expect that armed conflict will take place in a globalized arena in the foreseeable future.

8.3. Sustainable Development and Environmental Security

Currently the most discussed therapies against maldevelopment and environmental conflicts are the normative concepts of *sustainable development* and *environmental security*. The *Rio Declaration* stated explicitly that war actually prevents sustainable development and that peace, development, and environmental protection are therefore interwoven and indivisible.[1] The Rio statement notwithstanding, recent discussions concerning the two concepts mentioned are somewhat dissatisfying because they occur in two different

1 See principles 7, 14, and 27 of the *Rio Declaration*, 1992.

communities on two distinct tracks (*see: 2*). Each community - mainly development sociology and security studies - is confronted with countless attempts at defining its terms in light of standards set in its own discipline. Some scholars even maintain that both notions are mutually exclusive. Others disagree by suggesting that they have one and the same meaning, while still others simply define their notion all-encompassingly trying to integrate the other. This endeavor is most often assumed by the opponent community to be an unfriendly takeover.

However, there have hardly been any conceptual thoughts about how to systematically relate sustainable development and environmental security to each other or about how to make both normative horizons congruent. This final section is dedicated to propose such a systematized approach to deal with the two concepts. It considered them to be neither the same nor completely different. Instead, I will show how they both interconnect and overlap.

As has been noted throughout this study, peace is challenged in many regions by unsustainable resource use. In most cases transformation is more related to the development dilemma than to the security dilemma; however, both dilemmas are closely intertwined. The issue arises how sustainability can be made operational by keeping environmental assets renewable above and beyond their regenerative abilities. And how peace can be maintained by cooperative regulation of conflicting interests concerning resource use. Technical concepts can hardly suffice as the answer. Socioeconomic, institutional, and demographic factors inevitably come into focus due to the ecological scarcity of degraded resources, which can only be overcome by politically induced social change. Social strategies for regulating environmentally caused scarcity conflicts must be linked with cooperative mechanisms for exchanging equivalents to make peace and sustainability simultaneously a means and an end of successful regulation.

8.3.1. THE DEVELOPMENT/SECURITY PRISM

As both environmental security and sustainable development are dealing with society-nature relationships, they may be seen as part of two different triangles. The first one, the *security triangle*, links the three *threat* dimensions: human security, ecosystem security, and national/international security; the second one, the *development triangle*, links the three *vulnerability* dimensions: human development, ecosystem development, and national/international development. If we connect the security and development triangles (arrows), we virtually end up with the shape of a prism (*see: fig. 8.1.*).

Along the arrows connecting the respective edges there are countless issues dealing with each of the three dimensions, which environmental security as well as sustainable development consist of. Some of the issues tend more to the security side, addressing the avoidance of threat to security; others tend more to the development side, aimed at overcoming vulnerability.

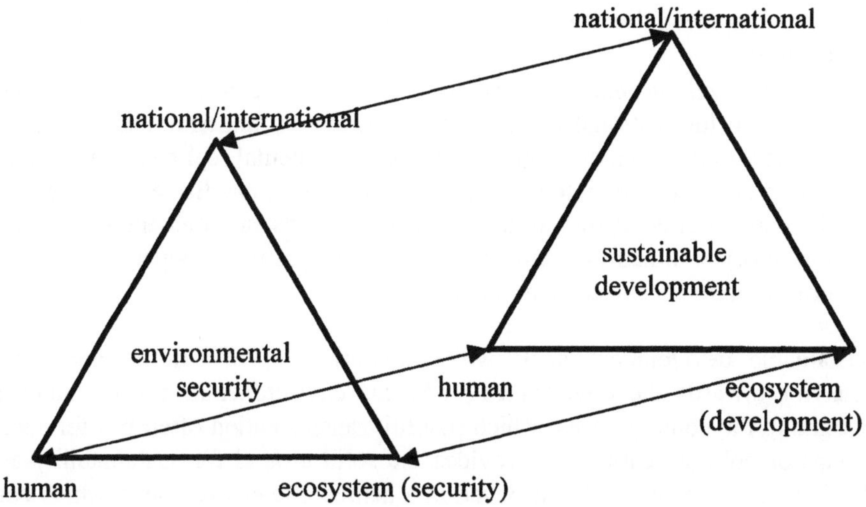

Figure 8.1. Development/security prism

Sustainable development focuses on the dilemma between economic growth and a way of resource use that respects the regeneration rate of renewable resources. Sustainability through qualified and *dynamic* development must be achieved at all levels of the society-nature relations: the level of scientific and technically sound approaches to ecosystem development, the level of adapted human development by mobilizing all aspects of human capital (from traditional knowledge to computer science) to overcome environmental degradation and resource depletion; and finally the level of national and international development through environmentally sound incentives and binding regulations, such as environmental legislation, regimes, and international treaties.

Environmental security is focused on the dilemma between securing and preserving societal and political institutions and leaving room for dynamic development. The major challenge is to secure each step achieved towards sustainable living in the sense that certain values should be *preserved* and not violated by economic growth and resource-use patterns; those values encompass

- *ecosystem security* consisting of behavioral, political, and legal measures to be taken in establishing an institutional framework in which sustainable development of an ecosystem can take place;

- *human or livelihood security* in a given habitat, including food security, safe access to freshwater and other resources needed, but also health, sanitation, etc. The human dimension encompasses *societal security* in a community or neighborhood, *e.g.*, not being blamed or policed all the time - or even displaced by force - only because of the daily need for scarce resources;
- *national and international security* deals with the environmentally induced threats to international relations, for instance, by undermining agreements on global commons or by questioning environmental regimes as a whole. It is also concerned with tendencies or events that jeopardize achievements in the domain of ecosystem or human security. Thus environmental security at this level is aimed at establishing legal and political cooperation in governing shared ecosystems by states.

Sustainable development deals with transformation of cultural and social behavior towards the environment, whereas environmental security aims at mitigating the consequences, which painful transformation often has for social groups or political entities. It provides the solid ground for sustainability and dynamic development to be achieved in a democratic environment, which contributes to prevent violent conflict and zero-sum (Rambo-type) games over access to scarce resources. Sustainable development is associated with *vulnerability* and its reduction, while environmental security is related to *threats* and their minimization; both correlate positively. Environmental security then provides the platform for both *mediating* environmentally induced conflicts (whether violent or not) and for political and legal *cooperation* (whether on a subnational, national, or international level).

Environmental security is an indicator if and how unsustainable development - which increases vulnerability - threatens social and political entities within the framework of society-nature relations. A lack of ecosystem development, for instance, can seriously affect behavioral, political, and legal measures and therefore provoke a perception of ecosystem insecurity. And unsustainable human development can seriously threaten livelihood security in all its major aspects. Degradation of the global commons can destabilize international cooperation and lead to increased pressure on resources through free riders, unilateral access to resources, and sharp competition. Each of the three edges of the twin-triangles forming the prism can be at stake, while unsustainability and insecurity in one of the corners affect one or both of the other two edges as well (*see: fig. 8.2., a, b, c*). The break-up of all three edges at the same time would create a situation of total insecurity and total unsustainability. Hence, sustainable development and environmental security are designed to enforce each other at every level of the societal-nature relations.

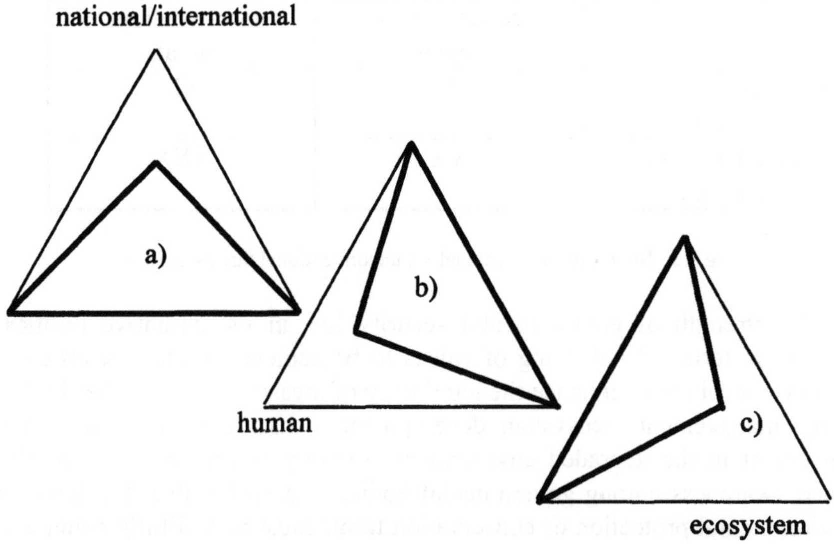

8.2. Each corner being threatened by insecurity and maldevelopment

Some critics consider the concept of environmental security being simply a tool for persuasion of the powerful defense sector of government. They argue that the defense community could misuse it as a tactical tool, for instance, in resource allocation struggles among ministries, or that it may see it as a bridge to preserve high defense budgets during minor military threats. Indeed, there is some evidence driven by experience that the defense/security community tends to give conventional answers to unconventional threats - dealing with symptoms rather than with causes (*see: fig. 8.3.*).

What is urgently needed instead are unconventional answers to unconventional threats. This will not happen if line-ministries such as the ministry of defense simply adopt environmental security as part of a broader concept of security or if development aid deals only with sustainability.

The approach presented here underlies the need to develop the twin concept as a powerful tool in itself in order to empower cross-sectional policy options within governments and administrations. Mainly the weaker sections - such as normally foreign aid and environmental ministries - might get a well-shaped tool to better communicate the related issue to the broad public.

	conventional threats	unconventional threats
conventional answers	X	XX
unconventional answers	XX	XXX

Figure 8.3. Matrix on conventional vs. unconventional threats/answers

Since the strength of environmental security lies in its normative political dimension in terms of balancing of values to be achieved, security considerations may contribute to enhance the legitimacy of measures taken in the field of resource management, ecosystem development, conservation of nature, and improvement in the degraded environment. Security considerations may also enhance awareness among governmental bodies that sustainable development in environmental protection or conservation terms must be skillfully communicated, mainly in regions where rural populations depend on scarce renewables, so that the people affected do not feel threatened. In the framework of good governance - and definitely not of an oppressive regime - the focus on environmental security may enhance transparency, participation, and thus negotiation or mediation of conflicts emerging from implementation of sustainable development as foreseen in the national environmental action plans decided in Rio in 1992.

Environmental security and sustainable development are designed as interconnected strategies to address poverty-related environmental concerns. These concerns are stemming from an offensive or even aggressive way of securing access to resources which is perceived as highly unjust by other actors. Since maldeveloped countries are agrarian societies in which 50 to 95 percent of the population depend heavily on renewable resources, internal violent conflicts and wars not only destroy the sociocultural network - such as traditional mechanisms to resolve resource conflicts (land, water, fisheries, etc.) by peaceful means. They also destroy the minimal ecological basis for subsistence farming, stockbreeding, and agriculture in general. Abandoned farms, decrease in the size of livestock, mined fields, refugees, etc. pose a major threat to human and ecosystem security, inhibiting sustainable living and transforming the ecoregion from a living space into an arena where struggles for scarce natural capital are about to escalate.

8.3.2. WHAT CAN BE LEARNED FROM THE BORANA SOLUTION?

How the twin concepts of environmental security and sustainable development strengthen each other will be shown by the example of a local conflict regulation dealing with all major aspects of the concept: threat minimization, enhancing livelihood and ecosystem security, as well as competing with vulnerability through more productive ways to use and distribute resources.

The *Borana solution* stands for settlement of an environmentally caused dispute at the subsidiary level of ethnic groups affected by land degradation, environmental scarcity, and drought. Neither central officials nor other third parties were involved as intermediaries. The incident occurred at the beginning of the 1990s in East Africa. No less than 14 ethnic groups and their representatives from delegations of elders sought acceptable regulation in a negotiating process for a sporadically escalating environmental conflict.[2]

In the center of the increasingly violent disputes - just as in their resolution - stood the people of Borana (or Borena). These were semi-nomads who grazed their herds in the border area of Northern Kenya and Southern Ethiopia in the arid and semi-arid countryside surrounding the Omo Valley. Under conditions of increasing scarcity, battles for grazing areas and watering spots as well as a rise in livestock poaching and plundering belonged to the region's challenges for survival.

The object of conflict was grazing property and access to watering spots for cattle of various neighboring semi-nomadic pastoralists belonging to several ethnic groups on Borana land. Thus in the background was a classic conflict over land use and property entitlement, as could be observed between traditional communities for centuries - and not just in this region. This land conflict had actually changed into an environmental conflict since the 1980s due to pressure of overuse, erratic rain, and continuing drought. The strong but regionally differing soil degradation had added a new dimension to the land conflict. The environmental problem - resulting from interaction between local overuse and regional impacts of climate change - contributed to a veritable security problem; heated-up conflict between the Borana, with their relatively good but also endangered land, and various ethnic groups, which fought for survival due to the drought.

Conflict Resolution Enhances Security and Sustainability

Beside the dispute settlement between former parties to the conflict, important rituals, symbolic acts, and steps towards confidence-building were worth noting

---

2 See in more detail Bächler (1996:376-396).

in this case of constructive conflict resolution. Especially the following aspects should be considered:

- Both sides expressed and advocated their immediate if not urgent interests (assertiveness), whereas both sides had the capacity to understand the other side's concerns and perspectives (empathy). First, this seemed to be a situation with two mutual exclusive goals: „private“ land tenure (Borana) vs. common land use (the others).
- However, against the background of the worsening environmental situation, both sides made a cooperative move to create value. Since subdivision of the already scarce land was out of the question, a kind of asymmetric solution had to be targeted; a tradeoff had to be found that would lead to a win-win situation because any win-lose situation would have induced a violent struggle over land.
- Indeed, the immediate material resolution of the conflict contained *exchange of equivalents*. The exchange consisted of confirming the Borana's property rights on their land, while the neighbors were guaranteed *user rights on the same land*. Thus it was agreed that the livestock herders were free to move with their cattle between the areas of the conflict parties, but only after the sorghum and corn harvests. With the freedom of movement and the access to the Borana land the great problem - simultaneous availability of water and grazing land, which must be made accessible to shepherds and their herds all year and within a determined time frame - could be solved permanently.
- Due to this success it was decided to expand the peace talks to further groups with which open or smoldering conflicts festered. Representatives of 14 peoples finally took part in talks and also accepted the agreed mechanisms for a lasting solution.
- Supporting measures for sustainable peace were agreed to. For one, a type of regional security and survival community was founded after finalizing the treaty through a ceremony, which was at the same time religious and secular. In a symbolic act weapons were destroyed by their producers. In addition, a neutral committee was formed of 40 members from five groups entrusted to monitor border offenses and violations of the agreement. Finally, projects were agreed to which would immediately be aimed at strengthening the socioeconomic and socioecological bases of the precarious societal-nature relationships in the region: founding of one community school for all children from the various ethnic groups, an agricultural training center as a contribution to intermediate as well as long-term food security for the semi-nomadic livestock breeders, a scientifically monitored animal-breeding facility as well as projects for improving water and health care in the region.

- With the cooperative approach to an urgent problem the creation of value - which includes the supporting measures - was tremendous, at least in comparison to a non-agreement and its possible consequences. The parties focused on interests and were able to broaden the perspective. To concentrate on the land issue was relatively easy afterwards. They not only agreed on land use and land tenure but were even able to introduce a component on how to get to more sustainable resource use without always undermining the basis for their own existence.

General Conclusions

The Borana solution allows broad generalization. It could be used for many conflicts now smoldering or already escalating in interiors of countries, such as between Fur farmers and nomads of Arab origin, the *dars* in the Jebel Marra mountains in Western Sudan or between Kyrgyzians and Usbeks in the Fergana Valley.

Applicability is somewhat difficult when the nation's central government intervenes or fuels further conflict issues by accommodating multinational consortia with large projects, such as dams or mining plans. Indeed, even in these cases the principle of exchange equivalents can be used as a specific form of tradeoff so that land, for instance, can serve as justifiable compensation. However, such negotiations are becoming lengthier and more complex, so that in cases in which external actors (states, multinational firms) intervene in an already precarious society-nature relationship, third parties must probably be consulted as mediators.

The Borana uphold the principle of subsidiarity. Affected actors act themselves without foreign influence. Despite conditions of acute shortage, they autonomously find their way to new regulation of land-use rights regarded as just. However, subsidiarity proves to be a barely sufficient condition in societies where poverty, marginality, and population growth accelerate social erosion processes due to ever more human beings exerting pressure on basic resources. The main reason for this is that, due to a shortage of economic alternatives and new methods for gaining energy, large shares of impoverished rural population in developing countries always have to rely on performing the same role and using the very same resources until they are fully degraded. Given the condition of uniformity, the threat is not only destruction of biological bases but even more the extremely vulnerable nature of a country's rural-dwelling majority: if 80 or 90 percent of the people depend upon the same farming activity, it can lead due to seasonal drought, fluctuations in precipitation patterns, or floods to the collapse of farming production and - as a result - famine and nearly inevitable migration and flight situations being seen as a last resort.

Thus to achieve sustainable solutions for both security and development, economic uniformity in rural sectors must be replaced by *socioeconomic diversity* adjusted to environmental conditions, population dynamics, and modernization attempts. Only if economic off-farm alternatives for the rural population are established - be it in new sectors of sustainable farming, in commerce, small rural industries, the service sector, or in administration of revalued small and provincial cities - can local production pressure on biological resources be reduced. And only in this way is there an intermediate and long-term chance to maintain biological wealth and renewable resources for future generations. To protect biodiversity socioeconomic diversity has to be strengthened.

Solution of the land issue in developing countries is enormously important. As already covered in Europe's early history, unclear property laws finally led to the disappearance of biological and human resources. Nationalizing land, legal uncertainty, and development by large farming producers pose a considerable potential for conflict, which may even increase through further degradation. Much more land must be privatized, so that small farmers may use the legal security to increase productivity, to test alternative methods, and to maintain genetic resources. Governments can play a considerable role in this process. Instead of policing environmentally dependent actors or enacting new laws with little consultation in advance, they can offer or at least facilitate an exchange of equivalents between the actors involved, according to the *formula*: land for reasonably paid work, resource use limitations for compensation payments or agricultural credit, ban on forest use for offering cheap alternative energy; after all, local actors know their own equivalency formula best.

However, these final thoughts are not terribly new, as the following passage from the North-South or Brandt Report illustrates:

> »Few threats to peace and survival of the human community are greater than those posed by the prospects of cumulative and irreversible degradation of the biosphere on which human life depends. In a global context, true security cannot be achieved by mounting buildup of weapons (defense in the narrow sense), but only by providing basic conditions for solving non-military problems which threaten them. Our survival depends not only on military balance, but on global cooperation to ensure a sustainable environment« (Brandt Report 1980).

* * *

CHAPTER 9

APPENDIX

Tables

Table 1:
List of African countries with possible correlations between desertification and violent conflicts

Location (HDI 1996)	conflict parties (actors)	key factors
Algeria (69)	Front Islamique du Salut FIS (the Faithful of the Sermon, the Brigades of God et al.) *vs.* government	o Sahara is the main ecoregion o desertification in the north o water scarcity through coastal urbanization o rural poverty o water conflict with Tunesia (Medjerda; Murzuk-aquifer with Libya?)
Chad (163)	Military faction, Revolutionary Forces of April 1st (1989) CSNPD (Committee of National Revival for Peace and Democracy) FNT (Front National Tchadien) MDD (-FANT) (Mouvement pour la Démocratie et le Dévelopment; Forces Armées Nationales du Tchad) Tuareg nomads vs. black farmers in the south (Sara, Massa)	o desertification, soil erosion o overuse of land, fuelwood (Central Chad, Tibesti) o shrinking Lake Chad

Ethiopia (168)	a) EPLF (Eritrean People's Liberation Front; until 1991) b) ALF (Afar Liberation Front, until 1991/92); c) TPLF (Tigray People's Liberation Front, until 1991/92); d) EPDM (Ethiopian People's Democratic Movement, later EPRDF, until 1991/92); e) EPRP (Ethiopian People's Revolutionary Party; until 1991/92), OLF (Oromo Liberation Front) *vs.* central (Tigray) government	o drought and soil erosion o rural livelihood insecurity o fuelwood scarcity o overuse (overgrazing) o rural migration, starvation
Kenya (128)	Social, ethnic, and religious tensions in the1980s (manifestations by the Muslim minority) and before the presidental elections in Dec. 1992; deadly quarrels in Nairobi in July 1990; Nov. 1991 until April 1992 ethnic clashes (Kalenjin tribes and others) in Western Kenya; ethnic clashes (Maasai), land conflicts and cattle razzias, triggered by President Arap-Moi.	o desertification and soil erosion in the north o overuse of scarce land o land-tenure conflicts o food crops vs. cash crops
Mali (171)	MPA (Mouvement Populaire de l'Azaouad) *vs.* government; Tensions and clashes between governmental troops and Tamaschek/Fulbe nomads 24 ethnic groups	o three ecoregions under erosion and overuse: o Sahara in the north (58%) o desertification in the semi-arid belt (30% of the terr.) o population pressure in the southern grasslands

Mauritania (149)	Traditional internal "north-south conflict" between Arab and black African tribes; conflict between lifestock breeders and farmers accentuates tensions; 1989 ethnic cleansing; Senegal Valley conflict with Senegal	o three ecoregions: Sahara, Senegal Valley and three mountain areas o desertification caused by overuse (only 0.2 % of territory are cultivated) o high rural exodus rate o urbanization 42% (Nouakchott) o unemployment 50%
Namibia (116)	Ovamboland people (smallholders and former workers of the disputed Rössing uran mine (British Rio Tinto Zink).	o Three ecoregions: Namib Desert along the coast line, Kalahari Desert in South-eastern N., arid and semi-arid central plateau o 80% of the population live in the north (Ovamboland) inducing high pressure on land o high rural exodus rate o water scarcity (no rivers)
Niger (174)	FLAA (Front de Libération de l'Aïr et l'Azawad); Tuareg nomads vs. central government; nomads vs. farmers (Oct. 1991 hundreds of Peulh and their cattle were killed by farmers); ethnic tensions/conflicts (Haussa 53%, Djerma-Sonrai 21.2%, Tuareg 10.4%)	o two ecoregions: Sahara and a 100 km arid/semi-arid belt in the south, 4% under cultivation o desertification and drought o lifestock was reduced by the 1984/85 drought (now 3.6 mio, in 1960 4.2 mio) o uran mining

Senegal (153)	MFDC (Mouvement des Forces Démocratiques de la Casamance); conflict between Diola and the (Muslim) Wolof (government); conflict with Mauritania in the Senegal Valley; farmers vs. nomads and lifestock breeders	o drought in the north caused migration of Wolof and Toucouleur into the south, introducing disputed cash crops o urbanization 40% o 23% of the population live in the coastal area (0.3% of terr.)
Somalia (172)	SNM (Somali National Movement) SPM (Somali Patriotic Movement) SSDF (Somali Salvation Democratic Front; active until 1991) USC (United Somali Congress; by Madhi, later by Aidhid) ongoing interclan clashes	o scarce natural resources in the highlands (low precipitation o large parts arid and semi- arid o desertification urges urbanization, mainly in Mogadishu
Sudan (146)	a) SPLA/SPLM (Sudan. People's Liberation Army/ Movement) Faction of SPLA/SPLM b) Fur vs. Dar tribes in Darfur, Jebel Marra,	o strong inter-ecoregional differentiations: Sahara Desert, arid, semi-arid, humid, and tropical climate in the south o desertification, drought, water and fuelwood scarcity in large areas o only 15% of potentially fertile land are under cultivation

Sources: see tab. 2.

Table 2:
Wars between 1/94 and 12/96, and HDI

Location	Type	HDI (Index)	Country Rank
1 Afghanistan	A-1	0.229 L	169
2 Algeria	A-2	0.746 M	069
3 Angola	A-2 / B -2	0.283 L	165
4 Azerbaijan vs. Armenia	C-2*	0.665/0.680 M	96 vs. 93
5 Bangladesh	AB-2	0.365 L	143
6 Bosnia-Hercegowina	BC-1		-
7 Burundi	A-1	0.282 L	166
8 Cambodia	A-1	0.325 L	156
9 Chad	AB-2	0.291 L	163
10 Colombia	A-2	0.840 H	049
11 Croatia	BC-2		-
12 Djibouti	AB-2	0.287 L	164
13 Ecuador vs. Peru	C-2*	0.784/0.694 M	64 vs. 91
14 Egypt (al-Gama'a al-l.)	A-2	0.611 M	106
15 Georgia, Abchasia	B-2	0.645 M	101
16 Guatemala	A-2	0.580 M	112
17 India	B-2	0.435 L	135
18 Indonesia (East Timor)	B-2	0.641 M	102
19 Iraq (Curds + Shiits)	AB-1	0.599 M	109
20 Israel vs. Palestinians	B-2	0.908 H	024
21 Kenya (ethnic viol.)	B-2	0.473 L	128
22 Lebanon	ABC-2	0.664 M	097
23 Liberia	A-1	0.311 L	158
24 Mali	B-2	0.223 L	171
25 Mexico (Chiapas)	B-2	0.845 H	048
26 Myanmar	AB-2	0.451 L	133

27 Niger	B-2	0.204 L	174
28 Nigeria (Ogoni)	AB-2	0.400 L	137
29 Northern Ireland	B-2	0.919 H	019
30 Pakistan	B-2	0.442 L	134
31 Papua N., Bougainv.	B-2	0.504 N	126
32 Peru	A-2	0.694 M	091
33 Philippines	B-2 / A-2	0.665 M	095
34 Rep. South Africa	AB-2 / A-2	0.649 M	100
35 Russia (Cecenia)	B-2	0.804 H	057
36 Rwanda	AB-1	0.332 L	152
37 Senegal	B-2	0.331 L	153
38 Sierra Leone	A-1	0.219 L	173
39 Somalia	AB-1	0.221 L	172
40 Sri Lanka	AB-1	0.698 M	089
41 Sudan	AB-2	0.359 L	146
42 Tajikistan	A-2	0.616 M	105
43 Turkey	B-2	0.711 M	084
44 Uganda	A-2	0.326 L	155
45 Yemen	A-2	0.366 L	142

Sources: Jaehrling/Nissen/Piper (AKUF Hamburg) 1996:89-123; Smith 1994:65-81; Keesing's Archives: Keesing's Register of World Events 1994, 1995, 1996; Africa Confidential, Vol. 35-37; ENCOP 1996. All wars are registered that occurred in the period between 1/94 and 12/96 no matter if they started earlier or later than 1/94 and if they ended before or after 12/96.

Key:

A	= Anti-regime war/violent conf.	Human Development:	
B	= other internal war/vc	H	= High (> 0.800)
C	= interstate war	M	= Medium (0.500 - 0.799)
AB	= mixed war/vc	L	= Low (< 0.500)
1	= with intervention		
2	= without intervention		

Table 3:
Violent conflicts in the 1990s, and HDI

Location	Type	HDI (Index)	Country Rank
1 Argentina	B-2 austerity)	0.885 H	030
2 Bolivia	B-1 (drugs)	0.548 M	111
3 Bangladesh	B-2 (unrest)	0.365 L	143
4 Bhutan	A-2 (minority)	0.307 L	159
5 Brazil	B-2 (unrest/riots)	0.796 M	058
6 Burkina Faso	A-2 (elections)	0.225 L	170
7 Cameroon*	A-2 (elections)	0.481 L	127
8 Central A.Rep.	A-2 (army mutiny)	0.355 L	148
9 Chile	B-2 (riots)	0.882 H	033
10 China (Tibet)	A-2 (riots)	0.609 H	108
11 Congo	A-2 (army mutiny)	0.517 M	125
12 Côte d'Ivoire	AB-2 (unrest)	0.357 L	147
13 Ecuador	AB-2 (riots)	0.764 M	064
14 El Salvador	A-2 (army, riots)	0.576 M	115
15 Ethiopia	A-2 (ethnopol.)	0.237 L	168
16 Eritrea**	AB-2 (army, riots)	-	-
17 Gabon	AB-2 (unrest)	0.557 M	120
18 Ghana	B-2 (tribal clash)	0.467 L	129
19 Haiti	A-1 (army, riots)	0.356 L	145
20 Honduras	B-2 (army, unrest)	0.576 M	114
21 Iran	A-1 (Kurds)	0.754 M	066
22 Kyrgyzstan	B-2 (ethnopol.)	0.663 M	099
23 Lesotho	A-2 (coup attempt)	0.464 L	130
24 Madagascar	A-1 (coup)	0.346 L	150
25 Malawi	AB-1 (riots)	0.321 L	157
26 Mali	A-2 (Tuareg)	0.223 L	171
27 Mauritania	B-2 (riots)	0.353 L	149

28 Mexico	B-2 (drugs)	0.845 H	048
29 Morocco	AB-2 (unrest)	0.534 M	123
30 Nicaragua	A-2 (guerilla)	0.568 M	117
31 Niger	A-2 (Tuareg)	0.204 L	174
32 Nigeria	AB-2 (riots)	0.400 L	137
33 Pakistan	B-2 (riots)	0.442 L	134
34 Romania	B-2 (protests)	0.738 M	074
35 Russian Fed.	A-2 (parliament)	0.804 H	057
36 Senegal	AB-2 (elections)	0.331 L	153
37 Serbia	AB-2 riots	-	-
38 Spain	A-2 (ETA)	0.933 H	010
39 Thailand	B-2 (army, riots)	0.832 H	052
40 Togo	A-2 (army, riots)	0.385 L	140
41 Tunisia	AB-2 (plot, riots)	0.727 M	078
42 Venezuela	A-2 (protests)	0.859 H	044
43 Zaire	A-2 (army mutiny)	0.371 L	141
44 Zambia	A-2 (coup, riots)	0.411 L	136

*Sources and key see: tab. 2. *Cameroon: there have also been clashes between Nigerian and Cameroonian forces in the disputed Bahassi peninsula in May 3-6, 1996. ** Eritrea is not listed in the HDI because of a lack of nationally aggregated data after independence in 1992.*

Table 4:
Low human development and violent conflicts/wars

Country	HDI Rank	Violent Conflict	War
1 Cameroon	127	x	
2 Kenya	128		x
3 Ghana	129	x	
4 Lesotho	130	x	
5 Equatorial Guinea	131		
6 Sao Tomé and Princ.	132		
7 Myanmar	133		x
8 Pakistan	134	x	
9 India	135		x
10 Zambia	136	x	
11 Nigeria	137	x	x
12 Lao People's Dem. R.	138		
13 Comores	139		
14 Togo	140	x	
15 Zaire	141	x	
16 Yemen	142		x
17 Bangladesh	143	x	x
18 Tanzania	144		
19 Haiti	145	x	
20 Sudan	146		x
21 Côte d'Ivoire	147	x	
22 Central African Rep.	148	x	
23 Mauritania	149	x	
24 Madagascar	150	x	
25 Nepal	151		
26 Rwanda	152		x

27 Senegal	153	x	x
28 Benin	154		
29 Uganda	155		x
30 Cambodia	156		x
31 Malawi	157	x	
32 Liberia	158		x
33 Bhutan	159	x	
34 Guinea	160	x	
35 Guinea-Bissau	161		
36 Gambia	162		
37 Chad	163		x
38 Djibouti	164		x
39 Angola	165		x
40 Burundi	166		x
41 Mozambique	167		
42 Ethiopia	168	x	
43 Eritrea	-	x	
44 Afghanistan	169		x
45 Burkina Faso	170	x	
46 Mali	171	x	x
47 Somalia	172		x
48 Sierra Leone	173		x
49 Niger	174	x	x
Total		23	21

Sources and key see: tab. 2.

Table 5:
HDI and distribution of violent conflicts/wars 1994 - 1997

1	H(57 countr.)		
10	-		
20	++		
30	- - (11)		
40	-		
50	-+-+		
57	+		
60	M (69 countr.)	+--	
70		+- -	
80		-+	
90		+++++++	
100		++++	
110		(31) + --+-+-	
120		- --	
126		-+	
130	L (48 Länder)		--++- -+
140			-- --+-++ (44 wars + vc)
150			----- ++-++
160			+-+--+++
170			-++-+- -
174			++ +-
	1 — 0.799	0.500	0.250 — 0
	high — medium	low	

(174 countries of HDI (UNDP) plus Eritrea, without Bosnia-Hercegovina/Croatia/former Yugoslavia)

+ war	high HDI	= 57 countries, 11 war-torn (19.3%)
- armed conflict		(incl. CIS and Israel)
	medium HDI	= 69 countries, 31 war-torn (44.9%)
	low HDI	= 48 countries, 44 war-torn (85.4%)

Table 6:
Human development vs. deprivation. 20 countries with high, medium and low HDI. 10 with violent conflict/war

HDI Rank, Country	access to safe water[1)] (%, 80-95) urban/rural		daily calories per capita[2)] (North = 100) 1965 / 1992		people in absolute poverty (%, 80-90)[1)] urban/rural		% of labor force agriculture, industry/service[2)] agricult. 1960	80	i + s 1990
High HDI	*no war or*		*violent*		*conflict*				
1 Canada	100	100	100	100	x	x		03	96
4 Netherlands	100	100	100	100	x	x		05	96
14 New Zealand	100	82	100	100	x	x		10	90
31 Costa Rica	88	99	84	93	24	34	51	26	74
56 Poland	94	82	100	100	x	x		27	73
Medium	*no war or*		*violent*		*conflict*				
64 Ecuador	82	55	67	83	40	65	59	33	67
71 Botswana	100	88	71	73	30	55	93	46	53
80 Ukraine	100	100	x	x	x	x		20	80
113 Mongolia	100	58	85	61	x	x	61	32	68
124 Zimbabwe	95	80	70	64	x	60	81	68	32
Medium	*with war or*		*violent*		*conflict*				
69 Algeria	85	55*	58	93	20	25	67	26	74
95 Philippines	93	77	66	73	40	64	64	46	54
105 Tajikistan	100	100	x	x	x	x		41	59
112 Guatemala	92	43	75	72	66	74	67	52	47
126 Papua N.G.	84	17	58	84	10	75	90	79	21
Low HDI	*with war or*		*violent*		*conflict*				
133 Myanmar	39	39	72	83	x	40	81	73	27
134 Pakistan	77	52	61	74	26	29	61	52	49
137 Nigeria	63	26	77	68	21	51	73	43	57
145 Haiti	37	23	71	55	65	80	80	68	32
152 Rwanda	84	67	59	58	30	90	95	92	08

Sources: [1)]Fig. drawn from World Resources 1996-97:152-3. [2)]Fig. drawn from Human Development Report 1996. Figures with a star (n) are drawn from Social Indicators of Development (World Bank) 1996; these figures are not totally compatible with the figures of World Resources Institute and Human Development Report because they are based on different time spans.*

Table 7:
Trends in demography and labor force. 20 countries with high, medium and low HDI. 10 with violent conflict/war

HDI Rank, Country	average annual population change in %[1] 80-85/90-95/00-05			density, population per sq. km[2] 80-85/89-94		average annual growth of labor force in %[1] 81-90/91-00		food production per capita 1993 79-81=100
High HDI		*no war or*		*violent*		*conflict*		
1 Canada	1.1	1.2	0.9	2.6	2.9	1.5	1.0	100*
4 Netherlands	0.5	0.7	0.3	388	409	1.3	0.5	108*
14 New Zealand	0.8	1.2	0.8	12	13	1.3	1.3	96*
31 Costa Rica	2.9	2.4	1.8	52	63	3.1	2.6	104
56 Poland	0.9	0.1	0.3	119	123	0.2	0.9	97*
Medium		*no war or*		*violent*		*conflict*		
64 Ecuador	2.7	2.2	1.7	32	39	2.9	2.8	110
71 Botswana	3.5	3.1	2.7	1.8	2.4	3.7	3.2	69
80 Ukraine	0.4	0.1	0.2	84	86	0.2	0.1	-
113 Mongolia	2.8	2.0	1.9	1.2	1.5	2.9	2.9	63
124 Zimbabwe	3.3	2.6	2.0	21	27	3.2	2.0	78
Medium		*with war or*		*violent*		*conflict*		
69 Algeria	3.1	2.3	2.0	9.1	11	4.1	3.5	119
95 Philippines	2.5	2.1	1.8	182	219	2.3	2.3	88
105 Tajikistan	2.8	2.9	2.5	32	39	3.2	3.2	-
112 Guatemala	2.8	2.9	2.7	73	92	2.9	3.4	94
126 Papua N.G.	2.2	2.3	2.1	7.4	8.8	1.8	1.9	103
Low HDI		*with war or*		*violent*		*conflict*		
133 Myanmar	2.1	2.1	1.9	55	66	1.7	1.8	107
134 Pakistan	3.7	2.8	2.7	121	154	3.6	3.0	118
137 Nigeria	2.9	3.0	2.7	90	114	2.2	2.4	129
145 Haiti	1.8	2.0	2.1	211	247	1.5	1.6	67
152 Rwanda	3.2	2-6	2.5	230	287	2.9	2.6	70

Sources: [1]Figures drawn from World Resources 1996-97:152-3. [2]Figures drawn from Social Indicators of Development (World Bank) 1996; rounded by the author. [3]Figures drawn from Human Development Report 1996. Figures with a star (n) are drawn from Social Indicators of Development (World Bank) 1996; these figures are not totally compatible with the rest of the figures in the same column, because they are based on different time spans.*

Table 8:
Trends in resource use. 20 countries with high, medium and low HDI. 10 with violent conflict/war

HDI Rank, Country	land use (000.000 ha) cropland - and land use change in - % since 1981-83[1)]		pasture		forests	%	cropland hectares per capita[1)]		annual rate of deforest. In %[2)]
	91-93	↓	91-93	↓	91-93	↓	1983	1993	1980-89
High HDI	*no war or violent conflict*								
1 Canada	46.0	-1.3	28.0	-3.3	494.0	11.6	1.82	1.58	-
4 Netherl.	0.9	12.0	1.0	-10.1	0.3	15.9	0.06	0.06	0.31*
14 New Zeal.	3.8	9.4	13.6	-5.3	7.4	3.6	1.10	1.09	-
31 Costa Rica	0.5	3.5	2.3	9.1	1.5	-5.2	0.21	0.16	3.1
56 Poland	14.7	0.9	4.0	0.6	8.8	0.9	0.40	0.38	0.06*
Medium	*no war or violent conflict*								
64 Ecuador	3.0	20.2	2.1	0.0	15.6	0.6	0.29	0.28	2.3
71 Botswana	0.4	5.0	25.6	0.0	26.5	0.0	0.40	0.30	0.1
80 Ukraine	34.5	-2.9	7.5	6.6	10.3	36.4	0.70	0.67	0.26*
113 Mongolia	1.4	11.1	124.8	1.1	13.8	-9.4	0.73	0.60	0.87*
124 Zimbabwe	2.9	3.4	4.9	0.0	8.8	-7.4	0.36	0.27	0.4
Medium	*current war or violent conflict*								
69 Algeria	7.9	7.0	30.8	-3.2	4.0	-9.5	0.35	0.29	2.3
95 Philippines	9.2	3.8	1.3	17.1	13.6	13.4	0.17	0.14	1.5
105 Tajikistan	0.8	-7.2	3.5	1.1	0.5	7.8	0.21	0.15	-
112 Guatema.	1.8	2.3	2.5	91.8	5.3	17.4	0.24	0.19	2.0
126 Papua NG	0.4	11.1	0.1	-17.3	42.0	0.0	0.11	0.10	0.1
Low HDI	*current war or violent conflict*								
133 Myanmar	10.1	0.1	0.4	0.7	32.4	0.9	0.28	0.23	2.1
134 Pakistan	22.9	12.4	5.0	0.0	3.5	14.5	0.21	0.16	0.4
137 Nigeria	32.4	6.1	40.0	0.0	11.4	-20.3	0.39	0.31	2.7
145 Haiti	0.9	1.3	0.5	-2.3	0.1	0.0	0.16	0.13	3.8
152 Rwanda	1.2	8.5	0.5	-15.2	0.6	-4.8	0.19	0.15	2.3

Sources: [1)]Figures drawn from World Resources 1996-97:152-3. [2)]Figures drawn from Human Development Report 1996. Figures with a star (n) are drawn from Social Indicators of Development (World Bank) 1996; these are not totally compatible with others in the same list because they are based on different time spans.*

Table 9:
Politico-institutional trends. 20 countries with high, medium and low HDI. 10 with violent conflict/war

HDI Rank, Country	type of regime 1990-95				change of regime 1990-95 through:		MILEX as % of combined education and health[1)]		political participation women in government at ministerial level[1)] 1995
	democracy		multiparty						
	yes	no	yes	no	war, coup	elections	1960	90/91	
High HDI	*no war or*				*violent*		*conflict*		
1 Canada	X		X			X	66	15	19.2
4 Netherl.	X		X			X	67	22	26.3
14 New Zeal.	X		X			X	29	16	7.4
31 Costa Rica	X		X			X	17	5	14.8
56 Poland	X		X			X	41	30	6.3
Medium	*no war or*				*violent*		*conflict*		
64 Ecuador	X		X			X	104	26	6.7
71 Botswana	X		X			X	-	22	0.0
80 Ukraine	X		X			X	-	-	0.0
113 Mongolia	X		X			X	-	-	0.0
124 Zimbabwe	X			X		X	-	66	3.0
Medium	*with war or*				*violent*		*conflict*		
69 Algeria		X		X	X		31	11	0.0
95 Philippines	X		X			X	44	41	8.3
105 Tajikistan		X		X	X		-	-	6.9
112 Guatema.	X		X			X*	45	31	18.8
126 Papua NG	X		X				-	41	0.0
Low HDI	*with war or*				*violent*		*conflict*		
133 Myanmar		X		X	X		241	222	0.0
134 Pakistan	X		X**			X	393	125	3.7
137 Nigeria		X		X	X		11	33	3.7
145 Haiti		X		X	X		100	30	17.4
152 Rwanda		X		X	X		-	25	7.7

*Sources: [1)] Figures drawn from HDR 1996: *In 1993 the court prevented President Serrano from imposing a new dictatorship. **Parties are not represented in the Consultative Council of 60 heads, all chosen by the powerful president.*

Table 10:
Regional aggregates of HDI and violent conflict/war

	population 1993 (000'000 total)[1]	rural population (% of total)[1] 1960	1993	HDI[1] (0-1) & PFI[2] (1-100)	WTS[3] (% of all countr.) W	VC	internal. Displaced ('000; total)[4]	refugees ('000; total)[1]
Sub-Saharan Africa	522	85	70	0.379 35.6	13	19 45 states	9'495	5'439
Arab States	241	70	50	0.633 24.0	8	2 19 states	1'700	1'230
South Asia	1'272	83	73	0.444 58.2	5	4 9 states	1'600	4'558
East Asia	1'272	80	69	0.633 40.8	0	1 5 states	-	290
East Asia excl. China	75	74	26	0.877 49.5*	0	0 4 states	1'000	2
Southeast Asia/Pacific	472	82	68	0.646	5	1 15 states	0'055	122
Latin America/Carib.	460	50	27	0.824 54.5	5	11 33 states	1'280	125
Least DCs	552	91	78	0.331	21	23 49 states	-	5'220
All DCs	4'299	78	64	0.563	35	39 126 stat.	-	11'787
Eastern Europe/CIS	391	53	34	0.773	7	4 25 states	2'740	656
Industrial countries	1'209	39	27	0.909	10	5 51 states	3'005	3'793
World	5'508	66	56	0.746	45	44 177 stat.	20'875	15'580

*Sources: [1)]Figures drawn from Human Development Report 1996. [2)]Data are drawn from [3)]See previous figures (1-3). [4)]Figures drawn from World Refugee Survey 1996, U.S. Committee for Refugees, Washington D.C. 1996, p.6. Information on internal displacement is very fragmentary. The table here presents only reported estimates and excludes countries in which significant displacement exists, but reliable estimates of the number of persons displaced are unavailable. The total number of displaced is undoubtedly much higher; probably higher than the 25-30 million one may find in literature (if one would include China, Ethiopia, Nigeria, and other big countries that are documented so far), and certainly much lower than 500 million, a figure often quoted, but nobody really knows from which source it stems (the Red Cross, however, refuses to be the source of it). *Excluding South Korea. **Including Eastern Europe and CIS.*

Table 11:
Typology of environmental conflicts and conflict intensity. 40 cases totally

A	Intrastate	
	A.I.	ethnopolitical conflicts
	A.II.	center-periphery conflicts
	A. III.	regionalist migration/displacement conflicts
B	Intrastate with transboundary dimension	
	B.IV.	transboundary migration conflicts
	B.V.	demographically caused conflicts
C	International	
	C.VI.	water/river basins conflicts
	C. VII.	globalized conflicts arising from distant sources

A Intrastate
A.I. ethnopolitical conflicts

conflict type	country: region	environmental problems	parties involved	conflict intensity
A. I.1.	Ghana: northern region	Sahel drought, land scarcity; overuse of renewables, farmers vs. livestock breeders	Konkombas vs. Nanumbas and allied Dagomba and Gonja (chieftancy areas vs. non-chieftancy areas)	violent conflict at the threshold of war, underlaying traditional land conflict
A. I.2.	Niger: arid and semi-arid belt	Sahel drought (1972-74 and 82-84), erratic precipitation patterns, degraded space of living for nomads and livestock breeders	Niger government (and farmers) vs. Tuareg nomads	protracted violent conflict and periods of war

A. I.3.	Rwanda	overuse of land, soil erosion, deforestation, subsistence crisis (farmer vs. stockholders), high population density	former Bahutu government vs. invading FPR Batutsi rebels and political opposition	war with massacres and genocide
A.I.4.	Sudan: Jebel Marra in Darfur	Sahel drought, desertification, regional overgrazing, land-tenure and land-use conflict	Fur farmers vs. Arab nomads, third party interventions (Sudanese government /Libya	periods of war, skirmishes and violent clashes
A. I.5.	Sudan: north-south	subsistence crisis caused by mechanized farming in the most fertile areas	Sudanese government vs. SPLM/ SPLA forces of the south	war, intertribal violent conflict (south vs. south)
A.I.6.	Tajikistan: Amu Darya Basin	hydrological system of Vakhsh, Pyandsh, causing upstream-downstream conflicts	fractional divides Grono Bakakshan and Kurgan Tyube region	violent conflict of low intensity, repeated clashes

A	Intrastate	
	A.II.	center-periphery conflicts

A.II.7.	Botswana: Southern Okavango, Maun	Southern Okavango Integrated Water Development SOIWD: water scarcity, overgrazing, mining, unique ecosystem	government vs. local population (peasants, cattle breeders) and Greenpeace / Kalahari Conservation Society	peaceful dispute settlement, project canceled in 1992; new attempts not to be excluded

A.II.8.	Bougain-ville: Papua-Newgui-nea	impoverishment of space of living, land and river pollution through Panguna copper mine	government of PNG vs. BRA guerilla forces	war that led to a struggle for secession
A.II.9.	Brazil: Amazon Region	overuse of renewables, above all the land in Southern Para	landless migratory peasantry vs. small/large-scale farmers	no violent conflict, but highly explo-sive, socioecono-mic and cultural "battlefield"
A.II.10.	Brazil: Lower Amazon	overuse of fish re-sources as an effect of the decline of jute/ varzea agriculture	community fishers (ribeirin-hos) vs. trawlers (geleiros)	violent conflict (low-intensity clashes) attempts of "dialoguing the conflict"
A.II.11.	Brazil: Tapajos River	overstrain of sink capacity through mercury pollution	gold washers vs. local population	no violent conflict, but some violent incidents, (mer-cury still is mys-tery and myth)
A.II.12.	Chile: Alto Bio Bio	ENDESA dams project: 6 dams planned; traditional land tenure at stake	central govern-ment vs. margin-alized pehuenche (pequenos agriculturos)	violent conflict (low intensity), INGO interven-tion to mitigate the conflict
A.II.13.	India: Narmada	Sardar-Sarovar dam project: flooding, destruc-tion of rural space of living	central govern-ment vs. indigen-ous people (Adi-vasi) and INGOs	no violent conflict, but non-violent protests, political repres-sion; INGO intervention
A.II.14.	Nigeria: Niger Delta Region	environmental devastation caused by oil mining in a densely populated rural production area	central govern-ment & Dutch Shell vs. indigen-ous/ethnic groups (Ogoni) and INGOs	violent conflict: clashes between army and Ogoni, heavy repression, high risk of escalation

A.II.15.	Nigeria: northern Sahel Belt	complex system of dams causing subsistence crisis and migration, depletion of fish stock in Lake Nguru	central government vs. farmers & fishermen, with an ethno-political dimension	no violent conflict, but protests and violent reactions by the army and police forces against farmers
A.II.16.	Papua Newguinea: Ok Tredi, Mount Fubilan	copper and gold mine in the rain forest, pollution of the Fly River System, heavy loss of biodiversity	government of PNG & mining company vs. indigenous subsistence farmers (Wopkaimin)	no violent conflict, but social protests, manifestations, sabotage, police violence, with a risk of escalation
A. II.17.	Philippines: Mount Apo Region	destruction of traditional land-tenure systems, expansion of mining, large-scale deforestation, dams, nuclear plants, Mt. Apo Geothermal Plant	central government and new private land-owners vs. indigenous Igorot & Lumads et al.	no violent conflict, but struggles against logging, mining, the so-called “development aggression”
A. II.18.	Senegal: Casamance	subsistence crisis and destruction of land-tenure systems caused by cash crops, over-use, overfishing	central government vs. Diola population (and armed forces)	anti-regime war, violent clashes between ethnic groups, terrorist assaults, massacres
A.II.19.	Thailand: northeast	large-scale Eucalyptus plantations destroying traditional agriculture	central government & private enterprises vs. rural populations & Karen immigrants	no violent conflict, but social protest, burning of plantations, forced displacement

A	Intrastate A.III. regionalist migration/displacement conflicts

A.III.20.	Algeria	soil erosion, water and land scarcity, subsistence crisis, urbanization (bidonvilles)	government vs. FIS and marginalized rural and urban poor	violent conflict, coup d'état against FIS, politico-military repression, social unrest and civil strife
A.III.21.	Brazil: Tocantins River Basin	impoverishment of rural space of living caused by Tucurui hydroelectric plant	local ribeirinhos vs. immigrants, rural exodus	no violent conflict, but 20 years of protracted conflict with violent social and ethnopolitical clashes
A. III.22.	China: Northern Henan Province	long-distance migration ("environmental refugees"), high population pressure, loss of arable land by flood/drought	local rural producers vs. immigrants and displaced people	no violent conflict, but high risk of exacerbating ancient regional, provincial, ethnic animosities
A.III.23.	Namibia: Ovamboland	high population pressure and crisis of subsistence, regional overuse of fertile and marginal land	migrants (small-scale peasants) vs. urban population (axis Oshakati-Ongwediva-Ondangwa)	no violent conflict, but growing social tensions along the (semi-)urban axis

B	Intrastate with transboundary dimension B.IV. transboundary migration conflicts

B.IV.24.	Algeria: Southern Algeria vs. Mali and Niger	erratic rainfall, unusual drought (desertification), local overuse of land and water resources	central government vs. Tuareg immigrants to Wilayat Tamanrasset and Illizi	violent conflict: low intensity, skirmishes, criminal violence

B	Intrastate with transboundary dimension B.V. demographically caused conflicts

B.V.25.	Bangladesh vs. India: Assam Province	population pressure, subsistence crisis, “environmental refugees” and land degradation	Assam government and local people vs. Bengali immigrants	violent conflict, low-intensity, social and ethnopolitical clashes
B.V.26.	Bangladesh: Chittagong Hill Tract	population pressure, competition over land, subsistence crisis, displacement causing land degradation and deforestation	central government vs. “Shanti Bahini” (Chakmas)	guerilla war, counter-insurgency,ethnic clashes
B.V.27.	Indonesia: Java	population pressure on land: degradation, subsistence crisis, displacement (Transmigrasi)	central government vs. indigenous people (Irian Jaya)	no violent conflict, but growing social and political (regionalist) tensions

C	International C.VI. water/river basins conflict

C. VI.28.	Bangladesh vs. India: Ganges River Basin	Farakka barrage on the Ganges River: up-downstream conflict, flood vs. drought	Bengali government vs. Indian government	no violent conflict, asymmetry between big and small player, political tensions
C. VI.29.	Central Asia: Aral Sea Basin	shrinking Aral because of irrigation for cotton, soil erosion (salinization) and freshwater scarcity	Uzbek government vs. Kazachian & Turkmenian government & vs. local population	no violent conflict, threat of secession of Karakalpakia from Uzbekistan

C. VI.30.	Central Asia: Fergana Valley	shared irrigation systems: withdrawal from tributaries to Fergana Valley	Uzbek vs. Tajik population	violent conflict: ethnic clashes, high tension
C. VI.31.	China vs. Thailand /Laos Cambo-dia/Viet-nam: Mekong Basin	planned cascade of 15 dams by the upper riparian China, up-downstream conflict	Chinese government vs. governments of the lower riparians	no violent conflict, but growing inter-national tensions, future risk of escalation
C. VI.32.	Hungary vs. Slovakia: Danube Basin	Gabcikovo hydro-power station and diversion of Danube	Hungarian government vs. government of Slovakia & Hungarian minority	no violent conflict, but mili-tary threats and ethnopolitical tensions, settle-ment by the Int. Court of Justice
C. VI.33.	Kyrgyz-stan vs. Uzbeki-stan	upstream-down-stream conflict: hydrological system Naryn and Toktogul	Uzbek vs. Kyrgyz population in the Fergana Valley	violent conflict: violent ethnic clashes, high tensions
C. VI.34.	Lake Chad Basin	shrinking Lake Chad, persistent drought, high eva-poration, subsi-stence crisis, immigrants from Nigeria	Nigerian government vs. Cameroonian & Chadian government	no violent conflict, but tensions: water regime is disputed (there is a Basin Com-mission)
C. VI.35.	Mexico vs. USA: Colorado and Rio Grande Basin	upstream-down-stream conflict: pollution (fertili-zers) and overuse of water resource (withdrawal and salinization)	Mexican government vs. US government, regional actors (twin cities and rural producers)	no violent conflict: bilateral negotiations, bargaining, water regimes

C. VI.36.	Middle East: Euphra-tes-Tigris Basin	upstream-down-stream conflict caused by a huge dam project in Anatolia (GAP)	Turkish government vs. Syrian, Iranian and Iraki governments	no violent conflict, but mili-tary threats, periodically high international tensions
C. VI.37.	Middle East: Jordan Basin, Gaza Strip	absolute distribu-tion conflict: water scarcity (irriga-tion) and drought threatening small-scale farming	Israeli govern-ment vs. Syrian and Jordanese governments and Palestinians (Gaza)	no violent conflict, but part of Middle East wars
C. VI.38.	North-east Africa/ wider Horn: Nile Basin	upstream-down-stream conflict, absolute distribu-tion conflict in a drought-prone area depending on agriculture	governments of Zaire, Rwanda, Uganda, Kenya, Sudan, Ethiopia, and Egypt involved	no violent conflict, but high regional tensions as part of major historical conflicts in the wider Horn
C. VI.39.	Senegal vs. Mauri-tania: Senegal River Valley	dam project (OMVS) against persistent drought, desertification, expulsion of black farmers by Mauritania	Mauritanian governm. (Moors as white elite) vs. Senegalese government & black small-scale farmers	violent conflict: pogroms in 1989, transboundary fights between troops on both sides

C	International C.VII. global environmental conflicts

C. VII.40.	French Polyne-sia	environmental pollution and impoverishment of space of living caused by nuclear tests	French government vs. liberation movement and indigenous population	no violent conflict, but counter-insur-gency manifesta-tions, French navy vs. INGO (Greenpeace)

Table 12:
Three general patterns of environmental conflicts.
Basis: 40 case studies compiled in table 11

<table>
<tr><td>Type</td><td colspan="2">marginalization of poverty in marginal ecoregions</td><td colspan="2">creating (inter-) national "sacrifice areas" *mentioned twice</td><td colspan="2">degradation of international river basins (up-/downstream)</td></tr>
<tr><td>A I</td><td colspan="2">Ghana; Niger; Rwanda; Sudan (4, 5); Tajikistan</td><td colspan="2"></td><td colspan="2"></td></tr>
<tr><td>A II</td><td colspan="2">Brazil (9, 10, 11); Chile; Philippines; Senegal; Thailand</td><td colspan="2">Botswana; Bougainville; India; Nigeria; (14, 15); Philippines*; PNG</td><td colspan="2"></td></tr>
<tr><td>A III</td><td colspan="2">Algeria; China; Namibia</td><td colspan="2">Brazil (21)</td><td colspan="2"></td></tr>
<tr><td>B IV</td><td colspan="2">Southern Algeria</td><td colspan="2"></td><td colspan="2"></td></tr>
<tr><td>B V</td><td colspan="2">Bangladesh (25, 26) Indonesia</td><td colspan="2"></td><td colspan="2"></td></tr>
<tr><td>C VI</td><td colspan="2">Fergana Valley (Uzbek vs. Tajik pop.); Lake Chad Basin; Senegal Valley</td><td colspan="2">Aral Sea Basin</td><td colspan="2">Ganges Basin; Danube Basin; Mekong Basin; Colorado and Rio Grande Basin, Euphrates-Tigris Basin, Jordan Basin; Nile Basin, Senegal Valley*</td></tr>
<tr><td>C VII</td><td colspan="2"></td><td colspan="2">French Polynesia</td><td colspan="2"></td></tr>
<tr><td></td><td>↓</td><td>↓</td><td>↓</td><td>↓</td><td>↓</td><td>↓</td></tr>
<tr><td>T</td><td>security dilemma</td><td>development dilemma</td><td>security dilemma</td><td>development dilemma</td><td>security dilemma</td><td>development dilemma</td></tr>
<tr><td>A I</td><td></td><td>Ghana
Niger
Rwanda
Sudan
(4,5)
Tajikistan</td><td></td><td></td><td></td><td></td></tr>
</table>

A II		Brazil (9,10, 11) Chile Philip- pines Senegal Thailand		Botswana Bougain- ville, India; Nigeria (14,15) Philip- pines* PNG		
A III		Algeria China Namibia		Brazil (21)		
B IV	Southern Algeria	Southern Algeria*				
B V	Bangla- desh (25)	Bangla- desh (26) Indonesia				
C VI	Lake Chad Senegal Valley	Fergana Valley Senegal Valley*		Aral Sea Basin	Ganges Danube Mekong Colorado/Rio Grande Euphrates-Tigris Jordan Nile Senegal	
C VII			French Polyne- sia			

Table 13:
Single countries with environmental conflicts; correlated with HDI, PFI, and RAI

type & country	HDI[1)]	PF I [2)]	labor force: agr.	unemp % of total [3)]	resources[4)] avail. p. ca ha	CEU	RAI[5)]	conf. intens. no viol.	viol. conf.	war
A.I.1. Ghana	0.467	70	59	-	0.26	96	24.9		X	
A.I.2. Niger	0.204	72	90	-	0.42	38	15.9			X
A.I.3. Rwan	0.332	5	92	-	0.15	27	4.0			X
A.I.4. Sudan	0.359	4	69	-	0.49	65	31.8			X
A.I.5. Sudan				-						X
A.I.6. Tajiki	0.616	5	41	-	0.15	634	95.1			X
A.II.7. Bots	0.741	74	46	-	0.30	388	116.4	X		
A.II.8. Boug	0.504	65	79	-	0.10	238	23.8			X
A. II.9. Braz	0.796	73	23	6.2	0.31	666	206.5	X		
A.II.10. Bra				6.2					X	
A.II.11. Bra				6.2				X		
A.II.12. Chil	0.882	73	19	4.7	0.31	911	282.4		X	
A.II.13. Indi	0.435	63	64	-	0.19	242	45.9	X		
A.II.14. Nig	0.400	5	43	-	0.31	141	43.7		X	
A.II.15. Nig				-				X		
A.II.16. PN	0.504	65	79	-	0.10	238	23.8	X		
A.II.17. Phil	0.665	60	46	8.4	0.14	328	45.9	X		
A.II.18. Sen	0.331	39	77	-	0.30	115	34.5			X
A.II.19. Tha	0.832	61	64	1.5	0.36	678	244.8	X		
A.II.20. Alg	0.746	15	26	24	0.29	955	276.9		X	
A.II.21. Bra	0.796	73		6.2				X		
A.II.22. Chi	0.609	8	72	2.8	0.08	623	49.8	X		
A.II.23. Na	0.573	63	49	-	0.45	-	-	X		
B.IV.24. Al	0.746	15		24					X	
B.V.25. Ban	0.365	59	65	-	0.08	59	4.7		X	
B.V.26. Ban				-						X
B.V.27. Ind	0.641	44	55	-	0.16	321	39.3	X		

in comparison to:

Canada	0.951	84	5	9.5	1.58	7821	12357	X		
Germany	0.920	87	4	10.4	0.15	4170	625.5	X		
Switzerl.	0.926	87	6	4.2	0.07	3491	244.3	X		

Table 14:
River basin riparians with environmental conflicts; correlated with HDI, PFI, and RAI

type & country	HDI 1)	PFI 2)	annual water resources available 4)			annual withdr. Water resourc. 4)			conflict intensity		
			total km^3 (000)	ext. km^3	m^{3}pc	p.cap. m^3	%of total	%of agr	no vio	vio conf.	war
C.VI.28. *Gan*											
Banglad.	0.365	59	2357	1000	19.6	220	1	96	X		
India	0.435	63	2058	235	2.2	612	18	93			
C.VI.29. *Aral*											
Kazach.	0.740	40	169	56	9.9	2294	30	79	X		
Turkme.	0.695	5	72	69	17.6	6390	33	91			
Uzbeki.	0.679	11	129	98	5.7	4121	76	84			
C.VI.30. *Ferg*											
Tajikist.	0.616	5	101	48	16.6	2455	13	88		X	
Uzbeki.	0.679	11	129	98	5.7	4121	76	84			
C.VI.31. *Mek*											
China	0.609	8	2800	0	2.3	461	16	87	X		
Thailand	0.832	61	179	69	3.0	602	18	90			
Laos	0.400	26	270	-	55.3	259	0	82			
Cambod.	0.325	40	498	410	48.6	64	0	94			
Vietnam	0.523	23	376	-	5.0	414	8	93			
C.VI.32 *Dan*											
Hungary	0.855	82	120	114	11.9	661	6	36	X		
Slovakia	0.864	72	31	-	5.8	337	6	-			
C.VI.33. *Tok*											
Kyrgyzs.	0.663	19	61	0	13	2729	24	90		X	
Uzbeki.	0.679	11	129	98	5.7	4121	76	84			

C.VI.34.											
Cha											
Nigeria	0.400	5	280	59	2.5	41	1	54	X		
Camero.	0.481	24	268	0	20.3	38	0	35			
Chad	0.291	7	43	28	6.8	34	0	82			
C.VI.35.											
Col											
USA	0.940	84	2478	19	9.4	1870	19	42	X		
Mexico	0.845	47	357	-	3.8	899	22	86			
C.VI.36.											
Eup											
Turkey	0.711	54	193	7	3.1	585	17	57	X		
Syria	0.690	5	54	28	3.7	435	9	83			
Irak	0.599	3	109	66	5.3	4575	43	92			
Iran	0.754	16	118	-	1.7	1362	39	87			
C.VI.37.											
Jord											
Israel	0.908	66	2	0.5	0.4	408	86	79	X		
Syria	0.690	5	54	28	3.7	435	9	83			
Jordan	0.741	40	1.7	0.4	0.3	173	32	65			
C.VI.38.											
Nile											
Zaire	0.371	8	1019	84	23.2	10	0	23	X		
Rwanda	0.332	5	6	-	0.8	102	12	94			
Uganda	0.326	51	66	27	3.1	20	-	60			
Kenya	0.473	44	30	10	1.1	87	7	76			
Ethiopia	0.237	32	110	0	2.0	51	2	86			
Sudan	0.359	4	154	119	5.5	633	12	94			
Egypt	0.611	35	58	55	0.9	956	97	85			
C.VI.39.											
Sen											
Senegal	0.331	39	39	13	4.7	202	3	92		X	
Maurita.	0.353	15	11	11	5.0	923	14	92			
C.VII.40.											
Pac											
France	0.935	87	198	18	3.4	665	19	15	X		
Fr. Poly.	-	-	-	-	-	-	-	-			

Sources for tab. 13 and 14: [1)] Fig.: HDR 96. [2)] Fig.: Freedom House 94. [3)] Fig.: HDR and ILO, YB of Labour Statistics 1996. Fig. on unemployment questionable (inaccurate or manipulated data and methodological problems concerning def. of unemployment (ILO 1996:379). [4)] Fig.: World Resources 96/97. [5)] RAI = available land (ha) per capita times commercial. energy use (CEU).

Abbreviations

ANC	African National Congress (Rep. of South Africa)
APROSOMA	Association for the Social Promotion of the Masses (Rwanda)
BDI	Biodiversity Indicator
CDR	Coalition for the Defense of the Republic (Rwanda)
CIS	Community of Independent States
ECACP	Environmental Change and Acute Conflict Project
ECOMAN	Environmental Conflict Management Project
ENCOP	Environment and Conflicts Project
EU	European Union
EZLN	Ejército Zapatista para la Liberación National (Mexico)
FAO	Food and Agriculture Organization (UN Agency, Rome)
FAR	Armed Forces of Rwanda
FPR	Patriotic Front of Rwanda
FRW	Franc Rwandais
GDP	Gross Domestic Product
GNP	Gross National Product
GRID	Global Resources Information Data System (UNEP)
ha	hectares
HDI	Human Development Index (UNDP)
HDR	Human Development Report (UNDP)
ICA	International Coffee Agreement
ICP	International Comparison Programme
IGAD	Intergovernmental Authority on Development (Horn)
IIASA	International Institute for Applied System Analysis
INGO	International Non-Governmental Organization
IOHE	International Organization for Human-Ecology
IUCN	International Union for the Conservation of Nature
MILEX	Military expenditures
MSM	Social Movement of the Bahutu (Rwanda)
MT	metric tons
MRND	National Republican Movement for Development and Democracy
NCI	Natural Capital Index
NGO	Non-Governmental Organization
OAU	Organization of African Unity
OECD	Organization for Economic Cooperation and Development
OMVS	Organisation pour la mise en valeur du Fleuve Sénégal
PARMEHUTU	Party of the Bahutu Emancipation Movement (Rwanda)

PFI	Political Freedom Index
PPP	Purchasing Power Parity (US dollars)
PRI	Partido Revolucionario Institucional (ruling party, Mexico)
RAI	Resource Availability Index
RNA	Remaining Natural Areas Index
SOIWDP	Southern Okavango Integrated Water Development Project
UN	United Nations
UNAMIR	United Nations Assistance Mission for Rwanda
UNDHA	United Nations Department of Humanitarian Affairs
UNDP	United Nations Development Programme
UNEP	United Nations Environment Programme
UNHCR	United Nations High Commissioner for Refugees
UN/PACD	UN Plan of Action to Combat Desertification
US	United States of America
WRI	World Resources Institute (Washington D.C.)
WWF	World-Wide Found for Nature

List of Tables

List of Figures

CHAPTER 10

GLOSSARY

abakonde	(kinyarwanda) Holders of traditional *ubukonde* land rights in Rwanda *(> ubukonde).*
actor	Organized group of people (collective), which forms part of a political conflict formation (party to a conflict).
akazu	(kinyarwanda) Little house. Means the circle of in-laws (such as the free-trade elite) of the former president of Rwanda, Juvénal Habyarimana (1973 - 1994).
arena	Area that encompasses different conflict parties or actors with contradictory interests and preferences, which mobilize their cultural, social, and economic capital in view of playing an active role in policy design and receiving a significant part of the cake. Cake means the arena's natural capital or the arena's landscape with its available renewable resources *(> actor).*
bega clan	Dominant clan in Rwanda producing the king's wives.
caste	Social structure that transforms the horizontal and unconnected coexistence of ethnically segregated groups into a vertical social system of super- and subordination.
catalyst	Any factor that controls the rate or intensity and the duration of a conflict, once initiated. A catalyst might serve to lengthen a conflict if it stabilizes opportunities and preferences of violence in a given conflict. It might cause a conflict to become extremely violent. Ethical deterioration in a conflict can itself be a catalyst inducing more violence. Together with channeling, a catalyst can also transform reasons. Ethnic groups, for instance, which may have had an economic reason to fight each other may - as ethical deterioration goes on -

perceive differing ethnicity as a reason rather than the resources they fight for *(> channel).*

cause — A meaningful connection between independent (*explanans*) and dependent variable (*explanandum*) in terms of cause, in the sense of primary source and effect.

channel — A line of political, social, economic, or national cleavage. To cite a channel is to explain the social, political, economic, and/or cultural structures that cause individuals to fall into the groups they do. Channels are designed to form the group identity. They figure primarily in sub-intentional explanations (why actors have certain beliefs and desires) and supra-intentional explanation (why individual actions have certain collective or cumulative effects), whereas triggers and targets figure in explanation of intentional action *(> trigger; target).*

chief — Authority in the Rwandan kingdom(s). In the 19th century the king appointed land and pasture chiefs for each district. Several districts constituted a province headed by a provincial chief who appointed the hill chiefs. The latter collected prestations on one or two hills, forwarded part of them to the land and pasture chiefs who then forwarded the rest of the goods to the royal court. Besides the chiefs dealing with land prestations there was the social army chief *(umuheto)* collecting animals and luxury items on the basis of groups the *umuheto* was responsible for.

clan — A group of lineages in Rwanda (ubwoko) based on putative agnatic descent from an eponymous ancestor. They constitute a social category and not a descent group members can trace back. Clans have no leader, no political role, and no functions apart from providing social identity. Each clan - depending on the author there exist 10 to 18 clans in Rwanda - includes members of all three ethnic groups.

dar — Arabic expression meaning homeland; Darfur, for instance, is the homeland of the Fur in Western Sudan. A *dar* is the territory of ethnic groups or tribes they traditionally claimed to be their own. Each *dar* encompasses the institutional framework of local self-government, which may be distinct from the neighboring groups.

degradation — A technical term in order to describe introduction of tracer gases into the atmosphere, conversion of energy into heat (according to the law of entropy), and changes in soils through washing away, drying up, overgrazing, overuse, deforestation, etc. Degradation is exclusively used here as an indicator of the degree of environmental transformation.

desertification — A specific type of soil erosion indicating a process of sustained decline of biological productivity of arid and semi-arid land. Desertification has been defined in the UN Plan of Action to Combat Desertification (UN/PACD) in 1977 as «land degradation in arid, semi-arid and dry sub-humid areas resulting mainly from adverse human impacts«.

domination — Concept stemming from the critical theory of the early Frankfurt school indicating asymmetric relationships in two interrelated sociocultural domains. First, it addresses the exploitation and deterioration of non-human nature by human beings. Second, it deals with discriminatory human-human aspects of society-nature relationships.

ecoregion — Originally used in biogeography to explore the reasons for the spatial and temporal distribution of organisms and biological systems. In political geography and conflict analysis it is used as a functional unit to highlight geographically induced specifics in society-nature relationships in terms of production patterns, settlement, culture, etc.

ejido — Land-tenure system to provide small producers with a plot of land, introduced in 1917 as a consequence of the Mexican revolution.

environmental conflicts: Connotes environmentally caused violent conflict or war. Environmental conflicts manifest themselves as political, social, economic, ethnic, religious, or territorial conflicts, or conflicts over resources or national interests, or any other type of conflict. They are traditional conflicts induced by environmental degradation. Environmental conflicts are characterized by the principal importance of degradation in one or more of the following fields: overuse of renewable resources; overstrain of the environment's sink capacity (pollution); impoverishment of the space of living. Violence occurs if and when actor (a) discriminates against actor (b) or several actors (b, c,...) in terms of access to renewables the actors request or depend on. Environmental conflicts are rebellions against discrimination within the context of transforming society-nature relationship *(> environmental discrimination; renewables; society-nature relationships; transformation; violent conflict/war).*

environmental discrimination: Occurs when distinct actors - based on their international position and/or their social, ethnic, linguistic, religious, or regional identity - experience inequality through systematically restricted access to natural capital (productive renewable resources in the framework of the landscape) relative to other actors *(> actor; landscape).*

ethnopolitical conflict: (Violent) political conflict between distinct ethnic groups defined as psychological communities whose members share a distinguishable and continuous collective identity based on cultural features and way of life.

identity group: Consists of a group of people within states that shares common experience and cultural relationships. It organizes itself in order to defend its interests and it may be defined as: ethnic group, indigenous people, religious group, socially defined group, regionally or geographically bound group.

igikingi: (kinyarwanda) (pl. ibikingi) Domain in the formerly freely accessible pasture region, granted by a political authority, comprising pasture and agricultural land (Rwanda).

ingobyu — (kinyarwanda) Land-tenure system: the head of a lineage holds a common land right in an igikingi domain (Rwanda).

interahamwe — (kinyarwanda) Militia loyal to the Habyarimana Government, which - together with *impuzamugambi,* another militia - began the massacres after the presidential plane crash in Kigali in April 1994.

isambu — (kinyarwanda) Land-tenure system: the head of a nuclear family holds an individual land right in an igikingi domain (Rwanda).

jellaba — Class of wealthy Arab commercial traders in Sudan.

kgotla-meeting — Traditional regional parliament, which is led by a local chief. It plays a significant political role in Botswana's democracy concerning issues that affect people on a community level.

landscape — Biophysical system that provides human beings with its life support by its function of hosting the natural resources water, biomass, minerals, and energy.

maquiladora — Industries on the Mexican side of the US/Mexican border that consist of assembly plants. They receive rough or semi-finished products from the US, assemble them at cheaper costs than in the US, and re-export them back to the US market.

national sacrifice area: Peripheral rural arenas from which central governments - often with the help of multinational firms - draw profits through exploitation of local resources (mining) by deteriorating the environment and without transferring income back to the area.

nyiginya — (kinyarwanda) Historically predominant Batutsi dynasty. Lineage that provided the kings (umwami) of Central Rwanda. The central kingdom in the 19th and early 20th century unified Rwanda, which forms the basis of the modern republic *(> umwami).*

poor state performance: A lack of state outputs regarding civil and political rights, welfare expenditure, livelihood security, resource management, income, and job creation. The state may not produce good outputs for two different reasons. Firstly, the decisions and actions of the state are correct in terms of publicly stated legitimate goals, but their impact is not strong enough to reach the goals. Secondly, the rulers, although proclaiming that the state enhances the public interest, may pursue ends that are actually in its own interest.

reason: A combination of actions that are perceived by an actor as »historical problems« (traumata, history of oppression, injustice, former wars, etc.), which influence his preferences in a way he thinks justify the resort to violence historically. The reason could be a strong factor in mechanisms leading to the outbreak of violence. Reasons play the mighty background role, which shape normative behavior more than rational behavior.

resource availability index: Index composed of available fertile land per capita and the commercial energy use per capita in order to determine an actor's dependence on the natural resources of the landscape *(> actor; landscape).*

society-nature relationship: Used in the context of the anthropogenic transformation of the environment through human action in terms of industrial and agricultural activities and their impact on the landscape *(> landscape; transformation).*

syndrome: A running together; in medicine, it constitutes a number of symptoms occurring together and characterizing a specific disease. In the context of the study here a syndrome means a running together of socioecological key factors in a certain politico-geographical arena *(> arena).*

structural heterogeneity: Defined as simultaneity of interrelated but distinct modes of production being separated by completely different productivity rates. The internal disparities in Third World economies are seen as induced by external forces depending on domination of the world market and on interests of highly productive »centers« creating subcenters or modernized segments in developing countries. Those internal sectors not forming part of the dynamic economic modernization will either be integrated and subordinated to the interests of the modern sector or, if not, will be socioeconomically marginalized.

target — An actor's objective, aim, or goal. The target is what the conflict is about, at least in the eyes of the parties to the conflict. If we can define the target(s), we can explain the purpose of the conflict and can see why an actor prefers resort to violence.

transformation — Focused here on society-nature relationship. To transform is to effect fundamental change in the appearance of nature by the ability of human beings to change landscapes or physical environments at local, regional, and global scales. Chiefly during the past 300 years the scales, rates, and kinds of environmental change have been fundamentally altered as humanity has passed through an era of rapid population growth and the development of a fossil-fuel-based industrial society.
Transformation is used to indicate both the domination of non-human nature by human beings and the human-human aspects of society-nature relationships. Concerning human-nature relationship, domination can be conceptualized with the *degradation* of a certain landscape's renewable resources. Concerning the human-human relationship the concept of discrimination against actors in gaining access to productive natural capital serves to analyze the sociopolitical impacts of domination of nature *(> degradation; landscape; environmental conflict, society-nature relationship).*

transhumance — Seasonal shift of herds between major climatic and ecological zones. Pastoralists move from permanent lowland winter habitations to camps in summer mountain pastures or away from mountain villages to more mild winter coastal pastures.

trigger — Event which causes an action or state of affairs to become the most favored alternative in someone's feasibility set. In terms of violent conflict, a trigger causes an actor who previously preferred nonviolent solutions to a problem to favor violent action instead. A trigger must always be seen near the outbreak of a violent conflict or war under consideration. It is part of the cause, whereas reasons are, by contrast, the causes actors fight for. Triggers are the outcome of decisions that led to violence - even though the latter is not necessarily intended *(> cause).*

ubuhake

(kinyarwanda) Form of traditional clientship in Rwanda, often related to cattle 'leasing'. The system consists of a feudal contract between families aimed at strengthening the power of the leading family.

ubukonde

(kinyarwanda) Land held through right of occupation, not granted by a chief.

ubwoko

(kinyarwanda) Clan *(> clan).*

umuheto

(kinyarwanda) Specific form of clientship encompassing kinship groups, which contribute prestations to a particular chief.

umutware

(kinyarwanda) Chief *(> chief).*

umwami

(kinyarwanda) King in early Rwanda heading a group of clans with supra-clan authority but without political function. Following the Bahutu example the Batutsi in the 16th century established a kingdom with sacred features, the king having been the highest judge, chief of the army, and ruler over the entire land *(> umwami).*

urugo

(kinyarwanda) Nuclear family; forming the basis of a minor lineage *(inzu)*, a major lineage *(umuryango)* and - in some parts of Rwanda - of a subclan *(> ishyanga).*

violent conflict

Organized armed struggles of some duration (more than a one-day upheaval) between two or more collective actors with political goals. Violent conflicts are below the threshold of war but have a strong tendency towards escalating to war *(> environmental conflict; war).*

war

Armed, violent mass conflict following a planned strategy, encompassing the following three constitutive qualitative criteria: (1) it must be a conflict with a minimum of continuity (months rather than days); (2) there have to be central organizations on both sides (this could also be a paramilitary or guerilla force); (3) at least one of the war parties has to be a government with regular or at least government associated troops *(> environmental conflict; violent conflict).*

wilaya — Regional government's administrative hierarchic structure put in place in Sudan in 1982. It is made up of several provinces each subdivided into districts, which, in turn, are further divided into councils, and local councils.

CHAPTER 11

BIBLIOGRAPHY

Abdulai, Napoleon (ed.) (1994): Genocide in Rwanda, Background and Current Situation, London.

Addicks, Gerd (1977): Aussagefähigkeit wirtschaftlicher und sozialer Indikatoren, Dissertation des Wirtschafts- und Sozialwissenschaftlichen Fachbereichs der Universität Augsburg.

Africa Confidential (1994): Vol. 35, London.

Africa Confidential (1995): Vol. 36, London.

Africa Confidential (1996): Vol. 37, London.

African Rights (1994): Rwanda: Death, Despair and Defiance, London.

Afrika Jahrbuch 1991 (1992): Politik, Wirtschaft und Gesellschaft in Afrika südlich der Sahara, Leske + Budrich, Opladen.

Altner, Günther et al. (eds.) (1995): Jahrbuch Ökologie, Beck'sche Verlagsbuchhandlung, München.

Altvater, Elmar (1987): Ökologische und ökonomische Modalitäten von Zeit und Raum, in: Prokla, Vol. 17, No. 2, pp. 35-54.

--- (1992): Der Preis des Wohlstands oder Umweltplünderung und neue Welt(un)ordnung, Verlag Westfälisches Dampfboot, Münster.

American Association for the Advancement of Science (AAAS) (1992): Environmental Dimensions of Security, Proceedings from a AAAS Annual Meeting Symposium, 9 February 1992, Washington D.C.

Amin, Samir (1990): Maldevelopment: Anatomy of a Global Failure, UN University Press, Tokyo, London, New Jersey.

Anonymous (n.y.): Histoire et chronologie du Rwanda.

Ansprenger, Franz (1992): Politische Geschichte Afrikas im 20. Jahrhundert, Beck, München.

--- (1994): Konflikte in Afrika, in: Europa Archiv, Vol. 49, No. 20, pp. 571-578.

Ante, Ulrich (1981): Politische Geographie, Westermann, Braunschweig.

--- (1985): Zur Grundlegung des Gegenstandsbereiches der Politischen Geographie. Über das "Politische" in der Geographie, Stuttgart.

Ardayfio-Schandorf, Elizabeth (1995): Gender and Environmental Management in Africa. Some Theoretical and Methodological Dimensions, in: Swiss Peace Foundation (ed.): War against Women. The Impact of Violence on Gender Relations, Berne, pp. 3-15.

Arizpe, Lourdes, Stone, M., and Major, David (1994): Population and the Environment, Westview Press, Boulder CO.

Arnold, Guy (1991): Wars in the Third World since 1945, Cassell Publishers Limited, London.

Axell, Karin and Wallensteen, Peter (1994): Conflict Resolution and the End of the Cold War, 1989-93, in: Journal of Peace Research, Vol. 31, No. 3, pp. 333-349.

Bächler, Günther (1990): Kollaps und Konflikt. Zur Konfliktdimension globaler ökologischer Zerstörung, in: antimilitarismus information, No. 12, pp. 5-14.

--- (1990): Ökologische Sicherheit und Konflikt. Arbeitspapiere der Schweizerischen Friedensstiftung No. 5, Bern.

--- (1993): Conflict and Cooperation in the Light of Human-Ecological Transformation, ENCOP Occasional Paper No. 9, Swiss Federal Institute of Technology/Swiss Peace Foundation, Zurich, Berne.

--- (1993): Konflikt und Kooperation im Lichte globaler humanökologischer Transformation. ENCOP Occasional Paper No. 5, ETHZ/Schweizerische Friedensstiftung, Zürich, Bern.
--- (ed.) (1994): Umweltflüchtlinge. Das Konfliktpotential von morgen? agenda, Münster.
--- (1995): Desertification and Conflict. The Marginalization of Poverty and of Environmental Conflicts, in: J. Puigdefabregas and T. Mendizabal (eds.): Desertification and Migrations, Geoforma ediciones, Logrono, pp. 185-224.
--- (1995): The Anthropogenic Transformation of the Environment: A Source of War? In: Kurt R. Spillmann and Günther Bächler (eds.): Environmental Crisis: Regional Conflicts and Ways of Cooperation. Proceedings of an International Conference at Centro Stefano Franscini, Ascona/Switzerland, 2-7 October 1994. ENCOP Occasional Paper No. 14, Swiss Federal Institute of Technology/Swiss Peace Foundation, Zurich, Berne, pp. 11-27.
--- (1996): Wie sich umweltverursachte Konflikte friedlich regeln lassen, in: Dieter Senghaas (ed.): Frieden machen, Suhrkamp, Frankfurt/M.
--- (1997): Das Atatürk-Staudammprojekt am Euphrat-Tigris unter besonderer Berücksichtigung der sicherheitspolitischen Relevanz zwischen den Anliegerstaaten Türkei, Syrien und Irak, Bern (Gutachten für den Wissenschaftlichen Beirat Globale Umweltveränderungen (WB GU) der Bundesregierung, Bremerhaven).
--- et al. (1993): Umweltzerstörung: Krieg oder Kooperation? Ökologische Konflikte im internationalen System und Möglichkeiten der friedlichen Bearbeitung, agenda, Münster.
---, Böge, Volker, Klötzli, Stefan, Libiszewski, Stephan und Spillmann, Kurt R. (1996): Kriegsursache Umweltzerstörung. Ökologische Konflikte in der Dritten Welt und Wege ihrer friedlichen Bearbeitung (Bd. I), Verlag Rüegger, Zürich.
--- und Spillmann, Kurt R. (eds.) (1996): Kriegsursache Umweltzerstörung. Environmental Degradation as a Cause of War. Regional- und Länderstudien von Projektmitarbeitern. Regional and Country Studies of Research Fellows (Vol. II), Verlag Rüegger, Zürich.
--- (1996): Kriegsursache Umweltzerstörung. Environmental Degradation as a Cause of War. Länderstudien von externen Experten. Country Studies of External Experts (Vol. III), Verlag Rüegger, Zürich.
--- , Bellwald, Stefan, and Suliman, Mohamed (1996): Environmental Conflict Management. Approaches and Implementation in the Horn of Africa, Bern, Zürich, London.
Baier, S. (1976): Economic History and Development: Drought and the Sahelian Economics of Niger, in: African Economic History, Spring.
Bailey, Robert (1983): Delineation of Ecosystem Regions, in: Environmental Management, Vol. 7, No. 4, pp. 365-373.
--- and Hogg, Howard C. (1986): A World Ecoregions Map for Resource Reporting, in: Environmental Conservation, Vol. 13, No. 3, pp. 195-202.
Bart, François (1993): Montagnes d'Afrique. Terres paysanes. Le cas du Rwanda, Centre d'Etudes de Géographie Tropicale, Presses Universitaires, Bordeaux.
Bastian, Till (1991): Naturzerstörung: Die Quelle der künftigen Kriege, IPPNW, Wissenschaftliche Reihe No. 1, Heidesheim.
--- (1992): Städte, Flüchtlinge und Mangel. Wachstumsfixierte Welt-Wirtschaft und künftige Welt-Kriege, IPPNW, Wissenschaftliche Reihe No. 3, Heidesheim.
Bauerochse, Lothar (1993): Bauchspeck für den Frieden. Wie im Süden ein Streit mit traditionellen Mitteln beigelegt wurde, in: Der Überblick, No. 3, pp. 63-64.
Baxter, P.T.W. (1970): Stock Management and the Diffusion of Property Rights among the Boran, in: Institute of Ethiopian Studies (ed.): Proceedings of the Third International Conference of Ethiopian Studies, Addis Ababa, pp. 116-127.
Beaumont, Peter (1993): Drylands. Environmental Management and Development, Routledge, London.
Becker, Peter (1993): Rwanda, in: Dieter Nohlen and Franz Nuscheler (eds): Handbuch der Dritten Welt (Bd. 5). Ostafrika und Südafrika, Bonn, pp. 114-133.

Beckman, Bjorn (1984): Bakolori: Peasants versus State and Industry in Nigeria, in: Edward Goldsmith and Nicholas Hildyard (eds.): The Social and Environmental Effects of Large Dams, Cornwall: Quintrell & Co., Wadebridge, pp. 140-155.

Bencherifa, Abdellatif (1990): Demography and Cultural Ecology of the Atlas Mountains of Morocco: Some New Hypotheses, in: Bruno Messerli and Hans Hurni (eds.): African Mountains and Highlands. Problems and Perspectives, Walsworth Press, Marceline MS, pp. 369-378.

Bendrath, Ralf (1994): Die Diskussion um die "Überbevölkerung". Arbeitspapiere der Schweizerischen Friedensstiftung No. 19, Bern.

Bennet, Robert and Estall, Robert (eds.) (1991): Global Change and Challenge: Geography for the 1990's, Routledge, London, New York.

Bennett, Olivia (ed.) (1991): Greenwar: Environment and Conflict in the Sahel, Panos Publications Ltd., Budapest, London, Paris, Washington.

Bergesen, Helge O. and Parmann, Georg (eds.) (1993): Green Globe Yearbook 1993, Oxford University Press, New York.

--- (eds.) (1994): Green Globe Yearbook of International Co-operation on Environment and Development 1994, Oxford University Press, New York.

Bertram, Christoph (ed.) (1982): Third-World Conflict and International Security, Macmillan Press, London.

Beschorner, Natasha (1992): Water and Instability in the Middle East, Adelphi Paper No. 273, IISS, London.

Bézy, Fernand (1990): Rwanda 1962-1989. Bilan socio-économique d'un régime, Louvain-la-Neuve.

Billing, Peter (1992): Eskalation und Deeskalation internationaler Konflikte. Ein Konfliktmodell auf der Grundlage der empirischen Auswertung von 288 internationalen Konflikten seit 1945, Peter Lang, Frankfurt/M, Bern, New York, Paris.

Bilsborrow, Richard E. (1994): Rural Poverty, Migration and the Environment in Developing Countries. Three Case Studies, Development Economics, The World Bank, Washington D.C.

Blake, Gerald (1993): Transfrontier Collaboration: A Worldwide Survey, in: Arthur H. Westing (ed.): Transfrontier Reserves for Peace and Nature: A Contribution to Human Security, Nairobi, pp. 35-58.

Bliss, Frank (1989): Leben mit der Dürre. Alltag und Entwicklungsprobleme im Sahel am Beispiel der Provinz Darfur, Sudan, Bad Honnef.

Boeckh, Andreas (1992): Entwicklungstheorien: Eine Rückschau, in: Dieter Nohlen und Franz Nuscheler (eds.): Handbuch der Dritten Welt (Bd. 1). Grundprobleme, Theorien, Strategien, Bonn, pp. 110-130.

Böge, Volker (1992): Bougainville: A "Classical" Environmental Conflict? ENCOP Occasional Paper No. 3, Swiss Federal Institute of Technology/Swiss Peace Foundation, Zurich, Berne.

--- (1992): Proposal for an Analytical Framework to Grasp "Environmental Conflict". ENCOP Occasional Paper No. 1, Swiss Peace Foundation/Swiss Federal Institute of Technology, Zurich, Berne.

--- (1992): "Umweltflüchtlinge" als Ursache von Konflikten - das Konzept "ökologischer Sicherheit" als mögliche Antwort, in: Flucht und Migration in Europa. Arbeitspapiere des Instituts für Internationale Politik No. 20, Institut für Internationale Politik, Berlin, pp. 51-63.

--- (1993): Das Sardar-Sarovar-Projekt an der Narmada in Indien - Gegenstand ökologischen Konflikts. ENCOP Occasional Paper No. 8, Swiss Federal Institute of Technology/Swiss Peace Foundation, Zurich, Berne.

--- (1994): Umweltflüchtlinge als Ursache gewaltsamen Konfliktaustrags (Environmental Refugees as a Source of Violent Conflict), in: Günther Bächler (ed.): Umweltflüchtlinge. Das Konfliktpotential von morgen? agenda, Münster, pp. 164-169.

Boulding, K.E. (1978): Ecodynamics. A New Theory of Societal Evolution, Beverly Hills, London.
Boutros-Ghali, Boutros (1992): Agenda für den Frieden. Bericht des UN-Generalsekretärs an den Sicherheitsrat, in: Europa Archiv, No. 24, pp. D657-D673.
--- (1994): An Agenda for Development: Report of the Secretary-General A/48/49/665, 6 May 1994, in: An Agenda for Development - With Related Documents, United Nations Publication, New York.
Boyle, Alan (1994): The Convention on Biological Diversity, in: Luigi Campiglio et al. (ed.): The Environment after Rio. International Law and Economics, London, pp. 111-127.
Braeckman, Colette (1994): Rwanda. Histoire d'un génocide, Librairie Arthème Fayard, Paris.
Brandt Commission (1980): North-South: A Program for Survival. Report of the Independent Commission on International Develoment Issues, Pan Books, New York.
Braudel, Fernand (1995): The Mediterranean World in the Age of Philip II, Vol. 1, University of California Press, Berkeley, Los Angeles, London. First published in France under the title: La Méditerranée et le Monde Méditerranéen à l'Epoque de Philippe II, 3 Volumes, Paris 1949.
Brecke, Dieter (1997): Using Pattern Recognition to Identify Harbinger Configurations of Early Warning Indicators. Paper Prepared for Presentation at the ECPR Joint Sessions of Workshops, Berne, February 27 - March 4.
Breman, H. and De Wit, C.T. (1983): Rangeland productivity and exploitation in the Sahel, in: Science, No. 221, pp. 1341-1347.
Brimblecombe, Peter and Pfister, Christian (eds.) (1990): The Silent Countdown. Essays in European Environmental History, Springer, Berlin, Heidelberg.
Brock, Lothar (1991): Peace through Parks: The Environment on the Peace Research Agenda, in: Journal of Peace Research, Vol. 28, No. 4, pp. 407-423.
--- (1992): Die Dritte Welt in ihrem fünften Jahrzehnt, in: Aus Politik und Zeitgeschichte, No. 4, pp. 13-23.
--- (1992): Internationale Umweltpolitik: Zwischen Ahnung und Aufbruch, in: epd-Entwicklungspolitik, No. 6/7, pp. g-k.
--- (1992): Nord-Süd-Kontroversen in der internationalen Umweltpolitik: Von der taktischen Verknüpfung zur Integration von Umwelt und Entwicklung. HSFK Report No. 7, Frankfurt/M.
--- (1994): Ökologische Sicherheit. Zur Problematik einer naheliegenden Verknüpfung, in: Wolfgang Hein (ed.): Umbruch in der Weltgesellschaft, Deutsches Überseeinstitut, Hamburg, pp. 443-458.
--- und Hessler, Stephan (1992): Globaler Umweltschutz oder Öko-Imperialismus? Die ökologische Notwendigkeit neuer Nord-Süd-Beziehungen. Das Beispiel Amazonia, in: Berthold Meyer und Christian Wellmann (eds.): Umweltzerstörung: Kriegsfolge und Kriegsursache, Suhrkamp, Frankfurt/M, pp. 194-228.
Brogan, Patrick (1990): Die Unruhe der Welt: Die Enzyklopädie der Krisen und Konflikte unserer Zeit, Paul Zsolnay, Wien, Darmstadt.
Brooks, David B. and Lonergan, Stephen C. (1994): Watershed: The Role of Fresh Water in the Israeli-Palestinian Conflict, International Development Research Centre, Ottawa.
Brown, Janet W. (ed.) (1990): In the U.S. Interest: Resources, Growth, and Security in the Developing World, Westview Press, Boulder CO.
Brown, Lester R. (1977): Redefining National Security. Worldwatch Paper No. 14, Worldwatch Institute, Washington D.C.
--- (1988): The Changing World Food Prospect: The Nineties and Beyond. Worldwatch Paper No. 85, Worldwatch Institute, Washington D.C.
--- (1994): Who Will Feed China? In: World Watch, Vol. 7, No. 5, pp. 10-19.
--- and Kane, Hal (1994): Full House: Reassessing the Earth's Population Carrying Capacity, W.W. Norton & Co., New York, London.

--- and Wolf, Edward C. (1984): Soil Erosion: Quiet Crisis in the World Economy. Worldwatch Paper No. 60, Worldwatch Institute, Washington D.C.

--- (1985): Reversing Africa's Decline. Worldwatch Paper No. 65, Worldwatch Institute, Washington D.C.

---, Flavin, Christopher, and Postel, Sandra (1992): Zur Rettung des Planeten Erde. Strategien für eine ökologisch nachhaltige Weltwirtschaft, Fischer, Frankfurt/M.

--- et al. (1992): Zur Lage der Welt 1992. Worldwatch Institute Report, Fischer Taschenbuch, Frankfurt/M.

--- et al. (1993): State of the World 1993. A Worldwatch Institute Report on Progress toward a Sustainable Society, W.W. Norton & Co., New York, London.

--- (1994): State of the World 1994. A Worldwatch Institute Report on Progress toward a Sustainable Society, W.W. Norton & Co., New York, London.

--- (1995): State of the World 1995: A Worldwatch Institute Report on Progress toward a Sustainable Society, W.W. Norton & Co., New York.

--- (1996): State of the World 1996: A Worldwatch Institute Report on Progress toward a Sustainable Society, W.W. Norton & Co., New York.

Brown, Michael E. (ed.) (1996): The International Dimensions of Internal Conflict, MIT Press, Cambridge MA.

Brown, Neville (1989): Climate, Ecology and International Security, in: Survival, Vol. 31, No. 6, pp. 519-532.

--- (1994): Climate Change: A Threat to Peace, Vol. Conflict Studies 272, Research Institute for the Study of Conflict and Terrorism, London.

Brown, Sheryl J. and Schraub, Kimber M. (eds.) (1992): Resolving Third World Conflict: Challenges for a New Era, United States Institute of Peace Press, Washington D.C.

Brown Weiss, Edith (1989): In Fairness to Future Generations: International Law, Common Patrimony, and Intergenerational Equity, The United Nations University, Tokyo.

Bruckmeier, Karl (1994): Strategien globaler Umweltpolitik. "Umwelt und Entwicklung" in den Nord-Süd-Beziehungen, Westfälisches Dampfboot, Münster.

Bruene, Stefan (1994): Die französische Afrika-Politik: Ende einer Ära, in: Europa Archiv, Vol. 49, No. 20, pp. 587-594.

Bruhns, Barbro-Isabel und Kappel, Robert (eds.) (1992): Ökologische Zerstörung in Afrika und alternative Strategien. Bremer Afrika-Studien No. 1, Lit, Münster.

Bruno, Henri (1913): Le régime des eaux en droit musulman, Arthur Rousseau, Paris.

Brunnée, Jutta and Toope, Stephen J. (1995): Environmental Security and Freshwater Resources: A Case for International Ecosystem Law, in: Günther Handl (ed.): Yearbook of International Environmental Law, Vol. 5, 1994, Clarendon Press, Oxford.

Bruyninckx, Hans (1993): Environmental Security: A Conceptual Assessment. Paper Presented at the Second Conference of the European Peace Research Association, Budapest, 12-14 November.

Büttner, Sebastian und Simonis, Udo E. (1994): Wasser - ein globales Umweltproblem. Papiere des Wissenschaftszentrums Berlin, No. FS II, pp. 94-401.

Buschmann, Uwe (1986): Das Bodenrecht in Ruanda, edition herodot/Rader, Aachen.

Bush, Kenneth (1990): Climate Change, Global Security, and International Governance: A Summary of Proceedings of a Conference on Climate Change and Global Security, Ottawa, 11-12 April 1990, Working Paper No. 23, Ottawa.

Butts, Kent Hughes (ed.) (1994): Environmental Security. A DOD Partnership for Peace, Carlisle Barracks: Strategic Studies Institute, United States Army War College.

Byers, Bruce (1991): Ecoregions, State Sovereignity and Conflict, in: Bulletin of Peace Proposals, Vol. 22, No. 1, pp. 65-76.

Calliess, Jörg (ed.) (1994): Treiben Umweltprobleme in Gewaltkonflikte? Ökologische Konflikte im internationalen System und Möglichkeiten ihrer friedlichen Bearbeitung, Loccumer Protokolle No. 21, Evangelische Akademie, Rehburg-Loccum.
--- , Rüsen, Jörn und Striegnitz, Meinfried (eds.) (1989): Mensch und Umwelt in der Geschichte, Centaurus, Pfaffenweiler.
Campiglio, Luigi et al. (eds.) (1994): The Environment after Rio. International Law and Economics, Graham & Trotman/Martinus Nijhoff, London, Dordrecht, Boston.
Canadian Global Change Program Secretariat (1992): Environmental Degradation, Population Displacement and Global Security. An Overview of the Issues, Ottawa.
Caring for the Earth: A Strategy for Sustainable Living (1991), IUCN/UNEP/WWF, Gland, Switzerland.
Carino, Joji (1993): Local Solutions to Global Problems: Perspectives from the Cordillera Peoples' Movement, Berne (manus.).
Carrol, John E. (1988): International Environmental Diplomacy, Cambridge University Press, New York.
Carson, R. (1962): Silent Spring, Boston.
Cart, Henri-Philippe (1995): La coopération suisse au Rwanda ou les limites de l'aide extérieure, in: André Guichaoua (ed.): Les crises politiques au Burundi et au Rwanda (1993-1994), Paris, pp. 465-487.
Carter, L.J. (1977): Soil erosion: The Problem Persists despite the Billions Spent on It, in: Science, No. 196, pp. 409-411.
Choucri, Nazli (1992): Environment and Conflict: New Principles for Environmental Conduct, in: Disarmament, Vol. 15, No. 1, pp. 67-78.
--- (1993): Political Economy of the Global Environment, in: International Political Science Review, Vol. 14, No. 1, pp. 103-116.
Chrétien, Pierre (1992): La crise politique Rwandaise, in: Genève-Afrique, Vol. XXX, No. 2, pp. 121-140.
Clarke, Robin (1991): Water: The International Crisis, Earthscan Publications LTD, London.
Claus, Olaf (1996): "Batawana Say to Dredge a River is to Destroy It". The Case of the Southern Okavango Integrated Water Development Project, in: Bächler/Spillmann (eds.) (Vol. III), pp. 269-284.
Clausewitz, Carl (1976): On War. Ed. & transl. by Michael Howard and Peter Paret, Princetown University Press, Princetown.
Claussen, Eileen (1995): Environment and Security: The Challenge of Integration, in: Environmental Change and Security Project Report No. 1, Spring, pp. 40-43.
Clignet, Remi (1991): Conflict and Culture in Traditional Societies, in: Joseph V. Montville (ed.): Conflict and Peacemaking in Multiethnic Societies, Lexington Books, New York, pp. 65-81.
Cohen, M.N. (1977): The Food Crisis in Prehistory: Overpopulation and the Origins of Agriculture, New Haven.
Commission of Experts Established pursuant to Security Council Resolution 935 (1994), S/1994/1405, 9 December.
Commission on Global Governance (1995): Our Global Neighborhood. The Report of CGG, Oxford University Press, New York.
Cooley, John K. (1984): The War over Water, in: Foreign Policy, No. 54, pp. 3-26.
Coquery-Vidrovitch, Catherine (1992): Afrique Noire. Permanences et ruptures, L'Harmattan, Paris.
Corson, Walter H. (ed.) (1990): The Global Ecology Handbook, Beacon Press, Boston.
Cronon, William (1983): Changes in the Land. Indians, Colonists, and the Ecology of New England, Hill and Wang, New York.
Cummings, Barbara J. (1990): Dam the Rivers, Damn the People, Earthscan, London.

Cutter, Susan L. (1994): Exploiting, Conserving, and Preserving Natural Resources, in: Demko, George J. and Wood, William B. (eds.): Reordering the World. Geopolitical Perspectives on the 21st Century, Boulder CO, pp. 123-140.

Czekanowski, Jan (1917): Forschungen im Nil-Kongo-Zwischengebiet, Bd I, Ethnographie: Zwischenseengebiet; Mpororo, Rwanda, Leipzig (Wissenschaftliche Ergebnisse der Deutschen Zentral-Afrika-Expedition, Bd. 6, Erster Teil).

Czempiel, Ernst-Otto (1981): Internationale Politik: Ein Konfliktmodell, dtv, Paderborn.

Daase, Christopher (1992): Ökologische Sicherheit: Konzept oder Leerformel? In: Berthold Meyer und Christian Wellmann (eds.): Umweltzerstörung: Kriegsfolge und Kriegsursache, Suhrkamp, Frankfurt/M, pp. 21-52.

Dabelko, Geoffrey D. and Dabelko, David D. (1995): Environmental Security: Issues of Conflict and Redefinition, in: Environmental Change and Security Project Report, No. 1, Spring, pp. 3-13.

Daly, Herman E. (1991): Steady-State Economics, Island Press, Washington D.C.

Danielsson, Bengt (1993): Tests in French Polynesia and Political Protest, Berne (manus.).

Davidson, Basil (1992): The Black Man's Burden. Africa and the Curse of the Nation-State, Random House, New York.

Davidson, B. (1962): L'Afrique avant les Blancs - Découverte du passé oublié de l'Afrique, PUF, Paris.

Davis, Kingsley and Bernstam, Mikhail S. (1991): Resources, Environment, and Population: Present Knowledge, Future Options, in: Population and Development Review (Supplement to Vol. 16).

Day, David (1989): The Environmental Wars. Reports from the Front Lines, St. Martin's Press, New York.

De Lacger, Louis (1961): Le Ruanda, 2nd ed. Kabgayi, Rwanda.

Deleage, J.P. and Hemery, D. (1990): From Ecological History to World Ecology, in: P. Brimblecombe and C. Pfister (eds.): The Silent Countdown, Springer, Berlin et al., pp. 21-36.

Deng, Francis and Zartman, William I. (eds.) (1991): Conflict Resolution in Africa, Brookings, Washington D.C.

Denniston, Derek (1995a): High Priorities: Conserving Mountain Ecoystems and Cultures. Worldwatch Paper No. 123, Worldwatch Institute, Washington D.C.

--- (1995): Sustaining Mountain Peoples and Environments, in: Worldwatch Institute (ed.): State of the World: A Worldwatch Institute Report on Progress toward a Sustainable Society, W.W. Norton & Co., New York, pp. 38-57.

Dessler, David (1994): How to Sort Causes in the Study of Environmental Change and Violent Conflict, in: Nina Græger and Dan Smith (eds.): Environment, Poverty, Conflict, PRIO Report No. 2, Oslo, pp. 91-112.

Deudney, Daniel (1990): The Case Against Linking Environmental Degradation and National Security, in: Millennium: Journal of International Studies, Vol. 19, No. 3, pp. 461-476.

--- (1991): Environment and Security: Muddled Thinking, in: The Bulletin of the Atomic Scientists, Vol. 47, No. 3, pp. 22-28.

--- and Matthew, Richard (eds.) (1995): Contested Ground: Security and Conflict in the New Environmental Politics, Albany.

De Wall, Alexander (1989): Famine that Kills Darfur, Sudan 1984-1985, Oxford.

Die Herausforderung des Südens. Der Bericht der Südkommission. Über die Eigenverantwortung der Dritten Welt für dauerhafte Entwicklung (1991): Texte der Stiftung Entwicklung und Frieden, Bonn.

Diefenbacher, Hans und Ratsch, Ulrich (1992): Verelendung durch Naturzerstörung. Die politischen Grenzen der Wissenschaft, Fischer Sachbuch.

Diessenbacher, Hartmut (1994): Bürgerkrieg und Völkermord in Rwanda. Ethnischer Klassenkonflikt und Bevölkerungswachstum, in: Aus Politik und Zeitgeschichte, Beilage zur Wochenzeitung Das Parlament, B 31, pp. 14-23.

D'Hertefelt, Marcel (1971): Les clans du Rwanda ancien. Eléments d'ethno-sociologie et d'ethnohistoire, Annales, Musée Royal de l'Afrique Centrale: série in-8o, Sciences Humaines, Tervuren.

--- et al. (eds.) (1962): Les anciens royaumes de la zone interlacustre méridionale. Rwanda, Burundi, Buha, Musée Royal de l'Afrique Centrale, Tervuren.

--- et Coupez, Pierre (1964): La royauté sacrée de l'ancien Rwanda: texte, traduction et commentaire de son rituel, Annales, Musée Royale de l'Afrique Centrale: série in-80, Sciences Humaines. Tervuren.

Direktion für Entwicklungszusammenarbeit und Humanitäre Hilfe (DEH) (ed.) (1993): Aus unterschiedlichen Welten in eine gemeinsame Zukunft. Münchwiler Forum, DEH Informationsdienst, Bern

Dixon, John, Talbot, Lee M., and Le Moigne, Guy J.M. (1990): Dams and the Environment. Considerations in World Bank Projects, World Bank Technical Paper Nr. 110, The World Bank, Washington D.C.

Dokken, Karin (1995): Environment, Security and Regional Integration in West Africa, Oslo.

--- and Græger, Nina (1995): The Concept of Environmental Security - Political Slogan or Analytical Tool? Report Prepared for the Royal Ministry of Foreign Affairs, PRIO, Oslo.

Dregne, H.E. (1983): Desertification of Arid Lands, London.

--- (1992): Erosion and Soil Productivity, in: Asia, Journal of Soil and Water Conservation, Vol. 47, pp. 8-13.

Dunnigan, James F. and Bay, Austin (1991): A Quick & Dirty Guide to War. Briefings on Present and Potential Wars, Quill William Morrow, New York.

Durham, William H. (1979): Scarcity and Survival in Central America: Ecological Origins of the Soccer War, Stanford University Press, Stanford CA.

Durning, Alan B. (1992): Die Armutsfalle. Die Beziehung zwischen Armut und Umwelt; die Elendsspirale umdrehen, World Watch Papers No. 2, Wochenschau, Schwalbach.

Durth, Rainer (1993): Kooperation bei internationalen Oberlauf-Unterlauf-Problemen - ökonomische Aspekte. Diskussionspapier, Dezember 93, Graduiertenkolleg für Integrationsforschung Europakolleg Hamburg.

Earthscan (ed.) (1984): Environment and Conflict. Links between Ecological Decay, Environmental Bankruptcy and Political and Military Instability, Earthscan briefing document No. 40, Earthscan, London.

Easton, David (1965): A Systems Analysis of Political Life, Wiley, New York.

Eder, Klaus (1980): Die Entstehung staatlich organisierter Gesellschaften. Ein Beitrag zu einer Theorie sozialer Evolution, Suhrkamp, Frankfurt/M.

Edit No. 530/1 du 26. 5. 1961 relatif au régime ubukonde dans les territoires de Gisenyi et Ruhengeri, J.O. du 15.1.1992, 3-7-.

Edit du 25. 4. 1992 relatif aux associations sans but lucratif, J.O. du 15. 5. 1992.

Efinger, Manfred, Rittberger, Volker und Zürn, Michael (1988): Internationale Regime in den Ost-West-Beziehungen. Ein Beitrag zur Erforschung der friedlichen Behandlung internationaler Konflikte, Frankfurt/M.

Ehrensperger, Albrecht (1993): Transmigrasi - Staatlich gelenkte Umsiedlungen in Indonesien, ihre Ursachen und Folgen in ökologischer sowie konfliktueller Hinsicht, Bern (manus.).

Ehrlich, Paul R. und Ehrlich, Anne H. (1972): Bevölkerungswachstum und Umweltkrise, Frankfurt/M.

--- (1991): Healing the Planet. Strategies for Resolving the Environmental Crisis, Addison-Wesley Publishing Co., Reading.

--- et al. (1985): The Cold and the Dark: The World after Nuclear War, New York.

Eikenberg, Kathrin (1991): Kriege in Äthiopien, in: Jens Siegelberg (ed.): Die Kriege 1985 bis 1990. Analyse ihrer Ursachen, Lit, Hamburg, pp. 190-211.

El-Hinnawi, Essam (1985): Environmental Refugees, Nairobi (UNEP).

Elsenhans, Hartmut (1984): Rohstoffe und die Auseinandersetzung um Rohstoffe als Element der wirtschaftlichen Entwicklung in historischer Perspektive, in: Harald Müller und Detlef Puhl (eds.): Ressourcenpolitik. Konfliktpotentiale und Kooperationschancen bei der westlichen Rohstoffsicherung, Schriftenreihe der AFK No. 9, Frankfurt/M, pp. 3-39.

--- (1987): Nord-Süd-Beziehungen, W. Kohlhammer, Stuttgart.

Elster, Jon (1987): Subversion der Rationalität, Campus, New York.

--- (1989): Nuts and Bolts for the Social Sciences, Cambridge University Press, Cambridge MA.

Engelman, Robert and Koontz, Stephanie L. (1992): Population and the Environment: Impacts in the Developing World, Population Action International, Washington D.C.

Environmental Aftermath of the Gulf War, the (1992): A Report Prepared for the Committee on Environment and Public Works, Gulf Pollution Task Force by the Environment and Natural Resources Policy Division, American Law Division and the Science Policy Research Division of the Congressional Research Service, at the request of Joseph I. Lieberman and John H. Chafee, U.S. Government Printing Office, Washington D.C.

Environmental Degradation, Population Displacement and Global Security: An Overview of the Issues (1992), Canadian Global Change Program Secretariat, Ottawa.

Evers, H.D. und Schiel, T. (1988): Strategische Gruppen. Vergleichende Studien zu Staat, Bürokratie und Klassenbildung in der Dritten Welt, Reimer, Berlin.

Falkenmark, Malin (1986): Fresh Water as a Factor in Strategic Policy and Action, in: Arthur H. Westing (ed.): Global Resources and International Conflict. Environmental Factors in Strategic Policy and Action, Oxford University Press, Oxford, New York, pp. 85-113.

--- (1990): Global Water Issues Confronting Humanity, in: Journal of Peace Research, Vol. 27, No. 2, pp. 177-190.

--- (1994): Environment-related Conflicts: Taking a Water Cycle Perspective. Paper Presented at the International Conference on "Environmental Crisis - Regional Conflicts and Ways of Cooperation", Ascona/Switzerland, 3-7 October.

--- (1995): Eco-Conflicts - The Water Cycle Perspective, in: Kurt R. Spillmann und Günther Bächler (eds.): Environmental Crisis: Regional Conflicts and Ways of Cooperation. Proceedings of an International Conference at Centro Stefano Franscini, Ascona/Switzerland, 3-7 October, 1994. ENCOP Occasional Paper No. 14, Swiss Federal Institute of Technology/Swiss Peace Foundation, Zurich, Berne.

--- (1995): The Dangerous Spiral: Near-Future Risks for Water-Related Eco-Conflicts, in: International Committee of the Red Cross (ed.): Water and War. Symposium on Water in Armed Conflicts, Montreux, 21-23 November 1994, Geneva, pp. 7-10.

--- et al. (1980): Water and Society. Conflicts in Development, Vol. 2, Oxford.

--- , Lundqvist, Jan, and Widstrand, Carl (1990): Coping with Water Scarcity: Implications of Biomass Strategy for Communities and Policies, in: Water Resources Development, Vol. 6, No. 1, pp. 29-43.

--- and Subrapto, R. (1992): Population-Landscape Interactions in Development: A Water Perspective to Environmental Sustainability, in: Ambio, No. 1, pp. 31-36.

--- and Lindth, Gunnar (1993): Water and Economic Development, in: Peter H. Gleick (ed.): Water in Crisis. A Guide to the World's Fresh Water Resources, Oxford University Press, New York, pp. 80-91.

--- and Rockström, J. (1993): Curbing Rural Exodus from Tropical Drylands, in: Ambio, No. 7, pp. 427-437.

FAO (1991): Food and Agriculture Organization of the United Nations: World Soil Resources. An Explanatory Note on the FAO World Soil Resources Map at 1:25'000'000 scale. World Soil Resources Report No. 66, Rome.

Farès, Zahir (1992): Afrique et Démocratie. Espoir et Illusions, L'Harmattan, Paris.
Faure, Guy Olivier and Rubin, Jeffrey Z. (eds.) (1993): Culture and Negotiation. The Resolution of Water Disputes.
Feldbauer, Peter und Puhle, Hans-Jürgen (1992): Bauern im Widerstand. Agrarrebellionen und Revolutionen in Ländern der Dritten Welt und im vorindustriellen Europa, in: Historische Sozialkunde, Wien, No. 1 (Beiheft).
Ferdowsi, Mir A. (ed.) (1995): Die Welt der 90er Jahre. Das Ende der Illusion, Dietz Nachfolger GmbH, Bonn.
Flury, Manuel (1983): Krisen und Konflikte - Grundlagen. Ökologische, sozioökonomische und politisch-kulturelle Grundlagen zur Abschätzung von Krisen und Konflikten in Staaten insbesondere der Dritten Welt. Ein Beitrag zur entwicklungspolitischen Diskussion. Geographica Bernensia G 20, Geographisches Institut der Universität Bern.
--- (1992): Mit der Verwüstung leben, in: Entwicklung, Bern, April.
Founex-Report on Development and Environment. United Nations Conference on the Human Environment (1971), Founex, Switzerland.
Founex-Report on Development and Environment. Report and Working Papers of a Panel of Experts Convened by the Secretary-General of the United Nations. Conference on the Human Environment (Founex, Switzerland, 4-12 June 1971) (1972), Paris.
Franke, R. and Chasin, B. (1980): Seeds of Famine: Ecological Destruction and the Development Dilemma in the West African Sahel, Rowman & Allanheld, New Jersey.
Franzke, Michael (1994), in: Frankfurter Allgemeine Zeitung.
Freedom House (1994): The Comparative Survey of Freedom: 1994, Freedom Review, Vol. 25, No. 1, pp. 13-15.
Fritsch, Bruno (1993, 3. ed.): Mensch-Umwelt-Wissen: Evolutionsgeschichtliche Aspekte des Umweltproblems, Verlag der Fachvereine Zürich.
Gabriel, Oscar W. (ed.) (1993): Verstehen und Erklären von Konflikten. Beiträge zur nationalen und internationalen Politik, Minerva (Fachserie Politik), München.
Galaty, John G. and Johnson, Douglas L. (1990): The World of Pastoralism: Herding Systems in Comparative Perspective, The Guilford Press, New York.
--- (1990): Pastoral Systems in Global Perspective, in: John G. Galaty and Douglas L. Johnson (1990): The World of Pastoralism: Herding Systems in Comparative Perspective, The Guilford Press, New York, pp. 1-32.
Gallagher, Dennis (1994): Environment and Migration: A Security Problem? In: Günther Bächler (ed.): Umweltflüchtlinge. Das Konfliktpotential von morgen? agenda, Münster, pp. 65-72.
Gantzel, Klaus-Jürgen (1972): System und Akteur. Beiträge zur vergleichenden Kriegsursachenforschung, Düsseldorf.
--- (1981): Another Approach to a Theory on the Causes of War, in: Journal of Peace Research, Vol. 18, No. 1, pp. 39-56.
--- (1987): Tolstoi statt Clausewitz? Überlegungen zum Verhältnis von Staat und Krieg seit 1816 mittels statistischer Beobachtungen, in: Reiner Steinweg (ed.): Kriegsursachen, Friedensanalysen No. 21, Suhrkamp, Frankfurt/M.
--- (1988): Krieg in der Dritten Welt als Forschungsgegenstand. Eine Einführung zum Thema, in: Klaus-Jürgen Gantzel (ed.): Krieg in der Dritten Welt. Theoretische und methodische Probleme der Kriegsursachenforschung - Fallstudien, Studienreihe der AFK No. 12, Nomos, Baden-Baden, pp. 25-81.
--- (ed.) (1988b): Krieg in der Dritten Welt. Theoretische und methodische Probleme der Kriegsursachenforschung - Fallstudien, Studienreihe der AFK No. 12, Nomos, Baden-Baden.
---, Siegelberg, Jens und Schwinghammer, Torsten (eds.) (1992): Kriege der Welt: ein systematisches Register der kriegerischen Konflikte 1985-1992, Bonn.

Gasana, James K. (1995): La guerre, la paix et la démocratie au Rwanda, in: James K. Gasana (ed.) (1995): Les crises politiques au Burundi et au Rwanda (1993-1994). Analyses, faits et documents, Karthala, Paris, pp. 211-237.

--- et Nsengimana, Nkiko (1997): D'un génocide à l'autre: Plaidoyer pour une justice juste et pour le rétablissement dans l'autorité d'un courant démocratique au Rwanda, Project NOUER Nouvelle Espérance pour le Rwanda, Lausanne, Mai.

Gebremedhin, Naigzy (1997): Networking and Integrating the CONSDATA of the ECOMAN Project into Existing Information Systems. The Third ECOMAN Workshop, Asmara, Eritrea, 18-20 February.

--- et al. (1989): Environmental Security. A Report Contributing to the Concept of Comprehensive International Security, PRIO, Oslo.

Gerth, H.H. and Wright Mills, C. (eds.) (1958): From Max Weber: Essays in Sociology, New York.

Gettkant, Andreas (1996): Biologische Vielfalt. Jede Vertragspartei wird, soweit möglich und angebracht ..., in: Ökozid Journal, No. 11, pp. 8-14.

Giorgis, Anderberhan W. (1993): The Human and Ecological Consequences of the War in Eritrea, in: Terje Tvedt (ed.): Conflicts in the Horn of Africa: Human and Ecological Consequences of Warfare, EPOS, Uppsala, pp. 77-84.

Girardet, Herbert (1996): Das Zeitalter der Städte. Neue Wege für eine nachhaltige Stadtentwicklung, Deukalion, Holm.

Girgire, Aklu (1990): Official Opening, in: Bruno Messerli and Hans Hurni (eds.): African Mountains and Highlands. Problems and Perspectives, Walsworth Press, Marceline MS, pp. 3-4.

Giri, Jacques (1989): Le Sahel au XXIe siècle. Un essai de réflexion prospective sur les sociétés sahéliennes, Karthala, Paris.

Gizewski, Peter and Homer-Dixon, Thomas (1995): Urban Growth and Violence: Will the Future Resemble the Past? The Project on Environmental Population and Security, University College, Toronto.

Glaeser, Bernhard (ed.) (1989): Humanökologie. Grundlagen präventiver Politik, Westdeutscher Verlag, Opladen.

Glaser, Antoine et Smith, Stephen (1994): L'Afrique sans Africains. Le rêve blanc du continent noir, editions Stock, Paris.

Gleditsch, Nils P. (1992): Conversion and the Environment: An Overview, in: Nils P. Gleditsch (ed.): Conversion and the Environment, PRIO-Report No. 2, Oslo, pp. 23-44.

--- (1997): Armed Conflict and the Environment: A Critique of the Literature. Paper Presented to the NATO/CCMS Pilot Study Meeting in Ankara, 11-12 November 1996. Revised Version for the Meeting in Washington D.C., 21-22 January 1997, PRIO, Oslo.

Gleick, Peter H. (1990): Environment, Resources, and Security: Arenas for Conflict, Eras for Cooperation, Pacific Institute for Studies in Development, Environment and Security, Berkley CA.

--- (1991): Environment and Security: The Clear Connections, in: The Bulletin of the Atomic Scientists, Vol. 47, No. 3, pp. 17-21.

--- (ed.) (1993): Water in Crisis; A Guide to the World's Fresh Water Resources, Oxford University Press, New York.

--- (1996): Fresh Water: A Source of Conflict or Cooperation? A Survey of Present Developments, in: Bächler/Spillmann (eds.): Kriegsursache Umweltzerstörung. Environmental Degradation as a Cause of War. Länderstudien von externen Experten. Country Studies of External Experts, Vol. III, Verlag Rüegger, Zürich, pp. 1-25.

Godelier, M. (1984): L'idéel et le matériel, Fayard, Paris.

Görrissen, Thorsten (1993): Grenzüberschreitende Umweltprobleme in der internationalen Politik. Durchsetzung ökologischer Interessen unter den Bedingungen komplexer Interdependenz, Nomos, Baden Baden.

Goldstone, Jack A. (1992): Imminent Political Conflicts Arising from China's Environmental Crises, Occasional Paper Series of the Project on Environmental Change and Acute Conflict No. 2, Toronto, Cambridge MA.

Goodland, Robert et al. (eds.) (1992): Nach dem Brundtlandbericht: Umweltverträgliche wirtschaftliche Entwicklung, Deutsches Nationalkomitee für das UNESCO-Programm, Bonn.

Goodrich, Jill Wieder and Brecke, Peter (1996): The Paths from Environmental Change to Violent Conflict, Prepared for the Early Warning Conference and Workshop, November 14-16, at the University of Maryland.

Gore, Al (1993): Earth in the Balance. Ecology and the Human Spirit, Plume, New York.

Gorse, Jean Eugene and Steeds, David R. (1988): Desertification in the Sahelian and Sudanian Zones of West Africa, World Bank Technical Paper Number 61, The World Bank, Washington D.C.

Goudie, Andrew (1990, 3. ed.): The Human Impact on the Natural Environment, Oxford, Cambrigde.

Græger, Nina and Smith, Dan (eds.) (1994): Environment, Poverty, Conflict, PRIO Report No. 2, Oslo.

Green, Donald P. and Shapiro, Ian (1994): Pathologies of Rational Choice Theory. A Critique of Applications in Political Science, Yale University Press, New Haven, London.

Grosjean, Martin and Messerli, Bruno (1990): African Mountains and Highlands. Potential and Constraints, in: Bruno Messerli and Hans Hurni (eds.): African Mountains and Highlands. Problems and Perspectives, Walsworth Press, Marceline MS, pp. 38-48.

Guichaoua, André (1989): Destins paysans et politiques agraires en Afrique central, Tome 1: L'ordre paysan des hautes terres central du Burundi et du Rwanda, L'Harmattan, Paris.

--- (1992): The Problem of the Rwandese Refugees and the Banyarwanda Populations in the Great Lakes Region, Université de Lille (manus.).

--- (1995): La première génération: les réfugiés tutsi, trente ans après, in: André Guichaoua (ed.): Les crises politiques au Burundi et au Rwanda (1993-1994). Analyses, faits et documents, Karthala, Paris, pp. 339-348.

--- (ed.) (1995): Les crises politiques au Burundi et au Rwanda (1993-1994). Analyses, faits et documents, Karthala, Paris.

Gurr, Ted R. (1991): Theories of Political Violence and Revolution in the Third World, in: Francis M. Deng and William I. Zartman (eds.): Conflict Resolution in Africa, Washington D.C., pp. 153-189.

--- (1992): Why Minorities Rebel: A Global Analysis of Communal Mobilization and Conflict since 1945. Revised Draft - April 1992 - for Publication in the International Political Science Review.

--- (1993): Minorities at Risk: A Global View of Ethnopolitical Conflicts, U.S. Institute of Peace Press, Washington D.C.

--- and Harff, Barbara (1989): Victims of the State: Genocides, Policides and Group Repression since 1945, in: International Review of Victimology 1, No. 1, pp. 23-41.

--- (1994): Ethnic Conflict in World Politics, Westview Press, Boulder CO.

Haas, Peter M., Keohane, Robert O., and Levy, Marc A. (eds.) (1993): Institutions for the Earth, MIT Press, Cambridge MA.

Hanna, Susan and Munashinghe, Mohan (eds.) (1995): Property Rights and the Environment. Social and Ecological Issues. The Beijer International Institute of Ecological Economics and The World Bank, ESD Series, Washington D.C.

Harboth, Hans-Jürgen (1989): Dauerhafte Entwicklung (Sustainable Development). Zur Entstehung eines neuen ökologischen Konzepts, Berlin.

--- (1992): Die Diskussion um dauerhafte Entwicklung (Sustainable Development): Basis für eine umweltorientierte Weltentwicklungspolitik? In: Wolfgang Hein (ed.): Umweltorientierte Entwicklungspolitik, Schriften des Deutschen Überseeinstituts Hamburg No. 14 (2. ed.), Hamburg, pp. 37-62.

--- (1992): Armut und Umweltzerstörung in Entwicklungsländern, in: Herman Sautter (ed.): Entwicklung und Umwelt. Schriften des Vereins für Socialpolitik, Vol. 215, Duncker & Humblot, Berlin, pp. 41-72.

Hardin, Garrett (1968): The Tragedy of the Commons, in: Science, Vol. 162, No. 3859, pp. 1243-1248.

Harff, Barbara (1992): Recognizing Genocides and Policides, in: Helen Fein (ed.): Genocide Watch, Yale University Press, New Haven.

--- (1996): Early Warning of Humanitarian Crises: Sequential Models and the Role of Accelerators, Prepared for the Workshop on Risk Assessment and Crisis Early Warning, Nov. 14-16, University of Maryland at College Park.

Harir, Sharif (1993): Militarization of Conflict, Displacement and the Legitimacy of the State: A Case from Dar Fur, Western Sudan, in: Terje Tvedt (ed.): Conflicts in the Horn of Africa: Human and Ecological Consequences of Warfare, EPOS, Uppsala, pp. 14-26.

Harroy, Jean-Paul (1984): Rwanda. De la Féodalité à la Démocratie 1955 - 1962, Hayez, Bruxelles, Académie des Sciences d'Outre-Mer, Paris.

Hassan, Shaukat (1991): Environmental Issues and Security in South Asia, Adelphi Paper No. 262, London.

--- (1992): Environmental Sources of Conflict in the South Asian Subcontinent, in: Disarmament, Vol. 15, No. 1, pp. 79-95.

Hauff, Volker (ed.) (1987): Unsere gemeinsame Zukunft: Der Brundtland-Bericht der Weltkommission für Umwelt und Entwicklung, Eggenkamp, Greven.

Hauser, Jürg A. (1990): Bevölkerungs- und Umweltprobleme der Dritten Welt (2 Bde.), UTB, Bern, Stuttgart.

Haynes, Jeff (1996): Third World Politics. A Concise Introduction, Blackwell, Oxford.

Hein, Wolfgang (ed.) (1992): Umweltorientierte Entwicklungspolitik. Schriften des Deutschen Überseeinstituts Hamburg No. 14 (2 ed.), Hamburg.

Hellden, U. (1988): Desertification Monitoring. Is the Desert Encroaching? In: Desertification Control Bulletin (UNEP Nairobi), No. 17, pp. 8-12.

Helman, Udi (1990): Environment and the National Interest: An Analytical Survey of the Literature, in: The Washington Quarterly, Vol. 13, No. 4, pp. 193-206.

Henrion, Paul (1994): "Behind Rwanda's Disaster. A Plot by Extremist Bahutu?" In: International Herald Tribune, Monday, 14 November.

Hjort af Ornäs, Anders (ed.) (1992): Security in African Drylands. Research, Development and Policy, Uppsala.

--- and Krokfors, Christer (1992): Environment and International Security, in: Sverre Lodgaard and Anders Hjort af Ornäs (eds.): The Environment and International Security, PRIO-Report No. 3, Oslo, pp. 9-18.

--- and Salih, M.A. Mohamed (eds.) (1989): Ecology and Politics. Environmental Stress and Security in Africa, Scandinavian Institute of African Studies, Uppsala.

Hoagland, Sarah and Conbere, Susan (1991): Environmental Stress and National Security, Center for Global Change, Maryland.

Hoffmann, M.P. (1980): Prehistoric Ecological Crises, in: L.J. Bilsky (ed.): Historical Ecology, Port Washington, London, pp. 33-41.

Hofmeier, Rolf (ed.) (1993): Afrika Jahrbuch 1992. Politik, Wirtschaft und Gesellschaft in Afrika südlich der Sahara, Leske + Budrich, Opladen.

--- , Tetzlaff, Rainer und Wegemund, Regina (eds.) (1992): Afrika - Überleben in einer ökologisch gefährdeten Umwelt, Lit, Münster.

Holst, Johan Jorge (1989): Security and the Environment: A Preliminary Exploration, in: Bulletin of Peace Proposals, Vol. 20, No. 2, pp. 123-128.

Homer-Dixon, Thomas F. (1991): Environmental Change and Economic Decline, in: International Studies Notes, Vol. 16, No. 1, pp. 18-23.

--- (1991): On the Threshold: Environmental Change as Causes of Acute Conflict, in: International Security, Vol. 16, No. 1, pp. 76-116.

--- (1994): Across the Threshold: Empirical Evidence on Environmental Scarcities as Causes of Violent Conflict, in: International Security, Vol. 19, No. 1, pp. 5-40.

--- (1994): Environmental Scarcity and Intergroup Conflict, in: Michael T. Klare and Daniel C. Thomas (eds.): World Security. Challenges for a New Century, St. Martin's Press, New York, pp. 290-313.

--- (1994): Environmental Scarcities and Violent Conflict: Evidence from Cases, in: International Security, Vol. 19, No. 1, pp. 5-40.

--- (1995): Strategies for Studying Causation in Complex Ecological Political Systems. The Project on Environmental Population and Security, University College, Toronto.

--- , Boutwell, Jeffrey H., and Rathjens, George W. (1993): Environmental Change and Violent Conflict, in: Scientific American, February, pp. 38-45.

Horowitz, Donald L. (1985): Ethnic Groups in Conflict, University of California Press, London et al.

Hottinger, Arnold (1992): Wasser als Konfliktstoff: Eine Existenzfrage für Staaten des Nahen Ostens, in: Europa Archiv, No. 6, pp. 153-163.

Howard, Philip and Homer-Dixon, Thomas (1995): Environmental Scarcity and Violent Conflict: The Case of Chiapas, Mexico, University of Toronto.

Huber, Joseph (1989): Eine sozialwissenschaftliche Interpretation der Humanökoloie, in: Bernhard Glaeser (ed.): Humanökologie. Grundlagen präventiver Politik, Westdeutscher Verlag, Opladen.

Human Rights Watch (1995): World Report 1996, London et al.

Human Rights Watch/Africa (1996): Behind the Red Line. Political Repression in Sudan, New York et al.

Huntington, Samuel P. (1968): Political Order in Changing Societies, Yale University Press, New Haven, London.

Hurni, Hans (1990): Degradation and Conservation of Soil Resources in the Ethiopian Highlands, in: Bruno Messerli and Hans Hurni (eds.): African Mountains and Highlands. Problems and Perspectives, Walsworth Press, Marceline MS, pp. 51-63.

--- (1990): The First Workshop on African Mountains and Highlands: A Summary, in: Bruno Messerli and Hans Hurni (eds.): African Mountains and Highlands. Problems and Perspectives, Walsworth Press, Marceline MS, pp. 17-32.

Hutchison, Robert A. (ed.) (1991): Fighting for Survival: Insecurity, People and the Environment in the Horn of Africa, The World Conservation Union, Gland (Switzerland).

Imru, Mikael (1990): Environmental Security for the Horn of Africa, in: UNEP (ed.): Horn of Africa (Vol. I) (unpublished).

Ingham, Kenneth (1990): Politics in Modern Africa: The Uneven Tribal Dimension, Routledge, London, New York.

Institute of the Third World (ed.) (1990): "Third World Guide 91/92". Montevideo, pp. 479-480.

International Geosphere-Biosphere Programme (IGBP) (1990): The International Geosphere-Biosphere Programme: A Study of Global Change. The Initial Core Projects, IGPG Reports, Stockholm.

Intergovernmental Panel on Climate Change IPCC, Coastal Zone Management Subgroup (1992): Global Climate Change and the Rising of the Sea.

Intergovernmental Panel on Climate Change IPCC (1992): IPCC Supplement, World Metereological Organisation WMO, Geneva.

Ipsen, Knut (1990): Völkerrecht, Beck, München.
IISS (1994): "Strategic Comments", Rwanda's Rebels and Their Policies, 15th June.
IUCN; UNEP; WWF (1991): Caring for the Earth. A Strategy for Sustainable Living, Gland (Switzerland).
Jacobs, Jeffrey W. (1995): Mekong Committee History and Lessons for River Basin Development, in: The Geographical Journal, Vol. 161, No. 2, pp. 135-148.
Jacobson, Jodi L. (1988): Environmental Refugees: A Yardstick of Habitability, Worldwatch Paper No. 86, Washington D.C.
Jaehrling, Karen, Nissen, Astrid und Pieper, Dirk (1996): Das Kriegsgeschehen 1994/95. Ein kommentierter Überblick der Arbeitsgemeinschaft Kriegsursachenforschung Hamburg, in: Österreichisches Studienzentrum für Frieden und Konfliktlösung, Schweizerische Friedensstiftung (eds.): Theorie und Praxis ziviler Konfliktbearbeitung. Friedensbericht 1996. 1. Jahrbuch zur Konfliktlösung, Verlag Rüegger, Zürich, pp. 89-123.
Jahn, Thomas (1990): Das Problemverständnis sozialökologischer Forschung. Umrisse einer kritischen Theorie gesellschaftlicher Natur-Verhältnisse, in: E. Becker (ed.): Jahrbuch für sozial-ökologische Forschung, Frankfurt/M.
Jahrbuch Dritte Welt (1992): Daten, Übersichten, Analysen, Beck, München.
Jahrbuch Ökologie 1993 (1993): Beck, München.
Jazairy, Idriss, Alamgir, Mohiuddin, and Panuccio, Theresa (1992): The State of the World Rural Poverty. An Inquiry into its Causes and Consequences, IT Publications for IFAD, London.
Johnson, Stanley P. (ed.) (1993): The Earth Summit: The United Nations Conference on Environment and Development (UNCED), Graham & Trotman/Martinus Nijhoff, London.
--- (1995): The Politics of Population, Earthscan, Cairo 1994.
Jongman, A.J. and Schmid, A.P. (1994): Monitoring Human Rights. Manual for Assessing Country Performance, PIOOM, Leiden.
Kagame, Alexis (1952): Le Code des institutions politiques du Rwanda précolonial, Mémoire Institut Colonial Belge, présenté à la séance du 17 mars, Bruxelles.
--- (1959): La notion de génération appliqués à la généalogie dynastique et à l'histoire du Rwanda des Xe-XIe siècles à nos jours. Académie Royale des Sciences Coloniales, Classe des Sciences Morales et Politiques, mémoires in-80, n.s., tome 9, fasc. 5, Bruxelles.
--- (1972): Un abrégé de l'ethnohistoire du Rwanda, Butare.
Käkönen, Jyrki (1987): Natural Resources and Conflicts in the Changing International System. Three Studies on Imperialism, Avebury, Aldershot.
--- (ed.) (1992): Perspectives on Environmental Conflict and International Relations, Pinter Publ., London, New York.
--- (ed.) (1994): Green Security or Militarized Environment, Aldershot, Darthmouth.
Kälin, Walter and Moser, Rupert (eds.) (1993): Migrationen aus der Dritten Welt; Ursachen, Wirkungen, Handlungsmöglichkeiten, 3 ed., Haupt, Bern, Stuttgart, Wien.
Kandt, Richard (1921): Caput Nili. Eine empfindsame Reise zu den Quellen des Nils, Dietrich Reimer, Berlin.
Kates, Robert W., Turner II, B.L., and Clark, William C. (1990): The Great Transformation, in: B.L. Turner II et al. (eds.): The Earth As Transformed by Human Action: Global and Regional Changes in the Biosphere over the Past 300 Years, Cambridge University Press, Cambridge MA, pp. 1-18.
Kaufmann, Franz-Xaver (1973): Sicherheit als soziologisches und sozialpolitisches Problem, Ferdinand Enke Verlag, Stuttgart.
Keesing's Archives (1994): Keesing's Register of World Events 1994, Longman's, Harlow.
--- (1995): Keesing's Register of World Events 1995, Longman's, Harlow.
--- (1996): Keesing's Register of World Events 1996, Longman's, Harlow.
Kelman, Herbert C (1996): Negotiation as Interactive Problem Solving, in: International Negotiation 1, pp. 99-123.

Kende, István (1982): Kriege nach 1982. Militarpolitik Dokumentation No. 27, Frankfurt/M.
Kilian, Michael (1987): Umweltschutz durch Internationale Organisationen. Die Antwort des Völkerrechts auf die Krise der Umwelt? Duncker & Humblot, Berlin.
Klare, Michael T. (1994): World Security. Challenges for a New Century, St. Martin's Press, New York.
--- and Thomas, Daniel C. (1991): World Security. Trends & Challenges at Century's End, St. Martin's Press, New York.
Klötzli, Stefan (1992): Sustainable Development: A Disputed Concept. ENCOP Occasional Paper No. 2, Swiss Federal Institute of Technology/Swiss Peace Foundation, Zurich, Berne.
--- (1994): The Water and Soil Crisis in Central Asia - a Source for Future Conflicts. ENCOP Occasional Paper No. 11, Swiss Federal Institute of Technology/Swiss Peace Foundation, Zurich, Berne.
Kok, M.T.J. (1996): "Environmental Security and Sustainable Development". Proceedings International Workshop. Dutch National Research Programme on Global Air Pollution and Climate Change.
Körner, Peter (1993): Militante Konflikte im Schatten der Weltöffentlichkeit. Kriege und Gewalt in Afrika 1992, in: Österreichisches Studienzentrum für Frieden und Konfliktlösung, Schweizerische Friedensstiftung (eds.): Das neue Chaos der nationalen Egoismen. Friedensbericht 1993, Verlag Rüegger, pp. 91-102.
Krägenow, Timm (1996): "Joint Implementation" - Ein Beitrag zum Klimaschutz? In: Günther Altner et al. (eds.): Jahrbuch Ökologie, Beck, München.
Krusewitz, Knut (1985): Umweltkrieg: Militär, Oekologie und Gesellschaft, Athenaeum, Königstein/Ts.
Lachenmann, Gudrun (1984): Ökologie und Sozialstruktur im Sahel. Wechselwirkungen zwischen ökologischen und ökonomischen Veränderungen sowie soziostrukturellem und -kulturellem Wandel, in: Afrika Spectrum, No. 3, pp. 201-229.
--- (1990): Ökologische Krise und sozialer Wandel in afrikanischen Staaten, Spektrum No. 24, Breitenbach, Saarbrücken, Fort Lauderdale.
Lane, Jan-Erik and Ersson, Svante (1994): Comparative Politics. An Introduction and New Approach, Blackwell/Polity Press, Oxford, Cambridge.
Lang, Christoph I. (1995): Environmental Degradation in Kenya as a Cause of Political Conflict, Social Stress, and Ethnic Tensions. ENCOP Occasional Paper No. 12, Swiss Federal Institute of Technology/Swiss Peace Foundation, Zurich, Berne.
Lee, Richard B. and Devore, Irven (eds.) (1968/1987): Man the Hunter, Aldine de Gruyter, New York:
Leger Sivard, Ruth (1991): World Military and Social Expenditures 1991, World Priorities Inc., Washington D.C.
Leisinger, Klaus M. (1993): Hoffnung als Prinzip. Bevölkerungspolitik mit menschlichem Antlitz, Balance, Hannover.
--- (1994): Hoffnung als Prinzip. Bevölkerungspolitik mit menschlichem Antlitz (revised new edition), Balance, Hannover.
Lemarchand, René (1970): Rwanda and Burundi, Pall Mall, London.
--- (1994): Rwanda: The Rationality of Genocide, USAID/Abidjan (manus.).
Leonard, H.J. et al. (1989): Environment and the Poor. Development Strategies for a Common Agenda, Transaction Books, New Brunswick, Oxford.
--- (1989): Overview: Environment and the Poor. Development Strategies for a Common Agenda, in: Leonard et al., pp. 3-45.
Lesser, Ian O. (1989): Resources and Strategy. Vital Materials in International Conflict 1600 - the present, Macmillan, Hampshire, London.
Levy, Marc A. (1995): Is the Environment a National Security Issue, in: International Security, Vol. 20, No. 2, pp. 35-62.

--- (1995): Time for a Third Wave of Environment and Security Scholarship? In: Environmental Change and Security Project Report, No. 1, Spring, pp. 44-46.

Libiszewski, Stephan (1992): What is an Environmental Conflict? ENCOP Occasional Paper No. 1, Swiss Federal Institute of Technology/Swiss Peace Foundation, Zurich, Berne.

--- (1993): Umweltkonflikte - "die" Konfliktform im post-ideologischen Zeitalter? In: Wissenschaft und Frieden, No. 2, pp. 23-27.

--- (1995): Das Wasser im Nahost-Friedensprozess: Konfliktstrukturen und bisherige Vertragswerke unter wasserpolitischer Perspektive, in: Orient, Vol. 36, No. 4, pp. 625-648.

--- (1995): Water Disputes in the Jordan Basin Region and Their Role in the Resolution of the Arab-Israeli Conflict. ENCOP Occasional Paper No. 13, Swiss Federal Institute of Technology/Swiss Peace Foundation, Zurich, Berne.

--- (1996): Das Israelisch-Jordanische Wasserregime - ein Modell zur Lösung anderer Wasserverteilungskonflikte im Jordanbecken? In: Stephan Libiszewski und Manuel Schiffler (eds.): Wasserkonflikte und Wassermanagement im Jordanbecken. Arbeitspapier des Deutschen Instituts für Entwicklungspolitik, Berlin, pp. 73-88.

--- und Schiffler, Manuel (eds.) (1996): Wasserkonflikte und Wassermanagement im Jordanbecken. Arbeitspapier des Deutschen Instituts für Entwicklungspolitik, Berlin.

Lijphart, A. (1971): Comparative Politics and the Comparative Method, in: American Political Science Review, 65, pp. 682-93.

Lipschutz, Ronnie D. (1989): Global Resources and Environment: Arenas for Conflict, Opportunities for Cooperation. Paper Prepared for the East-West-Workshop on Common Security, Institute for Peace Research and Security Policy at the University of Hamburg, Hamburg, 4-5 December.

--- (1991): Wasn't the Future Wonderful? Resources, Environment, and the Emerging Myth of Global Sustainable Development, in: Colorado Journal of International Law and Policy, Vol. 2, No. 1, pp. 35-54.

--- and Holdren, John P. (1990): Crossing Borders: Resource Flows, the Global Environment, and International Security, in: Bulletin of Peace Proposals, Vol. 21, No. 2, pp. 121-133.

List, Martin and Rittberger, Volker (1992): Regime Theory and International Environmental Management, in: Andrew Hurrell and Benedict Klingsbury (eds.): The International Politics of the Environment. Actors, Interests, and Institutions, Oxford et al., pp. 85-109.

Lodgaard, Sverre (1992): Environment, Confidence-Building and Security, in: Sverre Lodgaard and Anders Hjort af Ornäs (eds.): The Environment and International Security, PRIO-Report No. 3, Oslo, pp. 19-30.

--- (1992a): Environmental Security, World Order, and Environmental Conflict Resolution, in: Nils Petter Gleditsch (ed.): Conversion and the Environment, PRIO-Report No. 2, Oslo, pp. 115-136.

--- and Hjort af Ornäs, Anders (eds.) (1992): The Environment and International Security, PRIO-Report No. 3, Oslo.

Loimeier, Roman (1986): "Nun schuldest Du uns 500.000 Francs...". Die Erschliessung des Senegaltales: Ökologische Konsequenzen eines Grossprojektes und bäuerlicher Widerstand, in: Peter E. Stüben (ed.): Nach uns die Sintflut. Staudämme, Entwicklungs"hilfe", Umweltzerstörung und Landraub, Focus, Giessen, pp. 119-141.

Luhmann, Niklas (1990): Ökologische Kommunikation: Kann die moderne Gesellschaft sich auf ökologische Gefährdungen einstellen? Westdeutscher Verlag, Opladen.

Lusigi, W.J. (1990): Networking in UNESCO's MAB Research- and Training Programmes. Aims at the Integrated Management of Ecosystems, in: M. Winiger, U. Wiesmann, and J.R. Rheker (eds.): Mount Kenya Area. Differentiation and Dynamics of a Tropical Mountain Ecosystem, University of Berne, Institute of Geography, Berne.

Mabbutt, J.A. (1984): A New Global Assessment of the Status and Trends of Desertification, in: Environmental Conservation, Vol. 11, No. 2, pp. 100-113.

Makinda, Samuel M. (1994): Die Tragödie in Ruanda, in: Europa Archiv, Vol. 49, No. 20, pp. 579-586.
Manifesto of the Bahutu (1957), in: F. Nkundabagenzi (1961): Rwanda politique, 1958-1960, Centre de Recherche et d'Information Socio-Politique, Brussels, pp. 21-22.
Maquet, Jacques J. (1954): Le système des relations sociales dans le Ruanda ancien, Musée Royal de l'Afrique Central, Tervuren.
--- (1961): The Premise of Inequality in Rwanda, London.
Markakis, John (1990): National and Class Conflict in the Horn of Africa, Cambridge University Press, Cambridge MA.
--- (ed.) (1993): Conflict and the Decline of Pastoralism in the Horn of Africa, MacMillan, Basingstoke.
Marsh, G.P. (1864): Man and Nature, New York.
Mathews, Jessica T. (1989): Redefining Security, in: Foreign Affairs, Vol. 68, No. 2, pp. 162-177.
--- (1991): The Environment and International Security, in: Michael T. Klare and Daniel C. Thomas (eds.): World Security, New York, pp. 362-380.
Matthew, Richard A. (1995): Environmental Security: Demystifying the Concept, Clarifying the Stakes, in: Environmental Change and Security Project Report, No. 1, Spring, pp. 14-23.
--- (1997): Rethinking Environmental Security. Paper Prepared for 1997 ISA Annual Meeting, Toronto.
Matthies, Volker (1988): Kriegsschauplatz Dritte Welt, Beck, München.
--- (ed.) (1992): Kreuzzug oder Dialog. Die Zukunft der Nord-Süd-Beziehungen, J.H.W. Dietz Nachf., Bonn.
--- (1994): Immer wieder Krieg? Eindämmen - beenden - verhüten? Schutz und Hilfe für die Menschen, Opladen.
--- (1995): Nicht mehr Krieg und noch nicht Frieden: Schlichtung und Wiederaufbau als Aufgaben der Entwicklungspolitik, in: Der Überblick, Vol. 31, No. 1, pp. 94-96.
--- (ed.) (1995): Vom Krieg zum Frieden: Kriegsbeendigung und Friedenskonsolidierung, Bremen.
McCabe, J.T. (1990): Turkana Pastoralism. A Case against the Tragedy of the Commons, in: Human Ecology, Vol. 18, pp. 81-103.
Meadows, Donella et al. (1972): The Limits to Growth. A Report for the Club of Rome's Project on the Predicament of Mankind, New York.
--- , Meadows, Dennis L., and Randers, Jorgen (1992): Die neuen Grenzen des Wachstums, Deutsche Verlags-Anstalt, Stuttgart.
Mendler, Martin und Schwegler-Rohmeis, Wolfgang (1989): Weder Drachentöter noch Sicherheitsingenieur. Bilanz und kritische Analyse der sozialwissenschaftlichen Kriegsursachenforschung. HSFK Forschungsbericht No. 3/1989, Hessische Stiftung Friedens- und Konfliktforschung, Frankfurt/M.
Mensching, Horst G. (1983): Die Verwüstung der Natur durch den Menschen in Historischer Zeit, in: H. Markl (ed.): Natur und Geschichte, München, Wien, pp. 147-170.
Menzel, Ulrich (1991): Geschichte der Entwicklungstheorie. Einführung und systematische Bibilographie, Deutsches Überseeinstitut, Hamburg.
--- (1992): Das Ende der Dritten Welt und das Scheitern der grossen Theorie, Suhrkamp, Frankfurt/M.
--- (1993, 2. Aufl.): Geschichte der Entwicklungstheorie. Einführung und systematische Bibliographie. Schriften des Deutschen Übersee-Instituts Hamburg, No. 18, Hamburg.
--- und Senghaas, Dieter (1986): Europas Entwicklung und die Dritte Welt. Eine Bestandesaufnahme, Suhrkamp, Frankfurt/M.
Merchant, Carolyn (1980): The Death of Nature. Women, Ecology and the Scientific Revolution, Harper Collins Publ., New York.

--- (1990): The Realm of Social Relations: Production, Reproduction, and Gender in Environmental Transformations, in: B.L. Turner II et al. (eds.): The Earth As Transformed by Human Action: Global and Regional Changes in the Biosphere over the Past 300 Years, Cambridge University Press, Cambridge MA, pp. 673-684.

--- (ed.) (1994): Ecology. Key Concepts in Critical Theory, Humanities Press, New Jersey.

Mertens, Heide (1993): Das Bevölkerungsproblem als Indikator ungleichgewichtiger Entwicklung, in: Handbuch der Dritten Welt, J.H.W. Dietz Nachf., Bonn.

Messerli, Bruno (1990): Die natürlichen Ressourcen - Grundlagen des Lebens und Überlebens in der Dritten Welt, in: E. Haselbach et al. (eds.): Dritte Welt und Naturwissenschaften, Universitätsverlag, Freiburg/CH, pp. 17-52.

--- et al. (1987): Umweltprobleme und Entwicklungszusammenarbeit: Entwicklungspolitik in weltweiter und langfristig ökologischer Sicht, Bern.

Meyer, Berthold und Wellmann, Christian (eds.) (1992): Umweltzerstörung: Kriegsfolge und Kriegsursache, Friedensanalysen No. 27, Suhrkamp, Frankfurt/M.

Meyer-Abich, Klaus M. (1984): Wege zum Frieden mit der Natur, Carl Hanser, München, Wien.

--- (1990): Aufstand für die Natur. Von der Umwelt zur Mitwelt, Carl Hanser, München, Wien.

Michler, Walter (1991): Weissbuch Afrika, J.H.W. Dietz Nachf., Bonn.

Midlarsky, Manus (1995): Environmental Influences on Democracy: Aridity, Warfare, and a Reversal of the Causal Arrow, in: Journal of Conflict Resolution, Vol. 39, pp. 224-262.

Mink, Stephen D. (1993): Poverty, Population, and the Environment, World Bank Disskussion Papers 189, Washington D.C.

Molt, Peter (1994): Ein Produkt der Kolonialherrschaft. Wie es in der einstigen "Schweiz Afrikas" zum Bürgerkrieg gekommen ist, in: Frankfurter Rundschau, Dokumentation, 20.6.

Molvaer, Reidulf K. (1989): Environmental Cooperation in the Horn of Africa, in: UNEP (ed.), Horn of Africa, Vol. I, Nairobi (unpublished).

--- (1990): Environmental Cooperation in the Horn of Africa. A UNEP Perspective, in: Bulletin of Peace Proposals, Vol. 21, No. 2, pp. 135-142.

--- (1991): Environmentally Induced Conflicts? A Discussion Based on Studies from the Horn of Africa, in: Bulletin of Peace Proposals, Vol. 22, No. 2, pp. 175-188.

--- (1992): Points for the Agenda of Studies in Environmental Security, in: Sverre Lodgaard and Anders Hjort af Ornäs (eds.): The Environment and International Security, PRIO-Report No. 3, Oslo, pp. 63-73.

--- (1994): Environmental Cooperation and Confidence Building in the Horn of Africa, Sage, London.

Moore, Barrington, Jr. (1966): Social Origins of Dictatorship and Democracy. Lord and Peasant in the Making of the Modern World, Beacon Press, Boston.

--- (1987): Soziale Ursprünge von Diktatur und Demokratie, Suhrkamp, Frankfurt/M.

Morgan, R.P.C. (1979): Soil Erosion, London.

Mortimore, M. (1989): Adapting to Drought: Farmers, Famines, and Desertification in West Africa, Cambridge.

Morton, Cf J. (1989): Ethnicity and Politics in Red Sea Province, Sudan, in: African Affairs, Vol. 88, No. 350, pp. 63-76.

Moser, Rupert (1994): Der geschichtliche Hintergrund der blutigen Konflikte in Rwanda und Burundi, in: Pressestelle (ed.), UNI PRESS 81, Berichte über Forschung und Wissenschaft an der Universität Bern, No. 81, Juni.

Moss, Richard H. (1993): Resource Scarcity and Environmental Security, in: SIPRI (ed.): SIPRI Yearbook, Stockholm, pp. 27-36.

Mounier, Frederic (1986): The Senegal River Scheme: Development for Whom? In: Edward Goldsmith and Nicholas Hildyard (eds.): The Social and Environmental Effects of Large Dams. Case Studies, Vol. 2, Penwell Ltd, Cornwall, pp. 109-119.

Moynihan, Daniel P. (1993): Pandaemonium: Ethnicity in International Politics, Oxford University Press, Oxford.
Müller, Friedemann (1992): Kategorisierung von Umweltproblemen als Konfliktfelder in den internationalen Beziehungen, in: Wolfgang Heydrich (ed.): Sicherheitspolitik Deutschlands, Nomos, Baden-Baden, pp. 501-507.
Müller, Harald (1986): Umwelt und gewaltsamer Konflikt. Umweltschäden und innerstaatliche Gewalt in der Dritten Welt, in: antimilitarismus information, No. 12, pp. III-170 - III-175.
--- (1990): Internationale Ressourcen- und Umweltproblematik, in: Manfred Knapp und Gert Krell (eds.): Einführung in die Internationale Politik, Studienbuch, München, pp. 351-382.
--- (1991): Welche Auswirkungen werden ökologische Probleme auf die Internationalen Beziehungen und auf Internationale Politik haben? Thesenpapier für das Experten-Kolloquium der Evangelischen Akademie Loccum "Weltsystem und Weltpolitik jenseits der Bipolarität" vom 21.-23.6.1991.
--- (1992): Umwelt und Konflikt, in: Berthold Meyer und Christian Wellmann (eds.): Umweltzerstörung: Kriegsfolge und Kriegsursache, Friedensanalysen No. 27, Suhrkamp, Frankfurt/M, pp. 72-99.
--- (1993): Die Chance der Kooperation. Regime in den internationalen Beziehungen, Wissenschaftliche Buchgesellschaft, Darmstadt.
Münch, Richard (1992): Die Struktur der Moderne, Suhrkamp, Frankfurt/M.
Murego, Donat (1975): La révolution rwandaise, 1959-1962, Institut des Sciences Politiques et Sociales, Louvain.
Mutter, Theo (1991): Rohstoffreichtum - Entwicklungsbasis oder Sackgasse? In: Peter J. Opitz (ed.): Grundprobleme der Entwicklungsländer, München, pp. 118-138.
Myers, Norman (1987): Not Far Afield: U.S. Interests and the Global Environment, Washington D.C.
--- (1987): Population, Environment and Conflict, in: Environmental Conservation, Vol. 14, No. 1, pp. 15-22.
--- (1988): Natural Resource Systems and Human Exploitation Systems. Physiobiotic and Ecological Linkages, World Bank Environment Department Working Paper No. 12, The World Bank, Washington D.C.
--- (1989): Environment and Security, in: Foreign Policy, No. 74, pp. 23-41.
--- (1989): Environmental Security. The Case of South Asia, in: International Environmental Affairs, Vol. 1, No. 2, pp. 138-154.
--- (1993): Ultimate Security. The Environmental Basis of Political Stability, W.W. Norton & Co., New York, London.
Myrdal, G. (1974): Ökonomische Theorie und unterentwickelte Regionen, Stuttgart.
Nahimana, Ferdinand (1993): Le Rwanda. Emergence d'un état, L'Harmattan, Paris.
Nardin, Terry and Mapel, David R. (eds.) (1992): Traditions of international ethics, Cambridge University Press, Cambridge.
Nations of the Earth. United Nations Conference on Environment and Development: National Reports Summaries (1992), Geneva.
Ndjimbi-Tshiende, Olivier (1992): Réciprocité-Coopération et le Système Palabrique Africain. Dissertationen Philosophische Reihe No. 10, EOS Verlag Erzabteil, St. Ottilien.
Nelson, Ridley (1990): Dryland Management. The Desertification Problem, The World Bank Technical Paper No. 116, The World Bank, Washington D.C.
Nembrini, Pier Giorgio (1995): Do Water Production and Treatment Facilities Need Greater Protection in Armed Conflicts? In: International Committee of the Red Cross (ed.): Water and War. Symposium on Water in Armed Conflicts, Montreux, 21-23 November 1994, Geneva, pp. 29-50.
Nennen, Heinz-U. (1991): Ökologie im Diskurs. Zu Grundfragen der Anthropologie und Ökologie und zur Ethik der Wissenschaften, Westdeutscher Verlag, Opladen.

Newbury, Catharine (1988): The Cohesion of Oppression. Clientship and Ethnicity in Rwanda, 1860-1960, Columbia University Press, New York.

Nicholson, Michael (1992): Rationality and the Analysis of International Conflict, Cambridge Studies in International Relations No. 19, Cambridge University Press, Cambridge.

Nkundabagenzi, F. (1961): Rwanda politique, 1958-1960, Centre de Recherche et d'Information Socio-Politique, Bruxelles.

Nkurikiyimfura, Jean-Népomucène (1994): Le gros bétail et la société rwandaise: évolution historique des XIIe-XIVe siècles à 1958, Paris.

Nohlen, Dieter (1994): Lexikon der Politik. Vol. 6: Internationale Beziehungen, München.

--- and Nuscheler, Franz (1992) (eds.): Handbuch der Dritten Welt. Vol. 1: Grundprobleme - Theorien - Strategien, J.H.W. Dietz Nachf., Bonn.

--- (1992): Was heisst Entwicklung? In: Dieter Nohlen und Franz Nuscheler (eds.): Handbuch der Dritten Welt. Vol. 1: Grundprobleme - Theorien - Strategien, J.H.W. Dietz Nachf., Bonn, pp. 55-75.

--- (1992): Indikatoren von Unterentwicklung und Entwicklung, in: Dieter Nohlen und Franz Nuscheler (eds.): Handbuch der Dritten Welt. Vol. 1: Grundprobleme - Theorien - Strategien, J.H.W. Dietz Nachf., Bonn, pp. 77-108.

Nordenfelt, Johan (1992): Environmental Destruction as Method of Warfare, in: Sverre Lodgaard and Anders Hjort af Ornäs (eds.): The Environment and International Security, PRIO-Report No. 3, Oslo, pp. 31-39.

North, Robert (1977): Toward a Framework for the Analysis of Scarcity and Conflict, in: International Studies Quarterly, Vol. 21, No. 4.

Nsengiyaremye, Dismas (1995): La transition démocratique au Rwanda (1989-1993), in: André Guichaoua (ed.): Les crises politiques au Burundi et au Rwanda (1993-1994), Paris, pp. 239-263.

Nuscheler, Franz (1992): Entwicklungspolitische Bilanz der 80er Jahre - Perspektiven für die 90er Jahre, in: Dieter Nohlen und Franz Nuscheler (eds.): Handbuch der Dritten Welt. Vol. 1: Grundprobleme - Theorien - Strategien, J.H.W. Dietz Nachf., Bonn pp. 156-178.

--- (ed.) (1994): Weniger Menschen durch weniger Armut? Bevölkerungswachstum - globale Krise und ethische Herausforderung, Pustet, Salzburg.

OECD (ed.) (1991): Environmental Indicators: A Preliminary Set, Paris.

--- (1992): Trends in International Migration. Continuous Reporting System on Migration.

--- (ed.) (1994): The State of the Environment, Paris.

Ogata, Sadako (1994): Environmental Refugees. Statement by the United Nations High Commissioner for Refugees, in: Günther Bächler (ed.): Environmental Refugees. A Potential of Future Conflicts, agenda, Münster, pp. 41-48.

Okidi, C.O. (1992): Environmental Stress and Conflicts in Africa: Case Studies of African International Drainage Basins. Draft Prepared for the Review Conference "Environmental Change and Acute Conflict" at the Brookings Institution, Washington D.C., 11-12 May.

Okoh, Peter B. (1994): Schutzpatronin der Ölkonzerne: Nigerias Regierung duldet die ökologische Zerstörung der Ölfördergebiete, in: Der Überblick, No. 1, März, pp. 10-14.

Oldeman, L.R., Hakkeling, R.T.A., and Sombroek, W.G. (1991): World Map of the Status of Human-Induced Soil Degradation. An Explanatory Note. Global Assessment of Soil Degradation GLASOD, International Soil Reference and Information Centre/UNEP, Den Haag.

Omara-Ojungu, P.H. (1990): Resource Management Implications of Ecological Dynamics in Mountainous Regions, in: Bruno Messerli and Hans Hurni (eds.): African Mountains and Highlands. Problems and Perspectives, Walsworth Press, Marceline MS, pp. 143.

Onishi, Akira (1996): FUGI Global Model as Global Early Warning System for Refugees, Soka University, Prepared for the Early Warning Conference and Workshop, November 14-16, at the University of Maryland.

Ophuls, William and Boyan, Stephen A. Jr. (1992): Ecology and the Politics of Scarcity Revisited, W.H. Freeman & Co., New York.
Opitz, Peter J. (1988): Das Weltflüchtlingsproblem. Ursachen und Folgen, Beck, München.
--- (ed.) (1990): Weltprobleme, Bundeszentrale für politische Bildung, Bonn.
--- (1991): Entwicklung im Zwielicht - Anmerkungen zur Struktur und zum Profil einer Dekade, in: Peter J. Opitz (ed.): Grundprobleme der Entwicklungsländer, München, pp. 7-44.
--- (ed.) (1991): Grundprobleme der Entwicklungsländer, Beck, München.
--- (1992): Droht der grosse Marsch 'gen Norden? Flüchtlingsströme und Völkerwanderungen, in: Volker Matthies (ed.): Kreuzzug oder Dialog. Die Zukunft der Nord-Süd-Beziehungen, Bonn, pp. 90-106.
Osman, Mohamed (1990): Verwüstung: Die Zerstörung von Kulturland am Beispiel des Sudan, edition CON, Bremen.
Ostrom, Elinor (1990): Governing the Commons: The Evolutions of Institutions for Collective Action, Cambridge.
OXFAM Poverty Report (The) (1995), Oxfam (UK and Ireland), Oxford.
Pagès, Albert (1933): Un royaume hamite au centre de l'Afrique, Institut Royal Colonial Belge, Bruxelles.
Paternostre de la Mairieu, Baudouin (1972): Le Rwanda, son effort de développement, ed. A. de Boeck, ed. Rwandaise, Bruxelles/Kigali.
Pearce, D., Barbier, E., and Markandya, A. (1990): Sustainable Development. Economics and Environment in the Third World, Hants, Vermont.
Percival, Valerie and Homer-Dixon, Thomas (1995): Environmental Scarcity and Violent Conflict: The Case of Rwanda. The Project on Environmental Population and Security, University College, Toronto.
Perner, Conradin (1993): "The Reward of Life is Death": Warfare and the Anyuak on the Ethiopian-Sudanese Border, in: Terje Tvedt (ed.): Conflicts in the Horn of Africa: Human and Ecological Consequences of Warfare, EPOS, Uppsala, pp. 125-142.
Peters, Matthias und Zeugin, Peter (1979): Sozialindikatorenforschung. Eine Einführung, Ferdinand Enke, Stuttgart.
Peterson, W. (1958): A General Typology of Migration, in: American Sociological Review, 23 (3), pp. 256-266.
Pethig, Rüdiger (ed.) (1992): Conflicts and Cooperation in Managing Environmental Resources, Springer, Berlin, Heidelberg, New York.
Pfister, Christian (1995): Das 1950er Syndrom. Der Weg in die Konsumgesellschaft, Publikation der Akademischen Kommission der Universität Bern, Bern.
--- (1992): Das 1950er Syndrom: Der Energieverbrauch unserer Zivilisation in historischer Perspektive, in: Natur und Mensch, Schweizerische Blätter für Natur- und Heimatschutz, pp. 1-6.
Pianta, Mario and Renner, Michael (1989): The State System and the Consequences for Environmental Degradation, in: IPRA Newsletter, Vol. 27, No. 1, pp. 8-17.
Pietras, Ziemowit Jacek and Pietras, Marek (eds.) (1991): International Ecological Security, Maria Curie-Sklodowska University Press, Lublin.
Pirages, Dennis C. (1991): The Greening of Peace Research, in: Journal of Peace Research, Vol. 28, No. 2, pp. 129-133.
--- (1991): Social Evolution and Ecological Security, in: Bulletin of Peace Proposals, Vol. 22, No. 3, pp. 329-334.
Planning Ministry (1992): Rapport Nationale du Rwanda, CNUED, Brésil, in: United Nations (ed.), Nations of the Earth Report, United Nations Conference on Environment and Development: National Reports Summaries, Vol. II, Geneva, pp. 265-270.
Polanyi, Karl (1995): The Great Transformation. Politische und ökonomische Ursprünge von Gesellschaften und Wirtschaftssystemen, Suhrkamp, Frankfurt/ M.

Pontenagel, Irm (ed.) (1993): Energy in Africa. Economic and Political Initiatives for Application of Renewable Energies in Developing Countries. Proceedings of the International Solar Energy Conference in Harare/Zimbabwe, 14-17 November 1991, Ponte Press, Bochum.

Porter, Gareth (1990): Post-Cold War Global Environment and Security, in: The Fletcher Forum of World Affairs, Vol. 14, No. 2, pp. 332-344.

--- (1995): Environmental Security as a National Security Issue, in: Current History, May, pp. 218-222.

--- and Brown, Janet W. (1991): Global Environmental Politics, Westview Press, Boulder CO.

Postel, Sandra (1992): Last Oasis: Facing Water Scarcity, W.W. Norton & Co., New York.

--- (1993): Water and Agriculture, in: Peter H. Gleick (ed.): Water in Crisis. A Guide of the World's Fresh Water Resources, Oxford University Press, New York, Oxford, pp. 57-66.

Pradervand, Pierre (1994): Afrika lebt. Die stille Revolution der afrikanischen Bauern, Pendo, Zürich.

Prins, Gwyn (ed.) (1993): Threats without Enemies. Facing Enviromental Insecurity, Earthscan, London.

PRIO and UNEP (eds.) (1989): Environmental Security. Report Contributing to the Concept of Comprehensive International Security, Oslo, Nairobi.

Quimpo, Nathan F. (1993): Environmental Conflict in the Philippines? Bern (manus.).

Ratsch, Ulrich (1991): Globale ökologische Konflikte, in: Gert Krell, Johannes Schwerdtfeger und Egon Bahr (eds.): Friedensgutachten 1991, Lit, Münster.

--- (1994): Wasser als knappe Ressource, in: Forschungsstätte der Evangelischen Studiengemeinschaft, Hessische Stiftung für Friedens- und Konfliktforschung, Institut für Friedensforschung und Sicherheitspolitik (eds.): Friedensgutachten 1994, Lit, Münster, pp. 130-145.

Renner, Michael (1989): National Security: The Economic and Environmental Dimensions, Worldwatch Paper No. 89, Worldwatch Institute, Washington D.C.

--- (1996): Fighting for Survival. Environmental Decline, Social Conflict and the New Age of Insecurity, The World Watch Environmental Alert Series, New York, London.

---, Pianta, Mario, and Franchi, Cinzia (1991): International Conflict and Environmental Degradation, in: Raimo Väyrynen (ed.): New Directions in Conflict Theory. Conflict Resolution and Conflict Transformation, London, pp. 108-128.

Reyntiens, Filip (1985): Pouvoir et droit au Rwanda, Musée Royal de l'Afrique Central, Tervuren.

Research Panel on Environment and Security of the Canadian Global Change Program (1996) (author: Steve Lonergan): Environment and Security: An Overview of Issues and Research Priorities for Canada, Canadian Global Change Program Technical Report No. 96-1, The Royal Society of Canada.

Rich, Bruce (1994): Mortgaging the Earth: The World Bank, Environmental Impoverishment, and the Crisis of Development, Beacon Press, Boston.

Richards, John F. (1990): Land Transformation, in: B.L. Turner II et al. (eds.): The Earth As Transformed by Human Action, Cambridge University Press, New York, pp. 163-178.

Richter, Roland (1992): Flüchtlingsfragen in Afrika. Zwangsmigrationen im Rahmen grösserer Entwicklungsprozesse, Nomos, Baden-Baden.

Rifkin, Jeremy (1991): Biosphere Politics: A New Consciousness for a New Century, New York.

River and Lake Basin Development (1990): Natural Resources/Water Series, No. 20, United Nations, New York.

Robinson, Nicholas A. (ed.) (1993): Agenda 21: Earth's Action Plan, IUCN Environmental Policy & Law Paper, No. 27, Oceana, New York.

Rodal, Berel (1994): The Environment and Changing Concepts of Security, in: Commentary - A Canadian Security Intelligence Service Publication, No. 47, August, pp. 1-7.

Rodenburg, Eric, Tunstall, Dan, and Bolhuis, Frederik van (1995): Environmental Indicators for Global Cooperation. Global Environmental Facility Working Paper Number 11, UNDP/UNEP/The World Bank, The World Bank, Washington D.C.
Romm, Joseph J. (1993): Defining National Security. The Nonmilitary Aspects: Council on Foreign Relations Press.
Rosenke, Werena und Siepelmeyer, Thomas (eds.) (1991): Afrika - der vergessene Kontinent? Zwischen selektiver Weltmarktintegration und ökologischen Katastrophen, unrast, Münster.
Roy, Beth (1994): Some Trouble with Cows. Making Sense of Social Conflict, University of California Press, Berkeley et al.
Rozanov, Boris G., Targulian, Viktor, and Orlov, D.S. (1990): Soils, in: B.L. Turner II (ed.): The Earth As Transformed by Human Action, Cambridge University Press, New York, pp. 203-214.
Ruhashyankiko, Nicodème (1977): Le droit foncier du Rwanda, Butare.
Rupesinghe, Kumar (ed.) (1992): Internal Conflict and Governance, St. Martin's Press, New York.
--- and Kuroda, Michiko (eds.) (1992): Early Warning and Conflict Resolution, St. Martin's Press, New York.
Russell, Sharon S., Jacobsen, Karen, and Stanley, William D. (1990): International Migration and Development in Sub-Saharan Africa: Overview, Vols. 1+2. World Bank Discussion Papers, African Technical Department Series No. 101/102, The World Bank, Washington D.C.
Russett, Bruce M. (1968): International Regions and the International Systems: A Study in Political Ecology, Rand McNally & Co., Chicago.
--- (1982): Security and the Resources Scramble: Will 1984 be like 1914? In: International Affairs, Vol. 58, pp. 42-58.
Sachs, Wolfgang (ed.) (1993): Global Ecology, Zed Books, London, New Jersey.
--- (1994): Globale Umweltpolitik im Schatten des Entwicklungsdenkens, in: Blätter für Deutsche und Internationale Politik, Vol. 39, No. 11, pp. 1367-1378.
Sadik, Nafis (1990): The State of the World Population 1990, United Nations Population Fund, New York.
Salih, M.A. Mohamed (1989): Political Coercion and the Limits of State Intervention, Sudan, in: Anders Hjort af Ornäs and M.A. Mohamed Salih (eds.): Ecology and Politics. Environmental Stress and Security in Africa, Scandinavian Institute of African Studies, Uppsala, pp. 101-115.
Sand, Peter H. (1990): Lessons Learned in Global Environmental Governance, Washington D.C.
Santerre, Frédérique (1996): Conflict, Security and the Environment, Graduate Institute of International Studies, Geneva.
Saro-Wiwa, Ken (1992): Genocide in Nigeria: The Ogoni Tragedy, Saros International, Port Harcourt et al.
Saucier, Jean-Francois (1974): The Patron-Client Relationship in Traditional and Contemporary Southern Rwanda, UMI Dissertation Services, Michigan.
Sautter Hermann (ed.) (1992): Entwicklung und Umwelt. Schriften des Vereins für Socialpolitik, Bd. 215, Duncker & Humblot, Berlin.
Schädle, Werner (1990): Wüsten wachsen. Der Einfluss des Menschen auf die trockenheissen Ökosysteme der Subtropen, in: Peter J. Opitz (ed.): Weltprobleme, Bundeszentrale für politische Bildung, Bonn, pp. 53-75.
Schildkrout, Enid (1979): The Ideology of Regionalism in Ghana, in: William A. Shack and Elliott P. Skinner (eds.): Strangers in African Societies, Berkeley, Los Angeles.
Schmeidl, Susanne (1995): From Root Cause Assessment to Preventive Diplomacy: Possibilities and Limitations of the Early Warning of Forced Migration, Dissertation, Ohio State University.

Schmid, Alex P. (1996): Indicator Development: Conceptual, Methodological and Data Issues Involved in Forecasting Conflict Escalation. Paper Prepared for the Early Warning Conference and Workshop: "Risk Assessment and Crisis Early Warning Systems", Center for International Development and Conflict Management, University of Maryland at College Park, 14-16 November.

Schmid, Josef (ed.) (1994): Bevölkerung - Umwelt - Entwicklung: Eine humanökologische Perspektive, Westdeutscher Verlag, Opladen.

--- (1994): Bevölkerung - Umwelt - Entwicklung: Forschungsrichtungen und aktuelle Argumentation, in: Josef Schmid (ed.): Bevölkerung - Umwelt - Entwicklung. Eine humanökologische Perspektive, Opladen, pp. 17-42.

Schmidt-Bleek, Friedrich und Klüting, Rainer (1994): Wieviel Umwelt braucht der Mensch? MIPS - Das Mass für ökologisches Wirtschaften, Birkhäuser, Berlin.

Schmieg, Evita (1995): Entwicklungszusammenarbeit als Beitrag zur Krisenprävention? Bundesministerium für Entwicklungszusammenarbeit, Bonn. Paper Presented at the 3rd International Workshop on "Peaceful Settlement of Conflict as a Joint Task for International Organisations, Governments and Civil Society", March 22-24, 1995, Evangelische Akademie Loccum.

Schmitz, Marc (ed.) (1992): Les conflits verts; la détérioration de l'environment, source de tension majeures, GRIP, Bruxelles.

Schrijver, Nico (1989): International Organization for Environmental Security, in: Bulletin of Peace Proposals, Vol. 20, No. 2, pp. 115-123.

Schubert, Gunter, Tetzlaff, Rainer und Vennewald, Werner (eds.) (1994): Demokratisierung und politischer Wandel: Theorie und Anwendung des Konzeptes der strategischen und konfliktfähigen Gruppen (SKOG), Lit, Münster.

Schweizerischer Bundesrat (1994): Leitbild Nord-Süd, Bern.

Schwind, Martin (1972): Allgemeine Staatengeographie, Walter de Gruyter, Berlin, New York.

Scoones, I. (1989): Economic and Ecological Carrying Capacity: Implications for Livestock Development in the Dryland Communal Areas of Zimbabwe, ODI Pastoral Network Paper No. 27b, London.

Semait, Bekure W. (1989): Ecological Stress and Political Conflict in Africa: The Case of Ethiopia, in: Anders Hjort af Ornäs and Mohamed Salih (eds.): Ecology and Politics. Environmental Stress and Security in Africa, Scandinavian Institute of African Studies, pp. 37-50 Uppsala.

Senghaas, Dieter (1974): Vorwort. Elemente einer Theorie des peripheren Kapitalismus, in: Dieter Senghaas (ed.): Peripherer Kapitalismus. Analysen über Abhängigkeit und Unterentwicklung, Suhrkamp, Frankfurt/M, pp. 7-36.

--- (1977): Weltwirtschaftsordnung und Entwicklungspolitik. Plädoyer für Dissoziation, Suhrkamp, Frankfurt/M.

--- (1982): Von Europa lernen. Entwicklungsgeschichtliche Betrachtungen, Suhrkamp, Frankfurt/M.

--- (1987): Die Entwicklungsproblematik. Überlegungen zum Stand der Diskussion, in: Aus Politik und Zeitgeschichte, Beilage zur Wochenzeitung Das Parlament, B 8, Februar, pp. 3-12.

--- (1988): Konfliktformationen im internationalen System, Suhrkamp, Frankfurt/M.

--- (1993): Ethnische Konflikte - Ursachen und Lösungswege, in: Spektrum der Wissenschaft, Juni, pp. 95-103.

--- (1993): Frieden als Zivilisierungsprojekt. Arbeitspapiere zu Problemen der Internationalen Politik und der Entwicklungsländerforschung No. 12, München.

--- (1994): Treiben Umweltprobleme in Gewaltkonflikte? Ökologische Konflikte im internationalen System und Möglichkeiten ihrer friedlichen Beqarbeitung, in: Jörg Calliess (ed.): Treiben Umweltprobleme in Gewaltkonflikte? Ökologische Konflikte im internationalen System und

Möglichkeiten ihrer friedlichen Bearbeitung. Loccumer Protokolle No. 21, Evangelische Akademie Loccum, Rehburg-Loccum, pp. 301-310.
--- und Menzel, Ulrich (1976): Multinationale Konzerne und Dritte Welt, in: Dieter Senghaas und Ulrich Menzel (eds.): Multinationale Konzerne und Dritte Welt, Opladen, pp. 47-66.
--- und Zürn, Michael (1992): Kernfragen für die Friedensforschung der Neunziger Jahre, in: Günther Bächler (ed.): Perspektiven: Friedens- und Konfliktforschung in Zeiten des Umbruchs, Verlag Rüegger, Zürich, pp. 379-389.
--- (1989) (ed.): Regionalkonflikte in der Dritten Welt. Autonomie und Fremdbestimmung, Nomos, Baden-Baden.
Shevardnadse, Edvard (1990): Ecology and Diplomacy, in: Environmental Policy and Law, Vol. 20, No. 1/2, pp. 20-24.
Shiva, Vandana (1991): Ecology and the Politics of Survival. Conflicts over Natural Resources in India, Sage, New Delhi, Tokyo.
--- (1994): Development, Ecology, and Women, in: Carolyn Merchant (ed.): Ecology. Key Concepts and Critical Theory, Humanities Press, New Jersey, pp. 272-280.
Siebert, Horst (ed.) (1991): Environmental Scarcity: The International Dimension, Tübingen.
Sieferle, Rolf P. (1984): Fortschrittsfeinde? Opposition gegen Technik und Industrie von der Romantik bis zur Gegenwart, Beck, München.
--- (ed.) (1988): Fortschritte der Naturzerstörung, Suhrkamp, Frankfurt/M.
Siegelberg, Jens (1990): Schritte zu einer Theorie des Krieges - Ein gesellschaftstheoretischer Entwurf für die vergleichende Kriegsursachenforschung. Arbeitspapiere der Forschungsstelle Kriege, Rüstung und Entwicklung im Institut für Politische Wissenschaft der Universität Hamburg, No. 42.
--- (ed.) (1991): Die Kriege 1985 bis 1990. Analyse ihrer Ursachen, Lit, Münster, Hamburg.
--- (1994): Kapitalismus und Krieg: Eine Theorie des Krieges in der Weltgesellschaft, Lit, Münster.
Silberbauer, George B. (1981): Hunter & Habitat in the Central Kalahari Desert, Cambridge University Press, Cambridge.
Simmons, P.J. (ed.) (1995): Environment and Security Debates: An Introduction. Report of the Environmental Change and Security Project No. 1, Spring, The Woodrow Wilson Center, Washington D.C.
Simonis, Udo E. (ed.) (1988): Präventive Umweltpolitik. WZB Reihe, Campus, Frankfurt/M, New York.
--- (1990): Lebensressource Wasser: Wasserknappheit und Wasserverschmutzung, in: Jahrbuch Dritte Welt 1991. Daten, Übersichten, Analysen, Beck'sche Reihe, München.
--- (ed.) (1991): Ökonomie und Ökologie: Auswege aus einem Konflikt, C.F. Müller, Karlsruhe.
--- (1992): Kooperation oder Konfrontation: Chancen einer globalen Klimapolitik, in: Aus Politik und Zeitgeschichte, No. 16, pp. 21-32.
--- und Weizsäcker, Ernst U. von (1990): Globale Umweltprobleme, in: Europa Archiv, No. 1, pp. 1-12.
SIPRI (1994): SIPRI Yearbook 1994, Oxford University Press, New York.
Sironneau, Jacques (1993): "L'eau ressource stratégique", Géopolitique, N° 43 (Autumn).
--- (1995): Is there an International Law on Water? Progress and Prospects for the Resolution of Conflicts over Water Use, in: International Committee of the Red Cross (ed.): Water and War. Symposium on Water in Armed Conflicts, Montreux, 21-23 November 1994, Geneva, pp. 53-71.
Sirven, P., Gotanegre, J.F. et Prioul, C. (1974): Géographie du Rwanda, Bruxelles, Kigali.
--- (1995): China's Environmental Refugees: Causes, Dimensions and Risks of an Emerging Problem, in: Kurt R. Spillmann and Günther Bächler (eds.): Environmental Crisis. Regional Conflicts and Ways of Cooperation, Report of the International Conference at Monte Verità, Ascona, Switzerland, 3-7 October 1994, Berne, Zurich 1995, pp. 75-92.

Small, Melvin and Singer, J. David (1982): Resort to Arms. International and Civil Wars. 1816-1980. 2. ed., Sage, Beverly Hills, London, New Delhi (1st ed.: Singer & Small: The Wages of War. 1816-1965. A Statistical Handbook. New York et al.).

Smith, Dan (1994): Dynamics of Contemporary Conflict: Consequences for Development Strategies, in: Nina Græger and Dan Smith (eds.): Environment, Poverty, Conflict, PRIO-Report No. 2, Oslo, pp. 47-90.

--- (1994): Dynamics of Contemporary Conflict: Consequences for Development Strategies. Paper for the PRIO Research Workshop on Peace and Conflict Implications of Environmental Sustainability.

Smock, David R. (ed.) (1993): Making War and Waging Peace: Foreign Intervention in Africa, United States Institute of Peace Press, Washington D.C.

Snow, Donald M. (1993): Distant Thunder. Third World Conflict and the New International Order, St. Martin's Press, New York.

Soroos, Marvin S. (1994): Global Change, Environmental Security, and the Prisoner's Dilemma, in: Journal of Peace Research, Vol. 31, No. 3, pp. 317-332.

Sottas, Beat (1996): Conflict Resolution in a Setting of Contradictory Regulations and Rationales: Approaches to Resource Management in the Ewaso Ng'iro Basin (Kenya), in: Proceedings of the Regional Workshop in Asmara, Eritrea, 12-16 February 1996, Environmental Conflict Management And Sustainable Development, Berne, Zurich, pp. 215-220.

Sperling, Louise, Galaty, John G., and Johnson, Douglas L. (1990): Cattle, Culture, and Economy: Dynamics in East African Pastoralism, in: John G. Galaty and Douglas L. Johnson: The World of Pastoralism: Herding Systems in Comparative Perspective, The Guilford Press, New York, pp. 69-98.

Spillmann, Kurt R. (1994): Eine andere Konfliktdimension im Nahen und Mittleren Osten: Wasser! In: Kurt R. Spillmann (ed.): Zeitgeschichtliche Hintergründe aktueller Konflikte III. Zürcher Beiträge zur Sicherheitspolitik und Konfliktforschung, No. 31, Forschungsstelle für Sicherheitspolitik und Konfliktanalyse, ETH Zürich, pp. 153-173.

--- und Bächler, Günther (eds.) (1995): Environmental Crisis: Regional Conflicts and Ways of Cooperation. Proceedings of an International Conference at Centro Stefano Franscini, Ascona/Switzerland, 2-7 October 1994. ENCOP Occasional Paper No. 14, Swiss Federal Institute of Technology/Swiss Peace Foundation, Zurich, Berne.

Spooner, B. (1987): The Paradoxes of Desertification, in: Desertification Control Bulletin, Vol. 15, pp. 40-45.

Staehelin, Fritz (1992): Bevölkerungswachstum und Umweltzerstörung aus der Sicht der Entwicklungszusammenarbeit. Referat zur Tagung der Schweizerischen Gesellschaft für Umweltschutz am 6. November 1992 in Rüschlikon.

Stahl, Karin (1992): Der Nord-Süd-Konflikt um "Umwelt und Entwicklung". Zwischenbilanz der Vorverhandlungen der UN-Konferenz über "Umwelt und Entwicklung", in: epd-Entwicklungspolitik, No. 1/2.

Starke, Linda (1990): Signs of Hope: Working Towards Our Common Future, Oxford University Press, New York.

Starr, Joyce R. (1991): Water Wars, in: Foreign Policy, No. 82, pp. 17-36.

--- and Stoll, Daniel C. (1987): U.S. Foreign Policy on Water Resources in the Middle East, Center for Strategic & International Studies, Washington D.C.

--- (ed.) (1988): The Politics of Scarcity: Water in the Middle East, Westview Press, Boulder CO.

Stiftung Entwicklung und Frieden. Ingomar Hauchler (ed.) (1995): Globale Trends 1996. Fakten, Analysen, Prognosen, Fischer, Frankfurt/M.

Swiss Priority Programme Environment, Eritrean Agency for the Environment, Center for Security Studies and Conflict Research, Swiss Peace Foundation Berne (ed.) (1996): Proceedings of the Regional Workshop on Environmental Conflict Management and Sustainable Development. Strengthening the Research Cooperation between South and North, Berne, Zurich.

Steiner, Dieter und Wisner, Ben (eds.) (1986): Humanökologie und Geographie. Zürcher Geographische Schriften, No. 28, Geographisches Institut ETHZ, Zürich.
Stern, Eric K. (1992): "High Politics" and Crisis: Military and Environmental Dimensions of Security, in: Sverre Lodgaard and Anders Hjort af Ornäs (eds.): The Environment and International Security, PRIO-Report No. 3, Oslo, pp. 85-103.
Stern, Paul C., Young, Oran R., and Druckman, Daniel (eds.) (1992): Global Environmental Change: Understanding the Human Dimensions, National Academy Press, Washington D.C.
Stone, Peter B. (1992): State of the World's Mountains. A Global Report, Zed Books, London.
Stryker Dirck J. (1989): Technology, Human Pressure, and Ecology in the Arid and Semi-Arid Tropics, in: H. Jeffrey Leonard and contributors: Environment and the Poor. Development Strategies for a Common Agenda, Transaction Books, New Brunswick, pp. 87-110.
Stüben, Peter E. (ed.) (1985): Kahlschlag im Paradies. Die Vernichtung der Regenwälder - Das Ende der Stammesvölker, Focus, Giessen.
--- (ed.) (1986): Nach uns die Sintflut: Staudämme - Entwicklungs"hilfe", Umweltzerstörung und Landraub. Ökozid, Jahrbuch für Ökologie und Bedrohte Kulturen, No. 2, Focus, Giessen.
--- und Thurn, Valentin (eds.) (1991): "WüstenErde": der Kampf gegen Durst, Dürre und Desertifikation. Ökozid, Jahrbuch für Ökologie und bedrohte Völker, No. 7, Focus, Giessen.
Studiengruppe (1996): Die Schweizerische Entwicklungszusammenarbeit in Rwanda. Bericht der vom EDA eingesetzten Studiengruppe, Bern.
Suhrke, Astri (1992): Environmental Degradation, Migration and Conflict. Case Study Prepared for the Review Conference "Environmental Change and Acute Conflict" at the Brookings Institution, Washington D.C., 11-12 May.
--- (1994): Environmental Refugees and Social Conflict. A Framework for Analysis, in: Günther Bächler (ed.): Umweltflüchtlinge. Das Konfliktpotential von morgen? agenda, Münster, pp. 83-100.
--- (1992): Civil War in Sudan: The Impact of Ecological Degradation. ENCOP Occasional Paper No. 4, Swiss Federal Institute of Technology/ Swiss Peace Foundation, Zurich, Berne.
--- (1993): Der afrikanische Kontinent ist von innen her bedroht. Krieg und Frieden in Afrika 1992, in: Österreichisches Studienzentrum für Frieden und Konfliktlösung, Schweizerische Friedensstiftung (eds.): Das neue Chaos der nationalen Egoismen. Friedensbericht 1993, Verlag Rüegger, Chur, Zürich, pp. 77-90.
--- (1994): War in Darfur: The Desert versus the Oasis Syndrome, in: Günther Bächler und Kurt R. Spillmann (eds.): Kriegsursache Umweltzerstörung. Environmental Degradation as a Cause of War, Verlag Rüegger, Chur, Zürich, pp. 145-180.
Suliman, Mohamed (ed.) (1990): Green-House Effect and Its Impact on Africa, Institute for African Alternatives IFAA, London.
--- (ed.) (1991): Alternative Development Strategies for Africa. Vol. 2: Environment - Women, Institute for African Alternatives IFAA, London.
Swain, Ashok (1993): Conflicts over Water; The Ganges Water Dispute, in: Security Dialogue, Vol. 24, No. 4, pp. 429-439.
--- (1993): Environment and Conflict: Analysing the Developing World, Report No. 37, Uppsala University, Department of Peace and Conflict Research, Uppsala.
Taylor, Charles Lewis and Jodice, David A. (1994): World Handbook of Political and Social Indicators, Yale University Press, New Haven, London.
The Challenge to the South. The Report of the South Commission (1990), Oxford University Press, New York.
The Refugee Policy Group (1992): Migration and the Environment, Washington D.C.
Thomas, Caroline (1992): The Environment in International Relations, Royal Institute of International Affairs, London.
--- and Howlett, Darryl (eds.) (1993): Resource Politics; Fresh Water and Regional Relations, Open University Press, Buckingham.

Thomas, David S.G. and Middleton, Nicholas J. (1992): World Atlas of Desertification, Edward Arnold.
--- (1994): Desertification: Exploding the Myth, John Wiley & Sons, Chichester.
Thomas, William L. (ed.) (1956): Man's Role in Changing the Face of Earth. Vol. 2, Chicago.
Timberlake, Lloyd (1990): Krisenkontinent Afrika: Der Umwelt-Bankrott. Ursachen und Abwendung, Peter Hammer, Wuppertal.
Tinker, Jon and Timberlake, Lloyd (1985): Environment and Conflict. Exploring the Link between Resources and Strife, in: World Press Review, No. April, pp. 29-31.
Tiruneh, Andargachew (1993): The Ethiopian Revolution 1974-1987. A Transformation from an Aristocratic to a Totalitarian Autocracy, Cambridge University Press, Cambridge MA.
Tolba, Mostafa K. (1992): Saving Our Planet: Challenges and Hopes, Chapman & Hall, London.
--- et al. (1989): Die Umwelt bewahren. Die globale Umweltkrise als neue Dimension von Sicherheitspolitik. Ökologische Argumente für eine Wende in der internationalen Politik, Stiftung Entwicklung und Frieden, Bonn-Bad Godesberg.
--- , El-Kholy, Osama A., El-Hinnawi, E. et al. (eds.) (1992): The World Environment 1972-1992. Two Decades of Challenge, UNEP Nairobi, Chapman & Hall, London.
Trolldalen, Jon M. (1992): International Environmental Conflict Resolution. The Role of the United Nations, UNIDIR, Oslo, Washington, Geneva, New York.
Tuchman Matthew, Jessica (1989): Redefining Security, in: Foreign Affairs, Vol. 68, No. 2, pp. 162-178.
Turner II, B.L. et al. (1990): The Earth As Transformed by Human Action: Global and Regional Changes in the Biosphere over the Past 300 Years, Cambridge University Press, Cambridge, New York.
Tvedt, Terje (ed.) (1993): Conflicts in the Horn of Africa: Human and Ecological Consequences of Warfare, Environmental Policy and Society, EPOS, Uppsala.
Ullmann, Richard H. (1993): Redifining Security, in: International Security, 8, pp. 129-153.
--- (1995): Redefining Security: The Human Dimension, in: Current History, May, pp. 229-236.
Ulrich, Rita (1989): Environment and Security in the Horn of Africa. A Report Prepared for the United Nations Environment Programme Peace, Security and the Environment Programme Regional Cooperation in the Horn of Africa, in: UNEP (ed.): Horn of Africa, Vol. I.
UN 1977: UN Plan of Action to Combat Desertification (UN/PACD).
UNCTAD (1992): Combating Global Warming. Study on a Global System of Tradeable Carbon Emission Entitlements, New York.
UNDP (1991): Human Development Report 1991, Oxford University Press, New York.
--- (1994): Human Development Report 1994, Oxford University Press, New York.
--- (1995): Human Development Report 1995, Oxford University Press, New York.
--- (1996): Human Development Report 1996, Oxford University Press, New York.
UNEP (1988): Environmental Perspective to the Year 2000 and Beyond, Nairobi.
--- (1991): The State of the World Environment, Nairobi.
--- (s.a.): The Aral Sea: Diagnostic Study for the Development of an Action Plan for the Conservation of the Aral Sea.
UNEP; UNCOD (1978): United Nations Conference on Desertification 29 August - 9 September 1977: Round up, Plan of Action and Resolutions, UN Publications, New York.
UNFPA (1991): Population, Resources and the Environment. The Critical Challenges, London.
--- (1992): The State of World Population, UNFPA, New York.
--- (1992): Women Population and the Environment, UNFPA, New York.
--- (1993): The State of World Population 1993, UNFPA, New York.
UNHCR (1993): The State of the World's Refugees: The Challenge of Protection, Penguin, New York.
United Nations (1991): World Economic Survey 1991. Current Trends and Policies in the World Ecomony, United Nations, New York.

--- (1992): Agenda 21: The United Nations Program of Action from Rio, UN Publications, New York.
--- (1992): The Global Partnership for Environment and Development. A Guide to Agenda 21, Geneva.
--- (1993): Statistical Yearbook, 38 ed., United Nations, New York.
--- (1993): Report of the United Nations Conference on Environment and Development. Rio de Janeiro, 3-14 June 1992. Vol. I: Resolutions Adopted by the Conference, New York.
--- (1993): Report of the United Nations Conference on Environment and Development. Rio de Janeiro, 3-14 June 1992. Vol. II: Proceedings of the Conference, New York.
--- (1993): Report of the United Nations Conference on Environment and Development. Rio de Janeiro, 3-14 June 1992. Vol. III: Statements Made by Heads of State or Government at the Summit Segment of the Conference, New York.
United Nations Research Institute for Social Development (UNRISD) (1995): Rebuilding War-Torn Societies. An Action-Research Project on Problems of International Assistance in Post-Conflict Situations, Geneva.
United Nations Security Council (1993): Report of the Secretary-General on Rwanda, S/26350, 24 August.
--- (1993): Resolution S/RES/812, 12 March.
--- (1993): Resolution S/RES/846, 22 June.
--- (1994): Resolution S/RES/935, 4 October.
U.S. Committee for Refugees USCR (1995): World Refugee Survey 1995, Washington D.C.
--- (1996): World Refugee Survey 1996. An Annual Assessment of Conditions Affecting Refugees, Asylum Seekers, and Internally Discplaced People, Washington D.C.
Uwimana, Musema (1982): Exposé sur la politique nationale de conservation, amélioration et fertilité des sols au Rwanda, in: Bulletin Agricole du Rwanda, Avril.
Vansina, Jan (1962): L'évolution du royaume rwanda des origines à 1900, Académie Royale des Sciences d'Outre-Mer: Classe des Sciences Morales et Politiques, mémoires in-80, n.s., tome 26, fasc. 2, Bruxelles.
Väyrynen, Raimo (ed.) (1991): New Directions in Conflict Theory. Conflict Resolution and Conflict Transformation, Sage, London.
Verschave, François-Xavier (1994): Complicité de génocide? La politique de la France au Rwanda, La Découverte, Paris.
Wadlow, René (1988): Bioregions and Transnational Zones, in: Transnational Perspectives, Vol. 14, pp. 6-8.
Waever, Ole (1993): Securitization and Desecuritization. Centre for Peace and Conflict Research Working Paper No. 5, Copenhagen.
Wallensteen, Peter (1992): Environmental Destruction and Serious Social Conflict: Developing a Research Design, in: Sverre Lodgaard and Anders Hjort af Ornäs (eds.): The Environment and International Security, PRIO-Report No. 3, Oslo, pp. 47-54.
--- and Axell, Karin (1994): Conflict Resolution and the End of the Cold War, 1989-1993, in: Journal of Peace Research, Vol. 31, No. 3, pp. 333-349.
Weiner, Myron (ed.) (1993): International Migration and Security, Westview Press, Boulder CO.
Weizsäcker, Carl Friedrich von (ed.) (1971): Kriegsfolgen und Kriegsverhütung, Hanser, München.
Weizsäcker, Ernst U. von (1974): Humanökologie und Umweltschutz. Studien zur Friedensforschung, No. 8, Klett/Kösel, Stuttgart.
--- (1990): Erdpolitik: Ökologische Realpolitik an der Schwelle zum Jahrhundert der Umwelt, Wissenschaftliche Buchgesellschaft, Darmstadt.
--- et al. (1989): Globale Umweltprobleme. Globale Umweltpolitik. Interdependenz. Materialien der Stiftung Entwicklung und Frieden, No. 3, Stiftung Entwicklung und Frieden, Bonn.

Weltbevölkerungskonferenz der UNO in Kairo im September 1994 - eine Dokumentation, die (1995), in: Internationale Politik und Gesellschaft, No. 5, pp. 73-128.
Weltkommission für Umwelt und Entwicklung (1987): Unsere gemeinsame Zukunft, Eggenkamp, Greven.
Werle, Otmar und Weichert, Karl-Heinz (1987): Ruanda. Ein landeskundliches Porträt, Görres, Koblenz.
Westing, Arthur H. (1976): Ecological Consequences of the Second Indochina War, Almqvist & Wiksell, Stockholm.
--- (1977): Weapons of Mass Destruction and the Environment, Taylor & Francis, London.
--- (1980): Warfare in a Fragile World. Military Impact on the Human Environment, Taylor & Francis, London.
--- (1986): Environmental Factors in Strategic Policy and Action: An Overview, in: Arthur H. Westing (ed.): Global Resources and International Conflict. Environmental Factors in Strategic Policy and Action, Oxford University Press, Oxford, New York.
--- (1988): Towards Non-Violent Conflict Resolution and Environmental Protection. A Synthesis, in: Arthur H. Westing (ed.): Cultural Norms, War and the Environment, Oxford.
--- (1989): Environmental Approaches to Regional Security, in: Arthur H. Westing (ed.): Comprehensive Security for the Baltic. An Environmental Approach, London, pp. 1-14.
--- (1989): The Environmental Component of Comprehensive Security, in: Bulletin of Peace Proposals, Vol. 20, No. 2, pp. 129-134.
--- (1989): Environmental Security for the Danube Basin, in: Environmental Conservation, No. 16, pp. 323-329.
--- (1993): Transfrontier Reserves for Peace and Nature. A Contribution to Human Security, Nairobi.
--- (ed.) (1984): Environmental Warfare. A Technical, Legal and Policy Appraisal, Taylor & Francis, London, Philadelphia.
--- (ed.) (1985): Explosive Remnants of War. Mitigating the Environmental Effects, Taylor & Francis, London.
--- (ed.) (1986): Global Resources and International Conflict. Environmental Factors in Strategic Policy and Action, Oxford University Press, Oxford, New York.
--- (ed.) (1988): Cultural Norms, War and the Environment, Oxford University Press, Stockholm.
--- (ed.) (1989): Comprehensive Security for the Baltic. An Environmental Approach, London.
--- (ed.) (1990): Environmental Hazards of War. Releasing Dangerous Forces in an Industrialized World, Sage, London.
Wiesenthal, Helmut (1987): Einleitung: Die Ratlosigkeit des homo oeconomicus, in: Jon Elster (ed.): Subversion der Rationalität, Campus, New York, pp. 7-19.
Wijkman, Anders und Timberlake, Lloyd (1986): Die Rache der Schöpfung. Naturkatastrophen: Verhängnis oder Menschenwerk? Piper, München, Zürich.
Winiger, M., Wiesmann, U., and Rheker, U. (eds.) (1990): Mount Kenya Area. Differentiation and Dynamics of a Tropical Mountain Ecosystem, Berne University, Institute of Geography, Berne.
Winnefeld, James A. and Morris, Mary E. (1994): Where Environmental Concerns and Security Strategies Meet: Green Conflict in Asia and the Middle East, Santa Monica CA.
Wirth, David A. (1989): Climate Chaos, in: Foreign Policy, No. 74, pp. 3-22.
Wissenschaftlicher Beirat der Bundesregierung, Globale Umweltveränderungen (1993): Welt im Wandel. Grundstruktur globaler Mensch-Umwelt-Beziehungen. Jahresgutachten 1993. Kurzfassung, Bonn.
WMO (1992): Climate Change. Environment and Development. World leaders' viewpoints, Geneva.
Wöhlcke, Manfred (1987): Umweltzerstörung in der Dritten Welt, Beck, München.

--- (1990): Umwelt- und Ressourcenschutz in der internationalen Entwicklungspolitik: Probleme und Zielkonflikte. Aktuelle Materialien zur Internationalen Politik, No. 23, Nomos, Baden-Baden.
--- (1991): Risiken aus dem "Süden". Neue Themen in den Nord-Süd-Beziehungen nach dem Ende des Ost-West-Konflikts, Stiftung Wissenschaft und Politik, Ebenhausen.
--- (1992): Gemeinsam in die Öko-Katastrophe? Umweltzerstörung und Ressourcenausbeutung, in: Volker Matthies (ed.): Kreuzzug oder Dialog. Die Zukunft der Nord-Süd-Beziehungen, Bonn, pp. 74-89.
--- (1992): Umweltflüchtlinge. Ursachen und Folgen, Beck, München.
--- (1993): Der ökologische Nord-Süd-Konflikt. Interessen, Argumente und Verantwortlichkeiten im Rahmen der internationalen Umweltpolitik, Beck, München.
Wolde-Mariam, Mesfin (1991): Suffering Under God's Environment. A Vertical Study of the Predicament of Peasants in North-Central Ethiopia, Walsworth Press, Marceline MS.
Wolf, Eric (1969): Peasant Wars of the Twentieth Century, New York.
World Bank (1983): Rwanda. Economic Memorandum, Recent Economic Developments and Current Policy Issues, Report No. 4059-RW, May 1983, The World Bank, Washington D.C.
--- (1986): Rwanda. Economic Memorandum, Recent Economic Developments and Current Policy Issues, Report No. 6191-RW, October, The World Bank, Washington D.C.
--- (1991): World Development Report 1991. The Challenge of Development, Oxford University Press, New York.
--- (1992): World Development Report 1992. Development and the Environment. Executive Summary, The World Bank, Washington D.C.
--- (1994): World Development Report 1994. Infrastructure for Development, Oxford University Press, New York.
--- (1995): World Development Report 1995. Workers in an Integrating World, Oxford University Press, New York.
--- (1995): Monitoring Environmental Progress. A Report on Work in Progress. ESD Series, Washington D.C.
--- (1996): Social Indicators of Development 1996, John Hopkins University Press, Baltimore, London.
--- (ed.): World Data on CD-ROM.
World Disasters Report 1993 (1993), Martinus Nijhoff, Dordrecht.
World Resources Institute (WRI) (1992): World Resources 1992-93: A Guide to the Global Environment. Toward Sustainable Development (1992), Oxford University Press, New York, Oxford.
--- (1994): World Resources 1994-95: A Guide to the Global Environment. People and the Environment. Resource Consumption. Population Growth. Women (1994), Oxford University Press, New York.
--- (1996): World Resources 1996-97: A Guide to the Global Environment. The Urban Environment (1996), Oxford University Press (a joint publication by WIR, UNEP, UNDP, the World Bank), New York.
World Soil Resources. An Explanatory Note on the FAO World Soil Resources Map at 1:25'000'000 scale (1991): World Soil Resources Report, No. 66, FAO, Rome.
Zartman, William I. (ed.) (1991): Resolving Regional Conflicts: International Perspectives. The Annals of the American Academy of Political and Social Science No. 518, Sage, Newbury Park, London, New Delhi.
--- (1992): International Environmental Negotiation: Challenges for Analysis and Practice, in: Negotiation Journal, Vol. 8, No. 2, pp. 113-123.
Zegeye, A. (1995): Environmental Insecurity. Dilemmas of Survival in the Horn of Africa.
Zürn, Michael (1992): Interessen und Institutionen in der internationalen Politik. Grundlegung und Anwendungen des situationsstrukturellen Ansatzes, Dissertation, Opladen.

--- (1994): Globale Gefährdungen - Auf dem Wege zu einer Weltrisikogesellschaft? In: Jörg Calliess (ed.): Treiben Umweltprobleme in Gewaltkonflikte? Ökologische Konflikte im internationalen System und Möglichkeiten ihrer friedlichen Bearbeitung. Loccumer Protokolle No. 21, Evangelische Akademie Loccum, Rehburg-Loccum, pp. 229-262.

Social Indicators Research Series

1. V. Møller (ed.): *Quality of Life in South Africa.* 1997 ISBN 0-7923-4797-8
2. G. Baechler: *Violence Through Environmental Discrimination.* Causes, Rwanda Arena, and Conflict Model. 1999 ISBN 0-7923-5495-8

KLUWER ACADEMIC PUBLISHERS – DORDRECHT / LONDON / BOSTON

MIX
Papier aus verantwortungsvollen Quellen
Paper from responsible sources
FSC® C105338

If you have any concerns about our products,
you can contact us on
ProductSafety@springernature.com

In case Publisher is established outside the EU,
the EU authorized representative is:
Springer Nature Customer Service Center GmbH
Europaplatz 3, 69115 Heidelberg, Germany

Printed by Libri Plureos GmbH
in Hamburg, Germany